PHARMACOGENOMICS

PHARMACOGENOMICS
SOCIAL, ETHICAL, AND
CLINICAL DIMENSIONS

EDITED BY

MARK A. ROTHSTEIN
Herbert F. Boehl Chair of Law and Medicine
Director, Institute for Bioethics, Health Policy and Law
University of Louisville, School of Medicine
Louisville, KY

WILEY-LISS

A JOHN WILEY & SONS, INC., PUBLICATION

Published by John Wiley & Sons, Inc., Hoboken, New Jersey.
Published simultaneously in Canada.

For general information on our other products and services please contact our Customer Care Department within the U.S. at 877-762-2974, outside the U.S. at 317-572-3993 or fax 31-572-4002.

Wiley also publishes its books in a variety of electronic formats. Some content that appears in print, however, may not be available in electronic format.

Library of Congress Cataloging-in-Publication Data:

Pharmacogenomics : social, ethical, and clinical dimensions /
    edited by Mark A. Rothstein.
      p. ; cm.
    Includes bibliographical references and index.
    ISBN 0-471-22769-2 (cloth)
    1. Pharmacogenomics. I. Rothstein, Mark A.
    [DNLM: 1. Pharmacogenetics. 2. Drug Design. 3. Ethics,
  Medical. 4. Legislation, Medical. 5. Public Policy. 6. Social
  Environment. QV 38 P53197 2003]
  RM301.3.G45P435 2003
  615' .7—dc21
2002012137

Printed in the United States of America

10 9 8 7 6 5 4 3 2 1

CONTENTS

V

FOREWORD

FRANCIS S. COLLINS, M.D., PH.D.

A century ago, with characteristic vision and eloquence, Sir William Osler penned these words: "Were man always, inside and outside, cast in the same mold, instead of differing from his fellow man, we would ere have reached some settled principles in our art." This variability and unpredictability of responses to medical intervention is familiar to every health care provider, and has been a cause of puzzlement and vexation to providers and patients throughout the entire history of medicine. Given an ideal situation where a diagnosis has been firmly established and rigorous evidence points to a validated pharmaceutical intervention, it is virtually never the case that 100% of the patients treated will enjoy an optimal outcome. Instead, some proportion (the responders) will achieve benefit, while another fraction will have no response (the non-responders), and a small proportion may suffer a significant side effect (the adverse responders). Pharmacogenomics offers the bright hope of being able to make prior predictions of these outcomes, allowing more informed choices and a better outcome.

Since Osler penned his lament, a certain amount of progress has been made in identifying the heritable basis of variable drug response, but it has been only modest. A few instances of dramatic adverse drug responses, such as malignant hyperthermia, or hemolytic anemia following chloroquine administration, have been tracked to their genetic roots by clever detective work. With the advent of additional tools and increasingly robust databases of genetic information, insights into more complex drug reactions (such as the role of Factor V Leiden in the development of deep vein thrombosis after estrogen administration) are beginning to be more discernable. But now, at the dawn of the 21st century, the powerful tools of genomics are finally assembled to make it possible to identify the polygenic basis of drug responses in a much wider variety of applications. Already a few examples are emerging, such as the recent elucidation of a significant genetic contribution to abacavir adverse response, recently mapped to the major histo-compatibility complex. The full flowering of pharmacogenomics will occur in the next decade, building upon the foundation of the sequence of the human genome and a detailed catalogue of human variation. In that regard, determining not only the common variants in the genome but how they are associated with their neighbors (the so-called Haplotype Map) will provide the kind of toolkit to carry out whole genome association studies to identify determinants of drug response that were only a distant dream a decade ago.

Despite some uncertainties, the clinical applications of pharmacoge-nomics are likely to be highly significant. In instances where the positive predictive value of a genetic test is quite high, there are compelling argu-ments to utilize such testing as soon as rigorous clinical data has been obtained. An example where this is already become the standard of care involves a variant in the TPMT gene. Homozygotes for this mutation will have a severe and even fatal reaction to the thiopurine class of drugs, the most commonly used of which is 6-mercaptopurine, a mainstay in the treat-ment of childhood acute lymphocytic leukemia. For many pharmacoge-nomics tests of the future, however, the positive predictive value may be only modest. But there may be good reasons even then to utilize genetic testing if several interventions are available and the test results may help to skew the odds towards a favorable outcome.

Five years ago pharmacogenomics was unfamiliar to most scientists and health care providers. Today we are witnessing an exponential increase in research on the topic, new journals are springing up, and observers in both the academic and private sector are predicting a substantial impact on clin-ical medicine in only a few years. In such a rapidly-changing environment, it is critical to stand back from the avalanche of scientific information and consider the broader view, with special attention towards the impact on society. In that regard, the birth of this new textbook, which considers social and ethical issues as well as scientific and clinical underpinnings, comes at an auspicious moment. In this remarkably broad and far-reaching work, editor Mark Rothstein and his distinguished list of contributors have laid out an impressive framework for the field. Beginning with a survey of public attitudes, and progressing through scientific, clinical, governmental, legal, economic, and societal issues, the text builds to a final provocative epilogue on consequences for public policy. If, as many predict, we are moving towards a "Dx; Rx" model for medicine, in which genetic testing routinely precedes the writing of prescriptions, this text should provide a thoughtful foundation upon which to build that future.

PREFACE

In 1998, when I began the initial research that culminated in this volume, the terms pharmacogenetics (the role of genetic variation in differential response to pharmaceuticals) and pharmacogenomics (the use of genomic technologies in assessing differential response to pharmaceuticals) were limited to the vocabularies of a few experts in genetics and pharmacology. Within a mere five years, however, several journals have been launched devoted exclusively to these topics, pharmacogenomics articles have begun appearing regularly in general scientific and medical journals, and even the popular press has begun to feature articles on the development of genome-specific medications.

A notable exception to the increasing focus on pharmacogenetics and pharmacogenomics is the literature dealing with the ethical, legal, and social implications of genetic and genomic research. Commentators on "ELSI" issues have only occasionally turned their attention to the range of emerging issues raised by pharmaceutical research and clinical applications based on human genetic variation. This volume attempts to reduce this scholarship deficit by presenting a multidisciplinary analysis of the scientific, clinical, economic, ethical, social, and legal implications of pharmacogenomics. It also includes, in Chapter 1, the results of the first comprehensive public opinion survey on pharmacogenomics and, in the Epilogue, a series of proposals to address the policy issues raised in the book's fifteen wide-ranging chapters.

Research for this book was funded by a grant from the National Institute of General Medical Sciences (lead institute), the National Human Genome Research Institute, and the National Institute for Environmental Health Sciences of the National Institutes of Health. I want to thank, in particular, Dr. Rochelle Long, of the NIGMS, for her commitment to the exploration of these issues as well as her support and substantive contributions.

The grant application was written and submitted while I was at the Health Law and Policy Institute of the University of Houston. My former colleague, Elaine Lisko, who left Houston in 1999, played an integral part in researching and drafting the grant application. Phyllis Griffin Epps, who remained in Houston after my departure in December 2000, continued to have a central role in the project, including the public opinion survey, editorial work on this volume, and taking the lead in several articles we coauthored on pharmacogenomics.

Telesurveys Research Associates of Houston conducted the public opinion survey. Rosie Zamora, Dick Jaffe, and Barry Petree demonstrated patience and good cheer through the numerous revisions of the survey instrument. Dr. Eun-Sul Lee of the University of Texas-Houston School of Public Health and Dr. Sharon P. Cooper, now at the Texas A&M University School of Rural Public Health, were the statistical consultants on the survey.

At the Institute for Bioethics, Health Policy and Law at the University of Louisville, my colleague Gabriela Alcalde helped me in conceptualizing the key issues and also took the lead in drafting two articles we coauthored on the topic. Dr. Mary Anderlik coauthored a paper with me on genetic research in which pharmacogenomics was a key element. Dr. Carl Hornung of the Department of Medicine contributed indispensable data analysis on the survey and coauthored Chapter 1 of this volume.

Dennis Holmgren, J.D. 2002 from the Louis D. Brandeis School of Law of the University of Louisville, provided valuable research assistance and generated the computer graphics for the figures in Chapter 1. Rob Wright, J.D. 2003, ably assisted with the final reference check.

Sue Rose was responsible for the manuscript preparation and formatting, which she did with her customary skill and equanimity through endless revisions. My long-time associate, Cathy Rupf, contributed expert grant administration and project planning at both the University of Houston and the University of Louisville. Judy Oller was responsible for the flawless financial management on the grant.

Last, but not least, I want to thank my wife Laura for leading me into challenging new professional adventures and providing the personal support to make them work.

Mark A. Rothstein
July 2002

INTRODUCTION: SCIENCE AND SOCIETY

Public Attitudes About Pharmacogenomics

Mark A. Rothstein, J.D. and Carlton A. Hornung, Ph.D., M.P.H.

I. Introduction

A substantial amount of public and private research is being conducted in pharmacogenomics, and there is great hope that new pharmaceutical products will be developed that decrease adverse drug reactions and improve the treatment of many serious diseases. Amid the scholarly debates about research methodology and ethics, regulatory strategies, clinical applications, and economic and legal effects, there has been relatively little discussion about the views of the public. Do members of the public support pharmacogenomic research? Do they understand the underlying science? Would they be willing to participate in research and, if so, under what conditions? What institutions do they most and least trust to conduct the research? Are they optimistic about applications of new therapeutics? Do they think pharmacogenomic research will benefit them? Do they think they would be able to afford the new medicines? Do they have concerns about the privacy and confidentiality of genetic information generated in the research or clinical stages? Are their attitudes influenced by demographic factors such as education, income, ethnicity, gender, and age?

This chapter reports some of the key findings of the first comprehensive public survey on pharmacogenomics. In several instances, the data generated by the survey challenge prior assumptions concerning public attitudes

Pharmacogenomics: Social, Ethical, and Clinical Dimensions, Edited by Mark A. Rothstein.
ISBN 0-471-22769-2   Copyright © 2003 Wiley-Liss, Inc.

about medical research, privacy, and health care. The data suggest additional areas for study as well as possible new approaches to policy development. In general, the data lead to the following three conclusions: (1) the public is generally interested in and concerned about genetics and pharmacogenomics; (2) many members of the public are confused about the nature and consequences of pharmacogenomics; and (3) responses to questions on these issues frequently varied on the basis of education, income, race and ethnicity, and age but infrequently on the basis of other demographic variables.

II. PRIOR RESEARCH

Although there have been no prior surveys on public attitudes toward pharmacogenomics research, there are a few surveys on genetics research and genetics in general. In 2000, the National Health Council released the results of its focus group study of genetic research (National Health Council, 2000). The participants recalled hearing, reading, or seeing information about medical research, but they did not specifically mention genetics research. When prompted about genetics, the knowledge level of most participants was described as "inaccurate or shallow," and the most common form of genetic research participants recalled was cloning. Over half of the participants said they would be willing to participate in genetic research by submitting a cheek swab or blood sample and their medical history. Confidentiality and possible discrimination headed the list of concerns about the possible uses of their genetic information, and moral concerns, such as "playing God," were raised about genetic research in general. Participants were comfortable with research conducted by university medical centers, but they were concerned about the possible influence of the profit motive in genetic research conducted by pharmaceutical companies. Some participants were reluctant to have the government play a role in genetic research because of a perceived inability to safeguard secrets. Participants considered voluntary health agencies to be "credible" sponsors of research, but the participants were concerned about the agencies' lack of expertise to conduct the research themselves.

In a nationwide interview survey conducted in the fall of 2001 by Peter D. Hart Research Associates, Inc., respondents were asked whether genetic research will result in medical treatments and cures for diseases. Forty percent said that it will almost certainly happen, and 53% said that it will probably happen (Peter D. Hart, 2001). When asked whether many serious diseases will be eradicated as a result of genetic research, 20% said that it will almost certainly happen and 54% said that it will probably happen. The responses on genetic discrimination are particularly interesting. When asked whether health insurance companies will use genetic information to deny people coverage if they are predisposed to diseases, 32% said that it will almost certainly happen and 47% said that it will probably happen. By contrast, when asked whether employers will use genetic information to discriminate against workers or job applicants who are predisposed to diseases, 16% said that it will almost certainly happen and 35% said that it will probably happen. As discussed below in this chapter, our survey did not detect a variance between public concerns about possible health insurance and employment uses of genetic information.

Another telephone interview survey was conducted by the *Los Angeles Times* in 2000. Respondents were asked, "Are you more inclined to think that [genetic] research will ultimately be beneficial, or harmful, to you and your family?" Overall, 25% said that it will be very beneficial, 34% said that it will be somewhat beneficial, 11% said that it will be somewhat harmful, 6% said that it will be very harmful, and 24% said that they did not know (*Los Angeles Times*, 2000). Men were more likely than women to believe that the research would be beneficial (64% vs. 54%), and those with a college degree or higher education level were more likely to consider the research to be beneficial than those without a college degree (75% vs. 54%). The correlation of optimism about genetic research with education level is in accord with our survey findings, but the significant gender difference is not.

Finally, another survey showed that the public has misperceptions about the incorporation of genetic research into clinical practice. A telephone interview survey conducted in 2000 by Rasmussen Research asked the following question: "Which is closest to the truth. . . . Genetic research has already improved several medical treatments, or using genetic research has mixed results for medical purposes?" Some 40% of respondents answered "already improved," 39% answered "mixed results," and 21% answered "not sure" (Rasmussen Research, 2000). As detailed below in this book, pharmacogenetics and pharmacogenomics are still in their infancy in terms of clinical applications, and, although recombinant DNA technology has been used in drug development, gene therapy must be described as a disappointment thus far. Consequently, the responses may reflect a lack of familiarity with the scientific progress (as supported by the high number of "don't know" answers), wishful thinking, or the effects of premature exuberance by scientists as reported in the popular media.

III. METHODOLOGY

Our survey consisted of randomly dialed telephone interviews of 1796 individuals across the country performed by Telesurveys Research Associates of Houston, Texas, under contract with the Institute for Bioethics, Health Policy and Law of the University of Louisville School of Medicine. The research was funded by the following three Institutes of the National Institutes of Health (NIH): National Institute of General Medical Sciences (lead institute),

National Human Genome Research Institute, and National Institute of Environmental Health Sciences. The interviews were conducted in the summer of 2001.

The survey instrument, drafted by the principal investigator and the survey contractor, went through 15 drafts and three rounds of peer review at the NIH. The instrument was pretested in 20 interviews for length (under 15 minutes) and clarity. The instrument was translated and back-translated by separate translators into Spanish, Mandarin and Cantonese Chinese, Vietnamese, and Korean. The instrument received approval from the Human Studies Committee at the University of Louisville. All interviewees gave oral consent at the beginning of the interview.

The overall sample size was designed at 1800 (with 1,796 completed). Oversampling was used to achieve a minimum subgroup sample size of 300 for whites, African Americans, Hispanics, and Asians. Race and ethnicity designations were based on self-identification. The investigators recognize that Hispanics and Asians are heterogeneous groups. The preferred sampling methodology would have used oversampling to include a sufficient number of Chinese Americans, Vietnamese Americans, Japanese Americans, Korean Americans, Filipino Americans, and other Asian subpopulations to detect important differences. Similarly, the preferred sampling methodology would have used oversampling for Mexican Americans, Cuban Americans, Puerto Rican Americans, and other Hispanic subpopulations. Native Americans also would have been included and sampled in sufficient numbers. Financial constraints, however, necessitated limiting the survey to four race and ethnicity categories.

Telephone interviews were conducted in English, Spanish, Chinese, and Korean. Up to five contact attempts were made for each telephone number at different times of the day. The response rate for residential calls where the call was answered (not counting businesses, FAX machines, or voice mail) was 84.6%. Area code selections in the 48 contiguous states resulted in the use of 24 area codes and 80 telephone exchanges. Although this is not a true random sample of the United States population, statistical inference on a national level is possible. The sample size yields a maximum margin of error of 2.3%, and the ethnic subgroup samples of 300 yield a statistical power of 0.90 for detecting differences of 6 percentage points in pairwise comparisons at an $\alpha = 0.05$ level.

The sampling frame was a three-stage, stratified cluster design with random digit dialing at the last stage. To accommodate the oversampling of racial/ethnic groups, a two-stage poststratification weighting was used to adjust the sample to the 2000 U.S. population. In the first stage, eight age groups were coded to correspond to the U.S. Census. The weight for stage 1 was calculated as the percentage of the U.S. population in each age category in each of the four census regions and divided by the comparable percentage in the sample. Weight 2 was then calculated as the percent distribution from the U.S. 2000 Census divided by the sample percent distribution of ethnicities weighted by weight 1. The final weight used in the analysis was calculated as the product of the two weights.

The survey contained 12 substantive questions, most with subparts, that asked about genetic testing, genetic research, access to genetic information, and prescription medications. The survey used the following 18 demographic variables: health status, prescription and nonprescription medicine usage, health of family members, size of household, age, education, regular use of computer, marital status, employment status, work in health care, residence in city or rural area, race/ethnicity, language spoken at home, country of birth, health insurance coverage, religion, income, and gender.

IV. KEY FINDINGS

The data contain a wealth of information. In this chapter we report the findings from five areas of inquiry: (1) willingness to participate in genetic research; (2) trust in various entities to perform genetic research; (3) perceptions of the affordability of pharmacogenomic-based medications; (4) concerns about the confidentiality of genetic information; and (5) stability of views during the interview.

A. WILLINGNESS TO PARTICIPATE IN GENETIC RESEARCH

One assumption frequently made by researchers and policy makers about individuals' willingness to participate in genetic research (as well as other forms of medical research) is that as the research involves greater disclosure of personal health information, individuals will be less likely to participate. Prior research supports this assumption (National Health Council, 2000, p. 18). We asked the following four questions (Questions 4A, 4B, 4C, 4D) related to this issue:

A. How likely would you be to participate in genetic research if you knew the results would be anonymous, meaning that nobody—not even the researchers—would know which test came from which person?

B. How likely would you be to participate in genetic research if it involved researchers reviewing your medical records in addition to a genetic test in order to explore possible links between genetic test results and health history?

C. How likely would you be to participate in genetic research if you knew your genetic test results, without your name or other identifying information, would be shared with other scientists nationwide for use in additional genetic research?

   D. How likely would you be to participate in genetic research if the pos-
      sibility existed to develop a treatment in the future based on your
      participation in the research study?

   For each of the four questions (asked in random order), respondents
were provided with the following answer options: very likely, somewhat
likely, somewhat unlikely, very unlikely, and don't know. Refusals also were
recorded.
   We anticipated that individuals' willingness to participate in genetic
research would vary according to the confidentiality of the research and
results and according to the potential benefit or outcome of the research. We
therefore expected the highest percentages of respondents saying that they
would be "very likely" to participate in research when the results would be
anonymous and when the development of a new treatment was a possible
outcome.
   About 2% of respondents said that they didn't know whether they would
participate in genetic research, and another 0.1% refused to answer these
questions. As expected, a large majority (79.8%) said they were very likely
(45.5%) or somewhat likely (34.3%) to participate in genetic research if their
participation meant the possibility of developing a new treatment. However,
when questions concerning the anonymity of the results were raised, the percentage of respondents indicating they would be "very likely" or "somewhat
likely" to participate changed considerably. If the results were anonymous,
so that not even the researchers knew an individual's test results, only 68.2%
said they would participate. Surprisingly, linking test results to the respondent's medical records or sharing test results without his or her name with
other scientists had little or no effect on likely participation rates. Sixty-eight
percent said they were at least somewhat likely to participate when research
involved reviewing their test results and medical records, whereas even more
(69.7%) were somewhat or very likely to participate when their test results
were shared with other scientists and other researchers.
   There are some significant differences according to race/ethnicity, education, and income of respondents. In general, the percentages of whites and
Asians who said they would be very likely or somewhat likely to participate
in genetic research are about 8 to 10 points higher ($p < 0.05$) than African
Americans and Hispanics.
   Higher levels of education are consistently associated with significantly
higher percentages saying they were very likely or somewhat likely to participate in genetic research ($p < 0.02$). There is about a 5% difference between
those with less than a high school education and those who graduated from
high school and another 5% difference between high school graduates and
those who attended or graduated from college. Finally, respondents with a
graduate degree were between 5% and 10% more likely to participate than
high school graduates. The exception to this occurred when respondents

were queried about research that linked test results with the patient's medical records ($p = 0.878$). Only about 60% of respondents with less than a high school education said they would participate in research if the results were anonymous or if the results were shared with other scientists. In contrast, 84.9% of those with a graduate or professional degree reported that they would participate if the possibility existed to develop a treatment in the future based on their participation.

The response to this question has important implications for research design. Anonymous genetic testing, genetic testing that is not linked with medical records, and genetic testing that is not shared with other researchers is less valuable scientifically. Many protocols, however, use these methods because of the perceived need to respond to the privacy and confidentiality concerns of potential participants. The data suggest that more informative data methods may be acceptable to potential participants, especially if the research can be fairly said to have the potential to lead to the development of treatments. Both altruism and self-interest probably combine to produce the strong support for research that may develop a treatment.

Of the various demographic factors, age is especially significant in the responses to this question. As indicated in Figure 1.1, for all age categories,

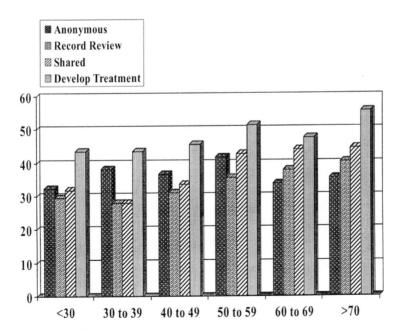

FIGURE 1.1. Percent Very Likely to Participate in Various Methods of Genetics Research (by Age).

testing that might lead to the development of treatments received the highest percentage of individuals very likely to participate. Of the other three categories of research, anonymous testing had the highest percentage for individuals below age 50, but anonymity became less important for respondents over age 50. For those age 60 and over, anonymity was the least important concern. Sharing of the information nationwide with other scientists, of concern to those under 50, was the second most likely condition for testing for those 50 and over. These data may be interpreted as suggesting that relatively younger respondents were more concerned about confidentiality, perhaps because they had the most to lose from disclosure in terms of insurance or employment discrimination. Another possible explanation is that older individuals, who were more likely to be retired or on Social Security, wanted to maximize the scientific value of their research participation and were not as fearful of the economic consequences of disclosure of genetic information.

B. TRUST IN VARIOUS ENTITIES THAT PERFORM GENETIC RESEARCH

Genetic research may be undertaken by various individuals and entities in the public and private sectors. We attempted to ascertain the level of trust for different entities by asking the following related questions (Questions 6A, 6B, 6C, and 6D):

- A. How much trust would you have in genetic research conducted by universities and medical schools?
- B. How much trust would you have in genetic research conducted by the federal government?
- C. How much trust would you have in genetic research conducted by drug companies?
- D. How much trust would you have in genetic research conducted by organizations like the American Cancer Society and the March of Dimes?

For each question (asked in random order) the respondents were provided with the following options: very likely, somewhat likely, very unlikely, and don't know. Refusals were also recorded.

The results confirmed our expectation that universities and medical schools (Question 6A) and organizations like the American Cancer Society or the March of Dimes (Question 6D) would have the greatest levels of trust (over 80%) whereas the federal government (Question 6B) and drug companies (Question 6C) would have the lowest levels of trust (approximately 50%). These findings are in accord with earlier focus group research (National Health Council, 2000, p. 19). The most interesting demographic variables for these questions were race/ethnicity and education.

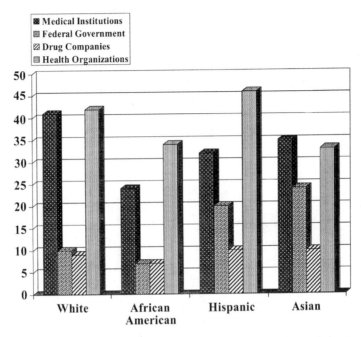

FIGURE 1.2. Percentage With "a Great Deal of Trust" in Various Organizations to Conduct Genetic Research (by Race/Ethnicity).

As indicated in Figure 1.2, whites had the highest level of trust for universities and medical institutions (medical institutions) (42%). Hispanics had the highest level of trust in organizations like the American Cancer Society or the March of Dimes (health organizations) (46%). Whites and Asians rated medical institutions and health organizations at about the same levels (41% and 42% for whites; 35% and 33% for Asians). African Americans and Hispanics had wider disparities between these categories (24% and 34% for African Americans; 32% and 46% for Hispanics). Whites and African Americans rated the federal government and drug companies approximately the same (10% and 9% for whites; 7% and 7% for African Americans). Hispanics and Asians, however, had much more trust in the federal government than in drug companies (20% and 10% for Hispanics; 23% and 10% for Asians). No other demographic variables have been identified to account for this greater level of trust in the federal government, including country of birth and language spoken at home. Two possible explanations are that the variation is related to cultural factors that need to be explored through qualitative research or that the variation is attributable to the heterogeneity of these ethnic categories and therefore additional, more targeted, survey research is needed.

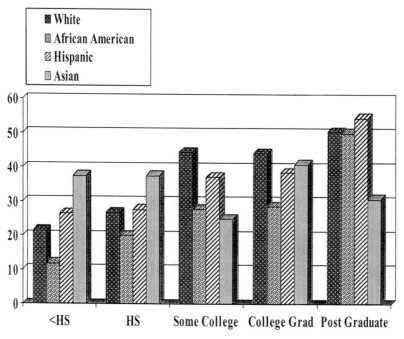

FIGURE 1.3. Percentage With "a Great Deal of Trust" in Medical Schools to Conduct Genetic Research (by race/ethnicity).

Adding education to race/ethnicity produces additional valuable information. As indicated in Figure 1.3, for universities and medical schools, trust levels increased with education for all race/ethnicity groups. Figure 1.4 indicates that for health organizations, the other research entities in which there was the most trust, education did not correlate with increased trust. Trust generally increased with education up until "some college" and then declined. For the groups with the least trust, the patterns are even more unusual. Figure 1.5 indicates that, with regard to the federal government, the trust level of whites declined modestly with education up to "some college" and then increased modestly. For Hispanics, trust was highest at the "some college" level. For African Americans, except for a modest spike at the "some college" level, trust in the federal government declined with education.

A different, but also unusual, pattern emerged with respect to drug companies. Figure 1.6 illustrates that "some college" was the great equalizer among the groups. For Hispanics, those who were not high school graduates and those who were college graduates had the most trust. Whites who were not high school graduates had the most trust, but the other education levels were about the same except for a slight increase among college

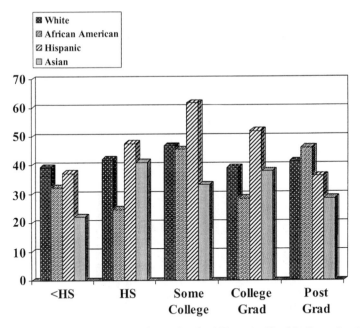

FIGURE 1.4. Percentage With "a Great Deal of Trust in Health Organizations" to Conduct Genetic Research (by race/ethnicity).

graduates. For African Americans, after "some college," trust declined precipitously with increased education in a pattern similar to African Americans' lack of trust for the federal government.

In an effort to determine the relative importance of demographic and socioeconomic factors as determinants of an individual's trust in these institutions and organizations to conduct genetic research, we created dummy variables (i.e., 0/1 coded indicator variables) for each category of race/ethnicity, gender, age, education, and income. The dummy variables were included in a logistic regression equation to predict the binary outcome "a great deal" or "some trust" for each institution. A forward selection procedure was used to identify statistically significant predictors.

Racial/ethnic group membership was expected to be an important factor in an individual's level of trust of these institutions. Controlling for gender, age, education, and income, African Americans were about 40% less likely than whites to trust universities; Asians and Hispanics were nearly twice as likely as whites to trust the federal government. However, race/ethnicity was not a factor in trust in the pharmaceutical industry nor in trust in health organizations.

Trust in universities and medical schools to conduct genetic research is most strongly associated with an individual's educational level. Race/

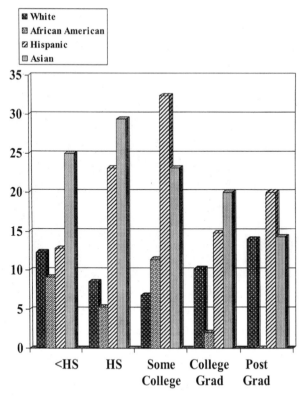

FIGURE 1.5. Percentage With "a Great Deal of Trust" in the Federal Government to Conduct Genetic Research (by Race/Ethnicity).

ethnicity, age, and income are not significant predictors of trust in universities and medical schools. In contrast, the proportion expressing trust in the pharmaceutical industry is most strongly related to income and to having a graduate degree. Race, age, and education are not as important. Finally, trust in health organizations to do genetic research is related to income, with greater numbers of individuals expressing trust at higher levels of income. It is noteworthy, however, that significantly fewer of those with less than a high school education and respondents older than age 70 express trust in health organizations.

These data demonstrate that concerns about conducting research in a culturally sensitive and responsive manner are well placed (Fisher, 1996; Foster et al., 1998; Greely, 1997). Perhaps the most striking feature of the data is the absence of trust in the federal government and drug companies by African Americans at the highest education levels. This may well be a legacy of Tuskegee and other historical instances of exploitative research of which

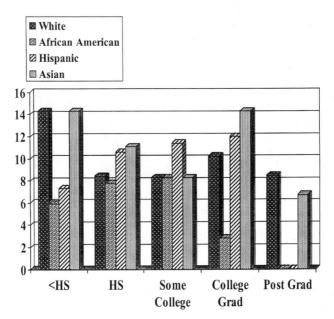

FIGURE 1.6. Percentage With "a Great Deal of Trust in Drug Companies" to Conduct Genetic Research (by race/ethnicity).

individuals with the most education are most likely to be aware (Jones, 1993; Reverby, 2000).

C. PERCEPTIONS ABOUT THE AFFORDABILITY OF GENOMIC-BASED MEDICINES

The cost implications of pharmacogenomics, both at the individual and societal levels, are the subject of considerable debate, as discussed in detail in Chapter 12 of this book. One possible scenario is that pharmacogenomic-based medications will be more expensive and therefore not as widely available as pharmaceutical products intended for wider distribution (Rai, 2001; Rothstein and Epps, 2001). We attempted to determine whether members of the public believed that they would be able to afford these new pharmaceutical products. To do so, we asked the following question (Question 9):

If medicines were developed that were matched to the genetic makeup of individuals, do you think that people of your income level could afford them?

The response options were yes, no, and don't know. Refusals also were recorded. Figure 1.7 plots the responses to the question by income level.

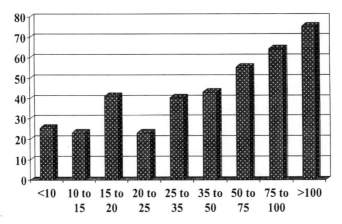

FIGURE 1.7. Percentage Believing They Could Afford Genetically Matched Drugs (by income).

These data are interesting in both relative and absolute terms. On the low income side, about 25% of individuals with incomes of less than $10,000 said that individuals of their income level could afford the new drugs. This represents a surprisingly high figure in light of the fact that it is unlikely that individuals with such low income could afford today's—presumably lower priced—medications. On the high income side, only 75% of individuals with incomes greater than $100,000 said that individuals of their income level could afford them, meaning that fully 25% of individuals in the highest income category believed that individuals of their income level could not afford pharmacogenomic drugs. In fact, not until the income category of $50,000–$75,000 do more than 50% of respondents believe that individuals of their income level could afford the new medicines. Of course, nobody, including the respondents, knows how much the new drugs will cost, and no specific prices were mentioned in the interviews. Not surprisingly, 17.2% of respondents said they did not know whether they could afford the new medicines.

The slope of the curve in Figure 1.7 also is noteworthy. The responses for income brackets of $15,000–$20,000 and $20,000–$25,000 seem incongruous. Superficially, either the former figure is too high or the latter figure is too low. One possible explanation is that the $20,000–$25,000 range represents the working poor, whose income is too high to qualify for Medicaid and similar need-based health care, but whose relatively low-paying jobs do not include any health benefits or do not include adequate health benefits for prescription drugs (Kuttner, 1999). Figure 1.8 supports this explanation. Individuals in the $15,000–$20,000 income range report higher levels of

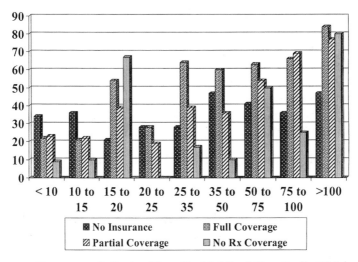

FIGURE 1.8. Percentage Believing They Could Afford Genetically Matched Drugs (by Insurance Coverage).

health coverage, including partial or full prescription drug coverage, than individuals with incomes in the $20,000–$25,000 range.

D. Concerns About the Confidentiality of Genetic Information

In a provocative article, Professor Sonia Suter raised the issue of whether genetic discrimination is largely a concern of the middle and upper classes.

> [G]enetic discrimination is principally a concern of the middle to upper classes, who have financial resources for testing and jobs and insurance they fear losing. This group of well-educated, well-off individuals has lobbied heavily for genetic legislation. . . . In short, genetics legislation becomes another middle-class entitlement (Suter, 2001).

Is she right? Although our survey was conducted before publication of Professor Suter's article, our research sheds light on her assertion. We attempted to determine the level and likely consequences of public concern about the disclosure of genetic information by asking the following questions (Questions 10A, 10B, and 10C):

A. If your employer could get the results of a genetic test that showed whether you were more likely to get sick in the future, what impact, if any, would this have on your willingness to take the test?

B. If your health insurance company could get the results of a genetic test that showed whether you were more likely to get sick in the future, what impact, if any, would this have on your willingness to take the test?

C. If your life insurance company could get the results of a genetic test that showed whether you were more likely to get sick in the future, what impact, if any, would this have on your willingness to take the test?

For each of the three questions (asked in random order), individuals were given the following choice of responses: much more likely, somewhat more likely, have no impact, somewhat less likely, much less likely, and don't know. Refusals also were recorded.

The data are quite revealing. To begin with, approximately 70% of respondents said that disclosure of test results to any of the third parties would make them less likely to take a genetic test. The responses were quite similar for employers, health insurers, and life insurers. This public view of equal concern for disclosure to employers, health insurers, and life insurers contrasts with the pattern of legislation enacted to protect the confidentiality of genetic information. Although there is a difference between privacy legislation and nondiscrimination legislation, they are related (Rothstein, 1997). Thus it is interesting to note that legislation prohibiting the use of genetic information in group health insurance has been enacted at the federal level (HIPAA, 2002); an overwhelming majority of states have enacted legislation prohibiting discrimination in individual health insurance (NCSL, 2002a); a little over half the states have enacted legislation prohibiting genetic discrimination in employment (NCSL, 2002b); and very little legislation has been enacted dealing with life insurance (NCSL, 2002c).

Figure 1.9 indicates the responses to Questions 10A, 10B, and 10C by race/ethnicity. Approximately 75% of whites said they would be less likely to take a genetic test if the results were disclosed, with a slightly higher percentage concerned about disclosure to life insurance companies. For African Americans, the overall percentage less likely to take a genetic test is approximately 60%, with a wider variation among categories, ranging from 63% for employers, to 60% for life insurers, and 55% for health insurers. For Hispanics, the less likely percentage falls below 50% for all categories: 49% for employers, and 42% for both health insurers and life insurers. For Asians, the widest variation is in the different "less likely" categories, ranging from 61% for life insurance to 42% for employers.

Another interesting feature of the data presented in Figure 1.9 is the surprisingly high percentage of individuals who actually said that disclosure of the results would make them *more likely* to undergo genetic testing. The percentage of individuals more likely to take a genetic test is approximately 10%

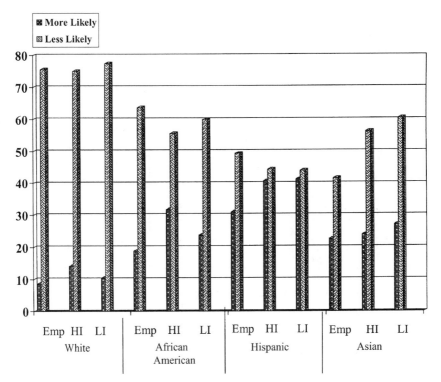

FIGURE 1.9. Percentage Willing to Undergo Genetic Testing if Results Were Available to Employers, Health Insurers, and Life Insurers (by race/ethnicity).

for whites, but it is much higher in other groups. For Hispanics, for both health and life insurance, disclosure of results was viewed as a positive factor by almost as many individuals as viewed it as a negative factor. We assume that respondents who viewed disclosure as a positive factor either did not understand the question or did not appreciate the possible negative consequences of disclosure.

Because of the high percentage of individuals who responded that having genetic test results shared with insurers and employers would be viewed as positive factors in deciding whether to undergo genetic testing, and the seeming correlation with race/ethnicity, we analyzed the influence of education and income of the responses. Figures 1.10 and 1.11 demonstrate clearly that for all groups, as education and income levels increase, disclosure of test results to employers, health insurers, and life insurers makes individuals less likely to be tested. Because race/ethnicity, education, and income are related, we used statistical analysis to determine whether it is possible to identify which of these variables is most predictive.

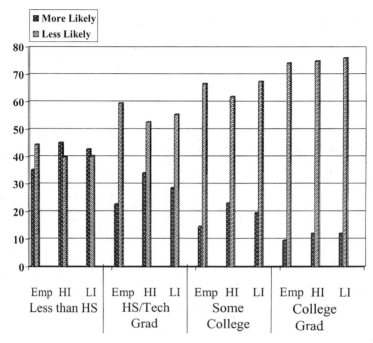

FIGURE 1.10. Percentage Willing to Undergo Genetic Testing (by education).

In the multivariate analysis, African Americans were nearly twice as likely as whites (OR = 1.88, p = 0.013), whereas Asians were 2.6 times (p = 0.001) and Hispanics 3.1 times ($p \leq 0.001$) as likely as whites to have such a test. These results suggest that confidentiality of genetic test results and possible discrimination by employers is not a salient concern of racial/ethnic minorities, but it is a concern among whites.

When asked about having a genetic test if the results were available to health insurers, the likelihood of having the test followed a similar pattern with a few notable exceptions. As was the case when the results would be available to employers, the likelihood of taking the test decreases with higher levels of education. However, when it comes to health insurers getting the results, the decrease in the likelihood of participating does not occur until an older age (i.e., above 50) and is also related to having a higher income and being male. Nevertheless, African Americans remain twice as likely (OR = 2.00; p < 0.001) and Hispanics 2.4 times (p > 0.001) as likely as whites to undergo genetic testing, again suggesting that the confidentiality and potential discrimination issues may be more important to whites. The decreased likelihood of men participating may be related to their being more

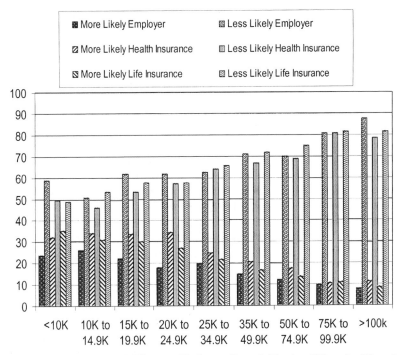

FIGURE 1.11. Percentage Willing to Undergo Genetic Testing if Results Were Available to Employers, Health Insurers, and Life Insurers (by income).

likely than women to carry health insurance coverage for their families and thus the heightened concern over the possible loss of coverage if insurance companies knew of a genetic link to disease. To test this further, we added a dummy indicator variable for those with full or partial health insurance coverage and re-ran the logistic regression analysis. The results for the demographic and socioeconomic variables were essentially unchanged and, as expected, having at least partial health insurance coverage was associated with an additional 25% decrease in the likelihood of reporting a willingness to have a genetic test when the results would be available to the health insurer.

The final question in this section of the interview inquired into the respondents' willingness to have a genetic test if the results were available to their life insurance company. Again, the likelihood of participation decreases with years of education and is significantly higher among racial/ethnic minorities compared with whites. What is noteworthy here is the finding that respondents who have total family incomes above $50,000 are less likely to have a genetic test if the results would be available to their life insurance company.

The pattern of results found here is surprising and at the same time internally consistent. What is noteworthy is the finding that white respondents, not members of racial/ethnic minority groups, are least likely to say that they would participate in genetic testing when the results of the test would be available to employers, health insurers, and life insurance companies. Besides the consistent finding for race/ethnicity, older age, particularly in the productive years between 40 and 60, as well as higher levels of education, are associated with a reduced likelihood of saying that they would participate. When results would be available to life insurance companies, individuals respond to participating in genetic testing in terms of their economic interests. Individuals with higher incomes are more likely to have life insurance and therefore have more to lose if life insurance companies have access to the results of genetic testing.

Another important demographic variable to consider is gender. Many of the state genetic nondiscrimination laws were enacted with lobbying support from breast cancer advocacy groups (Rothenberg et al., 1997). Does this fact suggest that women would be less likely than men to undergo genetic testing if the results could be obtained by employers, health insurers, and life insurers? Figure 1.12 indicates that this is not the case, and, in fact, men are slightly less likely than women to undergo genetic testing, especially where the results are disclosed to health insurers.

E. STABILITY OF VIEWS DURING THE INTERVIEW

We recognized that, despite the recent coverage of pharmacogenomic research in the media, the concept of genetic variability and pharmacogenomics may be new to most respondents. Therefore, we attempted to measure whether additional information about genetics and pharmacogenomics would affect the attitudes of individuals. To do this, we used a test-retest question. After obtaining informed consent, the first question (Question 1) we asked was:

> First, do you feel the study of genetic differences in the way individuals respond to specific medicines is. . . .

The options for answering were as follows: a good thing, more good than bad, equally good and bad, more bad than good, a bad thing, and don't know. Refusals were also noted.

During the course of the interview, a variety of information was given to respondents in framing and asking questions. This information included how genetic testing is performed and the purpose of genetic testing (Question 2), the use of genetic testing for prescribing appropriate medica-

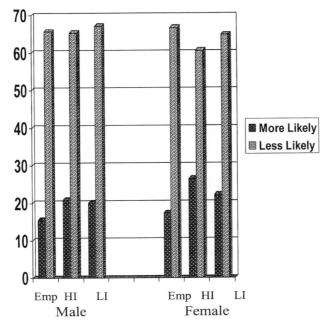

FIGURE 1.12. Percentage Willing to Undergo Genetic Testing if Results Were Available to Employers, Health Insurers, and Life Insurers (by gender).

tions (Question 3), genetic variation among individuals and groups of individuals (Question 7), and the use of genetic tests to make predictions about future health risks (Question 10). At the end of the substantive questions and before we asked for demographic information, we asked as Question 12 virtually the same question as we asked previously as Question 1:

> *Based on your prior knowledge and everything we have discussed in this interview, do you feel the study of genetic differences in the way individuals respond to specific medicines is. . . .*

The answer options were the same as for Question 1.

Figure 1.13 indicates that most of the change was in a positive direction. Individuals who initially answered that the study of genetic difference was a good thing were the most likely to repeat their answer on retest. Those who said it was a bad thing were most likely to change their answer. Figure 1.14 shows that African Americans were least likely to keep the same answer (54%), with 25% becoming more favorable and 19% becoming less favorable. Asians had the largest increase in positive versus negative, with 24% becom-

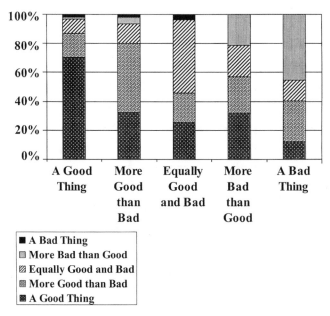

FIGURE 1.13. Response Shift During Interview in Views of Genetic Research.

ing more positive and 13% becoming more negative. Hispanics were the only group where the retest showed an increase in those who viewed the study of genetic difference less favorably, 31% to 17%.

More than 10% ($N = 183$) refused to answer (0.2%) or responded that they didn't know (9.9%) when asked the initial question of whether they felt the study of genetic differences in the way individuals respond to specific medicines is a good thing, more good than bad, equally good and bad, more bad than good, or a bad thing. These individuals were disproportionately female ($p = 0.046$), older ($p = 0.000$), and with less education ($p = 0.0003$) than those who gave an opinion about the value of genetics research.

Despite not expressing an initial opinion about whether research into genetic differences was a "good" or a "bad" thing, nearly all of these individuals expressed their opinion on the balance of questions about genetic research. In general, their opinions on those other questions did not differ significantly from other respondents. What is noteworthy is that 92.7% of individuals who did not have an opinion at the start of the interview expressed an opinion at the end of the interview, and more than 45% said that genetic research was a good thing (28.7%) or more good than bad (17%). Only 9.7% changed from "don't know" to a negative opinion of genetic research, and another 37.3% changed from "don't know" to perceiving genetic research as equally good and bad.

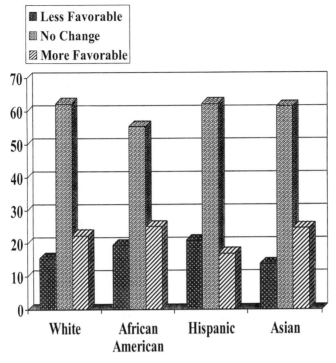

FIGURE 1.14. Response Shift During Interview on Views of Genetic Research (by race/ethnicity).

V. CONCLUSION

The survey revealed a public generally interested in genetics, optimistic about the prospects of genetic research to develop improved medications, and willing to participate in genetic research. For individuals who were unsure whether genetic research was a good or bad thing at the start of the interview, most were positive by the end of the interview. Many respondents however, also indicated a considerable degree of confusion about the possible consequences of third parties getting access to genetic information, caution about the roles of certain institutions engaged in genetic research, and a lack of confidence about whether they will be able to afford new pharmacogenomic-based medications.

Although numerous demographic variables were explored, individually and in combination, the only three variables with consistent statistical significance were race/ethnicity, income, and education. Age and gender were relevant to a lesser degree. Virtually none of the other variables, including health status, was of statistical significance.

The survey suggests the following four areas in need of further exploration and action: (1) increased public education about science and genetics; (2) greater cultural sensitivity in conducting research and providing genetic services; (3) continued reassessment of assumptions about public attitudes regarding research, confidentiality, discrimination, and related matters; and (4) additional research about consumer attitudes, including surveys in more ethnically homogeneous minority groups.

REFERENCES

Fisher, N.L., *Cultural and Ethnic Diversity: A Guide for Genetics Professionals*, Johns Hopkins University Press, Baltimore, 1997.

Foster, M.W. et al., "A Model Agreement for Genetic Research in Socially Identifiable Populations," *Am. J. Human Genetics* **63**, 696–702 (1998).

Greely, H.T., "The Control of Genetic Research: Involving the 'Groups Between,'" *Houston L. Rev.*, **33**, 1397–1430 (1997).

Peter D. Hart Research Associates, Inc., for the Academic Medicine Development Co. (2001), available at www.lasker_pcc_detail2.cfm?issue_type=medical_research&concern_genetics visited 4/25/02).

Jones, J.H., *Bad Blood*, Free Press, New York, 1993.

Kuttner, R., "The American Health Care System—Health Insurance Coverage," *N. Engl. J. Med.*, **340**, 163 (1999).

Los Angeles Times (2000), available at www.lasker_pcc_detail2.cfm?issue_type=medical_research&concern_genetics (visited 4/25/02).

NCSL 2002a, National Conference of State Legislatures, available at *www.ncsl.org/programs/health/genetics/ndiscrim.htm*

NCSL 2002b, National Conference of State Legislatures, available at *www.ncsl.org/programs/health/genetics/ndishlth.htm*

NCSL 2002c, National Conference of State Legislatures, available at *www.ncsl.org/programs/health/genetics/01life.htm*

National Health Council, *Perceptions of Genetics Research: A Focus Group Study* (2000).

Rai, A.K., "The Information Revolution Reaches Pharmaceuticals: Balancing Innovation Incentives, Cost, and Access in the Post-Genomic Era," *U. Illinois L. Rev.*, **2001**, 173–210 (2001).

Rasmussen Research (2000), available at www.lasker_pcc_detail2.cfm?issue_type=medical_research&concern_genetics (visited 4/25/02).

Reverby, S.H. (ed.), *Tuskegee's Truths: Rethinking the Tuskegee Syphilis Study*, University of North Carolina Press, Chapel Hill, NC, 2000.

Rothenberg, K. et al., "Genetic Information and the Workplace: Legislative Approaches and Policy Challenges," *Science*, **275**, 1755–1756 (1997).

Rothstein, M.A., (ed.), *Genetic Secrets: Protecting Privacy and Confidentiality in the Genetic Era*, Yale University Press, New Haven, CT, 1997.

Rothstein, M.A. and P.G. Epps, "Ethical and Legal Implications of Pharmacogenomics," *Nature Rev. Genet.*, **2**, 228–231 (2001).

Suter, S.M., "The Allure and Peril of Genetics Exceptionalism," *Wash. U. L. Q.*, **79**, 669–741, 721 (2001).

Pharmacogenomics: Pharmacology and Toxicology in the Genomics Era

Harvey W. Mohrenweiser, Ph.D.

I. History and Overview of the Genome Project

A. Initiation of the Genome Project

The biological and biomedical research world has undergone major changes in the last three decades. The development of molecular biological techniques that allowed the study of the basic features of the genetic material of cells changed the focus of the questions that could be asked from observations of consequences to studies directed toward defining the fundamental mechanisms of cellular biology. The availability of these techniques supported a revolution in biological research with the initiation of the Human Genome Project (HGP) in the mid- to late 1980s. The HGP was one of the first ventures of biology into "managed" high-visibility "big science." The genome effort did not emphasize specific questions nor did it address specific hypotheses, but instead emphasized large-scale, goal-driven, global data generation. This has also been called "industrial scale" data generation. The focus on large-scale, global data generation was coupled with the expectation that the data would be rapidly placed in the public domain. The HGP and its major funding agencies, the National Institutes of Health (NIH) and

Pharmacogenomics: Social, Ethical, and Clinical Dimensions, Edited by Mark A. Rothstein.
ISBN 0-471-22769-2   Copyright © 2003 Wiley-Liss, Inc.

the Department of Energy (DOE) in the United States and the Wellcome Trust in the United Kingdom, encouraged technical advances, development of instrumentation, and new experimental approaches as required to attain previously defined goals. The funding agencies also "encouraged" interaction, if not collaboration, among the laboratories being funded. The HGP moved the frontiers of biological and biomedical research, but it was as much a new approach to research as it was new science. It involved strong management by the funding agencies and large investments in a limited number of laboratories, rather than the more traditional investigator-initiated, hypothesis-focused funding strategies that are still the focal point for most biological research. The rapidity of the revolution is apparent when one realizes that the first recombinant DNA cloning experiments were conducted in 1972–1974 and DNA sequencing methods were developed in 1975–1977. Automated sequencing instrumentation became available in 1983, and the polymerase chain reaction (PCR) amplification technique was described in 1985. An article by Friedrich (1996) includes a dateline of a number of the significant events in molecular biology and technology that were necessary for the sequencing of the human genome. It emphasizes the rapidity with which the instrumentation and technology were applied in the effort to sequence the genome. Three articles in the journal *Genomics* in 1989 provide interesting historical perspectives of the initiation of the HGP (Cook-Deegan, 1989; Watson and Jordan, 1989; Barnhart, 1989).

B. IMPACT OF THE GENOME PROJECT ON RESEARCH

The directed, goal-focused influx of funding for the HGP has resulted in the generation of vast quantities of data, much of it in publicly accessible databases. This availability became functional reality when scientists began to have real-time access to these data from their offices via the World Wide Web in 1993–1994. The availability of both the technologies and the data that have accumulated in the last decade is facilitating rapid advancements in many disciplines. The availability of vast quantities of largely unanalyzed data has also generated a need for people with new skills and new disciplines. The field of bioinformatics has grown from the need to handle, analyze, and utilize large volumes of data. The data are often analyzed by individuals and groups not involved in data generation. An additional change in the scientific community since the early 1980s is in the scale of the involvement of the private sector, as both collaborators and competitors with the academic community, in data generation and especially in data analysis.

    The rapid successes of the HGP have resulted in the proliferation of other "–omics" projects. Discussions now center on the need to initiate similar projects, with names such as functional genomics, proteomics, structural genomics, toxicogenomics, and pharmacogenomics. These second-generation "genome" initiatives are proposing to emulate the large-scale, global data generation strategies employed in the HGP to determine the structure of large numbers of proteins (structural genomics), the physical interaction of all proteins (proteomics), and the function of all genes and their protein products (functional genomics). These –omics efforts emphasize application of high-throughput technologies to generate data to meet specific goals. As with the HGP, it is generally expected that the raw data from these new initiatives (if funding is from the public sector) will be rapidly deposited in databases for subsequent and additional analysis by other investigators.

C. PHARMACOGENOMICS

For the purposes of this discussion, pharmacologic research is very broadly defined and considered to include three areas. The first area is drug discovery or development. The second area is often described as pharmacogenetics. This is the study of differences in efficacy or toxicity of drugs and treatment protocols in different individuals and is increasingly focusing on understanding the genetic basis and biological mechanisms for hypo- and hyperresponse to pharmacologic agents. The third area, disease prevention, is not often included in discussions of pharmacogenetics and pharmacogenomics, but disease prevention is the ultimate treatment. As with individual variation in response to pharmacologic agents or drugs, there is also variation among individuals in the risk of disease following an exposure. Genetic susceptibility to disease is mechanistically closely coupled to pharmacogenetics and often involves the same sets of genes. The major difference between the two areas is the characteristics of the exposure. Drug exposures are well characterized and quantified, whereas exposures to pollutants are usually neither well characterized nor quantified. Studies of disease susceptibility have been included under such rubrics as environmental genomics and toxicogenomics rather than being considered a component of pharmacogenomics.

Pharmacogenomics can also be described as the application of high-throughput technology to generate data on a global scale at the interface of genetics, disease, toxicology, and pharmacology. An even more encompassing view would describe pharmacogenomics as an effort to (1) emulate the strategies of the HGP and (2) utilize the fruits of the HGP to reduce the incidence of disease and improve the practice of medicine, while reducing health care costs. As with the HGP, pharmacogenomics is generally a new approach to research and specifically data generation rather than new science. The questions and hypotheses of interest to pharmacology and toxicology have been with us for some time. This chapter focuses on describing the products of the HGP that are available for application and utilization in pharmacologic and toxicological research. It emphasizes the new tools that have been or are being added to our toolkit; other chapters emphasize the potential applications of the tools. The examples of utilization of information and technology of the genomics era were selected to present concepts and are not intended to be a review of any specific area.

II. ACCOMPLISHMENTS OF THE HUMAN GENOME PROJECT

A. PHYSICAL MAPS AND GENOMIC SEQUENCE

The pronouncements in early 2001 of the completion of the draft of the sequence of the human genome (Lander et al., 2001; Venter et al., 2001) highlighted the most visible of the accomplishments of the HGP. The sequence available at the end of 2001 covers over 95% of the genome. This is a significant accomplishment and a tremendous resource for other scientists, but it will still require significant continued effort to obtain high-quality sequence in the remaining regions of the genome, especially as many of the gaps have "interesting" features that make them difficult to clone and sequence. The analysis and annotation of the genome sequence will be a long-term effort. The major focus of the analysis thus far has been the search for open reading frames (ORFs) or coding regions, that is, the genes. The early computational analyses of the human genome predicted the existence of only 25,000–40,000 genes in the human genome (Lander et al., 2001), although this number has increased to 60,000–75,000 predicted genes in subsequent analyses (Hogenesch et al., 2001). This is only several times the number of genes identified during similar analysis of the genomes of organisms with simpler genomes that have been sequenced, such as *Drosophila* (the fruit fly) (Adams et al., 2000), *Arabidopsis* (a plant) (Bevan et al., 2001), and yeast (Goffeau et al., 1996). The analysis of expressed sequences suggests that 75,000–125,000 genes should be identified in the human genome (Liang et al., 2000). Irrespective of which estimate of gene number is closer to the ultimate truth, the number of genes encoded by the genome sequence is an underestimate of the cellular complexity. Many genes undergo differential splicing to generate a number of different transcripts from a single coding sequence or gene, and these transcripts often encode proteins with different functions. In addition, many proteins undergo modifications that may again change the functional characteristics of the protein. The number of different proteins is much larger than the number of genes. Although the sequence analysis is critical, it is only the first step in understanding the complexity of the functional machinery of the cell. At this stage of analysis, approximately 50% of the ORFs identified in the human genome encode proteins of unknown function. The proportion of genes with unknown function in human and other mammalian organisms is not dissimilar to the current state of analysis of the genomes of other organisms such as *Drosophila* and yeast. Gene identification and annotation remains a significant task in the effort to increase the utility of the genomic sequence for biological studies.

The next analytical challenge is the identification of the genomic regions responsible for regulation of gene expression. These are the sequence elements that, in response to specific signals, direct the timing of expression of genes during development and also the tissue and cellular localization of gene expression. Other sequence elements respond to signals to regulate the level of gene expression under different conditions or treatments. Differences in gene expression are associated with variation in response to drugs (Hustert et al., 2001). Disruption of appropriate gene expression is often a key early feature of many diseases. Understanding the processes involved in the regulation of gene expression, which is one aspect of the larger area of cellular communication and signaling, is a key to understanding disease processes. Having a catalog of all genes, along with knowledge of gene function and the mechanisms of regulation of gene expression, is expected to be a cornerstone for the development of new drugs and improved treatment protocols for intervening in disease processes.

B. DEVELOPMENT OF GENETIC MAPS

Genetic maps of the human genome have developed in parallel with the physical maps and many of the advances in molecular technology. Two genetic maps are being extensively employed today to localize the genetic component of disease phenotypes and characteristics to specific chromosomal regions. Tandem repeats of short stretches of repetitive sequence, for example, di- and trinucleotide repeats, are unstable, and thus extensive genetic heterogeneity has been generated during the replication of these regions over evolution. The microsatellite map was built before the availability of the draft sequence (Sheffield et al., 1995; Gyapay et al., 1996). The general strategy for building the microsatellite map was to identify clones containing more than 8–10 copies of the tandem repeat element(s). After the sequencing of a sufficient region flanking the repeat region to support the design of PCR primers, the extent of genetic variation in a sample of the population was estimated. The subset of loci exhibiting extensive polymorphic variation was then mapped to a chromosome, and the linear order was determined, relative to other genetic markers or loci. Many alleles, each with a different number of repeat elements, segregate at each of these loci; this highly informative genetic map is the platform for the rapid mapping of genetic traits to chromosomal regions (Weber and Broman, 2001). The microsatellite loci are most powerful for the mapping of traits or diseases segregating in large families. Sets of 200–400 well-characterized markers are being routinely utilized to scan the entire human genome in initial efforts to map the genes for specific diseases, phenotypes, and traits.

The availability of the sequence of the genome provides a platform for building even higher-resolution genetic maps, with the ultimate resolution being the single nucleotide. Thus recent attention has focused on building maps supported by differences among individuals in nucleotide sequence. Several strategies have been employed to identify single-nucleotide poly-morphisms (SNPs). The availability of the draft sequence of much of the human genome and rapid DNA sequencing technologies have made it fea-sible to screen or sequence a large number of different regions of the genome to identify sequence variation among unrelated individuals (Wang et al., 1998). The largest effort has been undertaken by the SNP Consortium, an effort funded by a number of pharmaceutical companies. This effort has identified some 1.42 million SNPs (Sachidanandam et al., 2001). These data have been deposited in a database (dbSNP) at the National Center for Biotechnology Information (NCBI). Most of these variants were identified by resequencing many different regions of the genomes of a small number of individuals. The individuals screened are from a resource selected to be a sampling of the gene pool of the U. S. population (Collins et al., 1998). A smaller number of the SNPs in this database were identified via searches of sequence databases, as described below. As only a small number of chromosomes were screened for variation, only the more highly poly-morphic variant alleles are generally well represented in this collection.

This extensive screen for variation confirmed that SNPs are the most common class of genetic variation. It is estimated that sequence differences exist at least every 1000–1500 nucleotides, with some calculations suggest-ing an even higher density of SNPs (Lai, 2001). This is an estimated 2–3 million sequence differences between the genomes of any two individuals. Although SNPs are closely spaced along the chromosome, each SNP consists of usually only two alleles, the wild-type, common, or most frequent and the variant or less common nucleotide. Note that wild-type and variant nucleotides are defined by relative allele frequencies and without regard to relative functionality or other attribute. The sequence in GenBank or other databases may be either wild type or variant, depending on the source of DNA sequenced. The microsatellite variants are less common than SNPs but have higher content of genetic informativeness, because 5–10 variant alleles exist at many of the microsatellite loci. Even these very high-density maps exhibit regional differences along the chromosomes in marker density and recombination rates or the relationship of physical or genomic distance to genetic distance (Yu et al., 2001; Broman et al., 1998). These differences between physical and genetic distances make it difficult to refine the local-ization of phenotypes in some regions of the genome. The molecular and biological basis for this organizational heterogeneity remains elusive.

C. CLASSES OF VARIATION ASSOCIATED WITH ELEVATED DISEASE RISK

These genetic maps have been the backbone for the highly successful endeavors of the molecular genetics community to identify highly penetrant disease genes segregating in "disease families" by positional cloning. In these families, specific variants at a locus are often associated with a more than 50- to 100-fold elevation in relative risk of disease, compared with other family members with the "normal" allele and the general population. Examples of disease genes identified by positional cloning include the breast cancer genes BRCA1 and BRCA2 (Miki et al., 1994; Tavtigian et al., 1995). These are examples where many different variants, but not all of the variants, in a gene are associated with high risks of cancer in the affected individual. This is also an example of the situation where genetic variation in either of two different genes is sufficient to place an individual at high risk of what was considered one disease, breast cancer. It is now known that the etiology of the disease and the biological characteristics of the tumors differ between BRCA1 and BRCA2 patients. Although the function of these genes is not completely understood, it is becoming clear that they have roles in ensuring that damaged DNA is repaired before the damaged DNA is replicated and transmitted to a (potentially precancerous) daughter cell. Understanding the genetics of breast cancer in these families has provided insight into the pathology of the disease, and knowledge of the function of the genes is providing information important to understanding the mechanisms and biological processes involved in tumor development. It is anticipated that this knowledge will result in more rational and more rapid drug design and treatment protocols for breast cancer.

A more complex example of the relationship of genetic variation to disease is the CACNL1A4 gene. This gene encodes a protein localized in the cell membrane and has a role in the transport of calcium into the cell. In this case, three different types of DNA sequence changes in a single gene are responsible for high risk of three different neurological conditions, including a specific class of migraine headache (Ophoff et al., 1996; 1998; 2001). Each class of variation disrupts the protein structure differently, and thus has a different impact on the ability of the protein to transport calcium and the degree of normal cellular function. An even higher level of complexity in the relationship of genetic variation to disease is observed for Alzheimer disease, where a subset of variants in one of three different genes, β-amyloid precursor and presenilin 1 and 2, places an individual at high risk of disease (Saunders, 2001; Sorbi et al., 2001). In addition, individuals with a variant allele at APOE, a gene involved in cholesterol metabolism, are at smaller but still significantly elevated risk of disease, even when the individuals have normal copies of all of the major Alzheimer disease genes. A fourth pattern appears to be emerging from the study of individuals with variant alleles of BRCA1 and BRCA2, in that variation at other genes, that is, the genetic background of the individual, influences the age-specific risk of breast cancer (Runnebaum et al., 2001; Wang et al., 2001). It is now known that specific variants in BRCA2 are associated with a high risk of early-onset breast cancer in some families, whereas in other families these same variants are being observed in elderly individuals who are apparently cancer free.

The availability of the sequence for much of the human genome is a major resource for positional cloning and identification of genes responsible for genetic diseases. In earlier days (really only 5 years ago), after a gene had been localized to a chromosomal region, it was usually necessary to build genetic and physical maps spanning the regions before focusing on gene identification. Reading the description of the cloning of the cystic fibrosis gene provides insight into the impact of the availability of the genomic sequence for much of the human genome (and high-resolution genetic maps described below) on gene identification and characterization (Riordan et al., 1985). Today, positional cloning efforts can often move rapidly from the mapping stage, which identifies the region of the genome containing the genetic variation (gene) associated with the phenotype or disease, to the stage of analyzing the most interesting candidate genes in the region. This can now often be accomplished by harvesting data available in public databases and without the need to build additional genetic maps or isolate and sequence DNA clones spanning a region of interest to identify potential candidate genes. The impact of these changes has been described in an article entitled "Leaving Kansas . . . Finding Genes in 1997" by Dr. Mary-Claire King (King, 1997), which was written before the draft sequence of the human genome was available.

The genetics, and thus the cloning, of the genes for these disease examples was relatively straightforward. The "disease families" were initially identified because of the large number of affected individuals expressing an obvious phenotype, for example, severe disease with early age of onset. Because of the rarity of the variant alleles causing a disease, only a single locus was responsible for the disease in each family. In the initial families studied in the effort to clone the gene, all or most individuals with the variant allele exhibited the disease phenotype. It was the availability of the large number of affected family members and clear evidence for genetic transmission that attracted researchers to these families. The identification and characterization of these genes has provided great insight into disease processes and has resulted in identification of genes and proteins that are anticipated to be potential targets for intervention in disease processes. However, the variants in these genes have only limited impact on the total burden of a disease in the population because of the relative rarity of these alleles. It is estimated that less than 5% of the disease burden for even such common diseases as breast cancer and Alzheimer disease are the result of variants in these "disease" genes.

D. GENETIC VARIATION AND PHARMACOLOGY

The same story plays out for the toxicity and efficacy of drugs, where a number of examples of a role for genetic variation in pharmacologic response have been identified. One example is the relationship of genetic variation in the enzyme thiopurine methyltransferase with the efficacy and toxicity of the thiopurine drugs (Weinshilboum, 2001). Individuals with reduced-activity variants of this enzyme are at elevated risk of drug-induced toxicity from standard treatment regimens. These same dosages have reduced impact on disease or efficacy in other individuals with hyperactive variants of this enzyme, as the drug is rapidly inactivated in these individuals. Other well-characterized examples of clinically relevant variation include the genes in the CYP3A family and CYP2D6, genes encoding proteins with roles in the activation and detoxification or inactivation of many drugs. Several recent reviews and commentaries include discussions of the current state of application of pharmacogenetics in the clinical laboratory and medical practice (Wolf et al., 2000; Kalow, 2001a; 2001b; McLeod and Evans, 2001; Shi et al., 2001; Zanger et al., 2001; Ingelman-Sundberg and Evans, 2001). A common aspect of the examples described is the involvement of a single gene in the activation and/or detoxification of the drug. A second feature is the existence of a relatively small number of common variants at each locus, each with a significant impact on the level of enzyme activity. Thus it is relatively straightforward to relate genotype to phenotype or individual response.

Progress in the study of the role of genetics in the incidence of complex diseases and traits, often characterized by complex patterns of inheritance (sometimes referred to as non-Mendelian inheritance) and gene-gene interaction, has been more limited. The same is true for diseases resulting from gene-environment (or exposure) interaction. In the successes described above, although variation in different genes may be associated with disease, the aberrant functioning of one gene was sufficient to place an individual at high risk of disease. This is not the situation for the more complex diseases and most of the disease burden in the population, where disease risk for each individual is defined by the combination of alleles inherited from parents at several different genes. The picture becomes even more complex when it involves variation in a number of genes and the level of exposure to disease-causing agents. In this context, the environmental component or exposure can be controlled (prescribed drug) or uncontrolled (pollution) or a lifestyle factor (smoking, diet). For the more simple Mendelian traits, individuals in a family can usually be divided into high risk and low risk on the basis of genetic variation at a single locus. For complex diseases, an almost continuous gradient of individual risk will exist within a family, and especially within a population. The individual risk will reflect genotypes at multiple loci and the exposure. In these cases, some individuals in the population having the "at-risk" genotype will not exhibit disease because they have not been exposed to levels of a drug or toxic agent above the threshold necessary to induce disease. This is the situation for the thiopurine toxicity described above. Most individuals with reduced enzyme activity in the population are not exposed to thiopurines and thus exhibit no readily obvious consequence of the genetic variation. For individuals with reduced enzyme activity, the toxicity is reduced to normal levels, but the efficacy of the drug is nearly normal when the dosage is reduced (Weinshilboum, 2001). The existence of individuals with disease but with neither obvious genetic risks nor exposure to compounds illustrates the difficulty of identifying both the exposure and genetic factors. Nonsmokers have lung cancer, just at a much lower rate than smokers, and some heavy, long-term smokers are disease free.

Diseases with complex patterns of inheritance, and where genetic variation comes into play after an exposure, account for the vast majority of the disease burden in the population. Examples of both the strategies and the problems involved in identifying genes with roles in these complex diseases can be appreciated from reading recent papers describing efforts to identify genes associated with risk for prostate cancer (Nwosu et al., 2001), type 2 diabetes, (Cox et al., 2001), and asthma (Xu et al., 2001). Cox (2001) outlines the challenges in moving from the identification of a chromosomal region associated with an elevated risk of disease via linkage mapping to the definitive estimate of risk associated with specific variants in specific genes. It should be noted that "common" refers to the relatively high incidence of the disease in the population and "complex" describes the pattern of inheritance of the genetic factors; neither term relates to the clinical characteristics of a disease. It should be apparent that progress is being made in understanding cellular biology and disease processes. However, even for the simpler situations, we have much to learn about the relationship of genetic variation to disease susceptibility.

The research products and information from the HGP are the reagents and tools for tackling these problems. High-density genetic maps are critical for identifying the chromosomal regions, and ultimately the genes, involved in susceptibility to diseases with complex patterns of inheritance and diseases resulting from the interaction of genetic variation and exposure. Continued development of new approaches to experimental design and data analysis—and in some instances entirely new paradigms for experimental approaches and data analysis—will be required to address these complex and data-intensive problems (Ponder, 2001; Risch, 2000). These are the problems that must be addressed and conquered if the much-discussed potential development of individualized disease treatment regimes is to become a reality for the common diseases affecting the population.

E. THE CANDIDATE GENE APPROACH FOR IDENTIFYING RISK GENES

An alternative to the genome-wide scanning strategy for identifying susceptibility genes is increasingly being utilized. This strategy, often referred to as a "candidate gene approach," screens selected sets of genes deemed potentially relevant to a pharmacologic response or disease susceptibility for variation in DNA sequence. The selection process utilizes current (and imperfect) knowledge of the function of genes and disease processes. Only the coding segments or exons and the regions expected to have roles in regulation of expression of these genes are usually screened for variation. Variants in these regions are expected to have a higher potential to impact gene expression or protein function than do variants in intronic and inter-genic regions.

Two strategies have been used to identify variants in known genes. The first strategy is the utilization of data already available in public databases. The Integrated Molecular Analysis of Genomes and their Expression (I.M.A.G.E.) Consortium was initiated in 1993 as a collaboration to identify all of the genes in the genome. The goal was to isolate and sequence all (or most) of the messenger RNAs (mRNA) or expressed sequences from many different tissues of an organism. In the context of cell function, the mRNAs can be considered the molds required for the process of transferring the blue-print (the functionally relevant component of the genome sequence) into wheels and doors (the proteins), which are subsequently combined to make cars (cells). The mRNA is known to be related to a segment of the DNA sequence, but the knowledge is not complete. The sequence obtained for each mRNA isolated was called an expressed sequence tag (EST) and was basically a "signature" for each gene. The goal was a catalog of the genes in an organism. The relative distribution and quantity of the individual mRNAs among different tissues provides insight into tissue-specific patterns of gene expression. Additional information can be obtained at <http://image.llnl.gov/image/html/igoals.shtml>. Because the tissues used for isolating the mRNA were obtained from different individuals, differences in the sequence of clones should reflect genetic variation in DNA sequence among individuals. The Cancer Gene Anatomy Project (cGAP), described below, is an effort similar to the I.M.A.G.E. project, except that its goal was to isolate expressed sequences from many different types of tumors. And because tumors were derived from different individuals, this was another source of sequence data to be screened for genetic variants (Clifford et al., 2000). Thus, by development of computer algorithms to search the sequence data files for the cDNA clones encoded by each gene and then screening each clone for deviations from the consensus sequence, potential SNPs could be identified. As these SNPs reside in the coding region of the genome, many of the sequence changes should result in amino acid sub-stitutions. This has been described as the *in silico* search for genetic variation (Grag et al., 1999; Beutow et al., 1999; Emahazion et al., 1999).

A second variant detection strategy involves directed resequencing of selected genomic regions from different individuals, usually between 50 and 100 unrelated individuals. Many of these efforts have screened a subset of the DNA Polymorphism Discovery Resource collection, a set of 450 samples selected by the NIH to be a sample of the ethnic diversity of the U.S. population (Collins et al., 1998). The DNA donors are anonymous but expected to be generally healthy individuals. Others have screened samples from individuals selected from diverse geographic regions of the world (Nickerson et al., 1998) or samples from individuals with a disease of specific interest (Hayward et al., 1999). These efforts are focused on identification of amino acid substitution variants and differ from the SNP Consortium effort, which had a goal of identifying the large number of SNPs required for building a high-density map and which treated all regions of the genome equally. It is estimated that fewer than 5% of the SNPs in the SNP Consortium collection are located in coding regions of the genome.

The candidate gene strategy emphasizes identification of amino acid substitution variants and other variations of potential functional relevance in genes believed to be have roles in the biology of a disease and/or expected to have a potential role in susceptibility to environmental exposure- or lifestyle factor-related disease. Examples of the disease outcomes that directed the selection of biological pathways and processes, and thus the genes included in these variation screening efforts, include cardiovascular disease, cancer, and asthma (Cambien et al., 1999; Cargill et al., 1999; Halushka et al., 1999; Shen et al., 1998). These studies to identify genetic variants have reported results from the screening of over 200 different genes. The results can be generalized as follows: (1) approximately three different amino acid substitution variants per gene were detected in the screening of 100–200 chromosomes from generally healthy individuals; (2) it is not uncommon to observe specific variants only in individuals with similar ethnic or geographic origins; (3) the average variant allele frequencies range from 3 to 6% in the different studies; (4) the genetic variation among individuals is extensive, and most individuals will exhibit complex genotypes when the multiple genes of a pathway are being studied; (5) over 60% of the substitutions involve the exchange of amino acid residues with dissimilar physical or chemical properties, suggesting that many of the substitutions should impact protein structure and function; and (6) the very large number of variants with individual allele frequencies of less than 5% account for at least 30% of the total variation among individuals. The initial goal of these variant identification efforts is the development of a catalog of variants. The entries in these catalogs can be found in a number of databases via links from NCBI. Not unlike the catalog of genes, the catalog of variants is available as a starting point for investigators in designing experiments to address specific hypotheses.

With the increasing availability of the data from these variation screens, the challenge is selecting the candidate genes most likely to be relevant for a disease and then documenting the subset of variants in a gene that are causally associated with an altered phenotype. Several approaches can be used to test for the potential functional relevance of a variant gene or protein. These variants can be characterized for biochemical activity (Hadi et al., 2000), but this is expensive given the large number of variants that continue to be identified. Thus some efforts have been initiated to predict the effect of amino acid substitutions on protein structure (Hadi et al., 2000) or residual protein function (Chasman and Adams, 2001), as a prelude to detailed functional analysis of a variant protein. The availability of the three-dimensional structures for a large number of different proteins would be a very useful resource for developing strategies for selection of the subset of specific variants impacting protein activity. It has been suggested that a Structural Genomics Project with a goal of obtaining the structure of most of the proteins encoded by the genome should be initiated. The impact of the variants on biological characteristics of cells with defined genotypes could be determined. Many of the cellular end points of highest interest to pharmacology would be associated with differences in toxicity or efficacy of a compound in cells of different genotypes. Ultimately, the association of genotype(s) with toxicity or efficacy of a drug or disease incidence could be directly estimated in population-based studies. These studies could involve only a few individuals, if the gene and the phenotype are well characterized and the impact on measured outcome is large. For many of the common diseases, the disease state is the consequence of the interaction of a number of the different variants existing at 5–10 genes and will require molecular epidemiology studies involving thousands of cases and controls.

This variation in search of function differs from the strategy that was commonly employed in the pregenome era, where the goal was usually to identify the genetic basis for a previously well-defined phenotypic characteristic, such as hyper- or hyporesponse to a therapeutic agent. This new genome-era strategy has been described as the genotype to phenotype strategy for estimating the relevance of genetic variation (Mohrenweiser and Jones, 1998). It is a reflection of the relative ease with which large quantities of preliminary data can be generated by a relatively small number of laboratories organized to focus on large-scale data generation. Successful utilization of this data requires that individual investigators have easy access to these databases.

III. COMPARATIVE GENOMICS

Although the "human" in "Human Genome Project" received virtually all the attention of the public discussion of the genome project, mapping and sequencing of the genomes of other organisms was occurring in the same time frame. The genomes of yeast, *Arabidopsis*, *Caenorhabditis elegans* (a worm), and *Drosophila* were generally complete in the late 1990s (Adams et al., 2000; Salanoubat et al., 2000; Wilson, 1999; Mewes et al., 1997). Draft sequences of the genomes of both the mouse and the rat should be completed in 2002. The availability of the sequence of the genomes of these and other organisms (e.g., zebra fish) provides resources for comparative genomics. The availability of these genomic sequences is extremely useful in the effort to determine the function of genes with currently unknown function, as it is easier to conduct initial experiments in model organisms than in humans and the function of most genes is similar in different organisms. The availability of genomic sequences from multiple organisms also will facilitate identification of noncoding sequences exhibiting minimal variation over evolutionary time. These regions have been suggested to be elements or sequences involved in regulation of gene expression (Cliften et al., 2001). Other applications that stem from the extensive databases on different organisms include development of better experimental models for the study of human disease and design and testing of new drugs. The characterization of the genomes of multiple organisms, with concurrent increases in understanding of their biology, increases the utility of these experimental models for pharmacology and toxicology research. Examples would include the development of transgenic mice in which specific genes have been inactivated (often referred to as gene knockouts) or modified so that the mice are models for Alzheimer disease (Chapman et al., 2001). Others are constructing mice with copies of variant DNA repair genes observed in screening human populations for use as models for the study of cancer susceptibility and aging (http://www.niehs.nih.gov/envgenom/mouse.htm).

IV. STUDIES OF GENE EXPRESSION

Localization of the tissue or cellular sites of gene expression is often an early experiment in the characterization of a newly discovered gene. Quantitation of the level and the tissue sites of gene expression during different stages of an organism's development helps in understanding the function of newly identified genes. The I.M.A.G.E. project described above was primarily focused on gene identification, with insight into gene expression being a secondary goal. The cGAP, in contrast, endeavored to systematically estimate the level of expression of a large number of genes in normal tissue and a large number of different precancer and cancer cells (<http://cgap.nci.nih.gov/>) (Strausberg et al., 2000). It is expected that extensive knowledge of the gene expression profiles will eventually lead to improved detection, diagnosis, and treatment of cancer.

A major advance in the technology for studies of gene expression has been development of the DNA "chip." DNA chips have short segments of DNA of known sequence attached to a surface, often glass or silicon, much like a computer chip. These 15–40 nucleotide sequences are designed to be complimentary to the sequence of known genes and thus, under specific experimental conditions, can quantitatively trap the expressed transcripts for a gene. Before development of the chip technology, quantitative analysis of gene expression usually focused on a few genes and a limited number of tissues or experimental conditions. It is now possible to have chips with 10,000–50,000 different unique sequences attached to a surface of less than the area of a coin. With the advent of the chip technology, investigators are now routinely monitoring the relative level of expression of thousands of different genes simultaneously in cells and tissues under different conditions and disease states (Lockhart and Winzeler, 2001). Applications of this technology have included searches for markers for early detection of Alzheimer disease (Pasinetti, 2001), markers for differentiating among classes of tumors (Brenton et al., 2001; Liotta and Petricoin, 2000) and differences in the expression profile between breast cancers derived from BRCA1 and BRCA2 patients (Hedenfalk et al., 2001).

The National Institute of Environmental Health Sciences (NIEHS) has recently initiated a coordinated, multidisciplinary research program, the National Center for Toxicogenomics (NCT), with the goal of obtaining large amounts of gene expression data relevant to toxicology (<http://www.niehs.nih.gov/nct/home.htm>). Toxicogenomics is a scientific field that strives to elucidate how the entire genome is involved in biological responses of organisms exposed to environmental toxicants/ stressors. The Environmental Genome Project is another NIEHS sponsored initiative (<http://www.niehs.nih.gov/envgenom/home.htm>). This project focuses on the role of genetic variation in the individual response or risk of disease following exposure to environmentally relevant agents. Similar research in the area of individual variation in response to pharmacologic agents can be reviewed through the Pharmacogenetics Research Network (<http://www.nigms.nih.gov/pharmacogenetics/index.html>).

The rapidly expanding databases from large-scale gene expression and protein characterization studies will have important roles in the analysis of the functional relevance of genetic variation in disease but will also undoubtedly uncover additional variation. The data from both gene and protein expression studies will be key in the effort to identify the elements in the genome involved in the regulation of gene expression. These data are also critical for gaining insight into the complex metabolic and communication networks of the cell, and especially the differences among healthy and non-healthy cells, tissues, and individuals. Efficient utilization of this vast array of genetic variation will require new tools for the study and analysis of complex genetic traits.

V. UTILIZATION OF THE FRUITS OF THE GENOME ERA IN PHARMACOGENOMICS

The potential applications of the technology and information resources of the HGP in pharmacologic research and medical practice have been extensively discussed (Collins and McKusick, 2001; Roses, 2001; McLeod and

Evans, 2001; Evans and Relling, 1999; Dean et al., 2001; Kalow, 2001a, 2001b; Shi et al., 2001) and are the focus of other chapters in this book. Identification of genes involved directly in disease processes and also susceptibility genes (actually, it is the susceptibility allele that is being identified) results in identification of potential cellular targets for pharmacologic intervention. This focuses the drug development activities toward drugs with potential to alter the expression of a gene or the function of a gene product with a key role in a disease process. An alternative strategy utilizes the available genomic sequence as a resource to be searched for new genes that could be potential targets for drug development, without prior knowledge of a relationship of the gene to a specific disease. The interest in the gene stems from characteristics of the sequence or a predicted biochemical or biological function and similarities to other genes with interesting functions. Although the availability of genetic variation in a target gene is a useful resource in characterization of the function of these genes and the potential relationship of a gene to disease, it is not required. As with the studies of the functional relevance of variation, the characterization of new genes will often start with the data available on expression patterns and protein characteristics. In either case, the goal is drug development and intervention in the disease process. The catalogs of known or predicted genes and variants expand the opportunities for both drug development strategies and the targets for intervention in disease processes. This can be considered the "genes and variants in search of a disease" strategy for drug discovery.

In addition to the identification of potential targets for drug development, identification of the basis for individual differences in response to pharmacologic agents provides the framework for the much-discussed potential to practice "individualized medicine." In the simpler cases described above, the lack of drug efficacy or the presence of toxic reactions to a drug in specific individuals will be due to the existence of a small number of variants in a few well-defined genes. Available molecular and biochemical technology, coupled with knowledge of the genetic basis for the aberrant response, makes it feasible to identify hyper- and hyporesponding individuals for a small number of agents. Unfortunately, individualizing treatment is currently feasible for only a small number of protocols. The problem of identifying the molecular basis for aberrant responses is more complex when the response is the consequence of genetic variation in a number of different genes, with each variant having a relatively small impact on the functional capacity of the pathway or process. It is also difficult to define the phenotype and to develop diagnostic genotyping tests for these conditions with complex patterns of inheritance. This is especially a problem when the deviation from the mean response is continuous and reflective of small contributions of multiple genes and multiple variants.

It has been suggested that knowledge of genetic variation will be useful information that can be incorporated at early stages of drug trials. This has potential for the common variation and simpler genetic traits, where effects might be observed in the few hundred individuals enrolled in early trials. Obviously, in the ideal case, sufficient a priori knowledge would be available so that genotype could be a component of the selection criteria for individuals to be enrolled in the clinical trial. For the more rare outcomes, which may result from rare variants or, more likely, from the complex interaction of several genes and genetic variants, it may still not be possible to observe adverse reactions until a drug is in widespread usage.

Significant progress continues to be made in cataloging and characterizing new genes and the genetic variation among individuals. Annotation of the sequence of the human genome, comparison of the human genome to genomes of other organisms, and profiling of the tissue and developmental stage of expression of genes are the current focus of genome-style efforts. The data expected from the other -omics projects (assuming that they become reality with a significant funding base and are as productive as the HGP) will add to the already large amount of data and information. This is the raw material to be utilized for the rapidly expanding field of pharmacogenomics. The challenge for multidisciplinary fields such as pharmacogenomics is increasingly not so much to generate new data but to efficiently utilize the vast amount of available information to address increasingly complex questions. This is increasingly akin to finding the bits of precious metal in a fast-flowing stream, and it sounds like the flow is even faster upstream. Although there will be more nuggets, it will be challenging to harvest them. Pharmacogenomics has the potential to revolutionize many aspects of the practice of medicine, including drug development and treatment protocols. Although the prospects are exciting, the scale of the impact of pharmacogenomics on medical practice and health care for large segments of the population is currently unknown.

ACKNOWLEDGMENT

This work was performed under the auspices of the U.S. Department of Energy by the University of California, Lawrence Livermore National Laboratory under contract No. W-7405-Eng-48.

REFERENCES

Adams, M.D., et al., "The Genome Sequence of *Drosophila melanogaster*," *Science* **287**, 2185–2195 (2000).

Barnhart, B.J., "The Department of Energy (DOE) Human Genome Initiative," *Genomics*, **5**, 657–660 (1989).

Bevan, M., et al., "Sequence and Analysis of the *Arabidopsis* Genome," *Curr. Opin. Plant Biol.*, **4**, 105–110 (2001).

Brenton, J.D., et al., "Molecular Profiling of Breast Cancer: Portraits but Not Physiognomy," *Breast Cancer Res.*, **3**, 77–80 (2001).

Broman, K.W., et al., "Comprehensive Human Genetic Maps: Individual and Sex-Specific Variation in Recombination," *Am. J. Hum. Genet.*, **63**, 861–869 (1998).

Buetow, K.H., et al., "Reliable Identification of Large Numbers of Candidate SNPs from Public EST Data," *Nat. Genet.*, **21**, 323–325 (1999).

Cambien, F., et al., "Sequence Diversity in 36 Candidate Genes for Cardiovascular Disorders," *Am. J. Hum. Genet.*, **65**, 183–191(1999).

Cargill, M., et al., "Characterization of Single-Nucleotide Polymorphisms in Coding Regions of Human Genes," *Nat. Genet.*, **22**, 231–238 (1999).

Chapman, P.F., et al., "Genes Models and Alzheimer's Disease," *Trends Genet.*, **17**, 254–261 (2001).

Chasman, D. and R.M. Adams, "Predicting the Functional Consequences of Non-Synonymous Single Nucleotide Polymorphisms: Structure-Based Assessment of Amino Acid Variation," *J. Mol. Biol.*, **307**, 683–706 (2001).

Clifford, R., et al., "Expression-Based Genetic/Physical Maps of Single Nucleotide Polymorphisms Identified by the Cancer Genome Anatomy Project," *Genome Res.*, **10**, 1259–1265 (2000).

Cliften, P.F., et al., "Surveying *Saccharomyces* Genomes to Identify Functional Elements by Comparative DNA Sequence Analysis," *Genome Res.*, **11**, 1175–1186 (2001).

Collins, F.S., et al., "A DNA Polymorphism Discovery Resource for Research on Human Genetic Variation," *Genome Res.*, **8**, 1229–1231 (1998).

Collins, F.S. and V.A. McKusick, "Implications of the HGP for Medical Science," *JAMA*, **285**, 540–544 (2001).

Cook Deegan, R.M., "The Alta Summit, December 1984," *Genomics*, **5**, 661–663 (1989).

Cox, N.J., "Challenges in Identifying Genetic Variation Affecting Susceptibility to Type 2 Diabetes: Examples from the Study of Calpain-10 Gene," *Human Mol. Genet.*, **10**, 2301–2305 (2001).

Cox, N.J., et al., "Seven Regions of the Genome Show Evidence of Linkage to Type 1 Diabetes in a Consensus of 767 Multiplex Families," *Am. J. Human Genet.*, **69**, 820–830 (2001).

Emahazion, T., et al., "Identification of 167 Polymorphisms in 88 Genes From Candidate Neurodegeneration Pathways," *Gene*, **238**, 315–324 (1999).

Evans, W.E. and M.V. Relling, "Pharmacogenomics: Translating Functional Genomics into Rational Therapeutics," *Science*, **286**, 487–491 (1999).

Dean, P.M., et al., "Industrial-Scale, Genomics-Based Drug Design and Discovery," *Trends Biotechnol.*, **19**, 288–292 (2001).

Friedrich, G.A., "Moving Beyond the Genome Projects," *Nat. Biotechnol.*, **14**, 1234–1237 (1996)

Garg, K., et al., "Identification of Candidate Coding Region Single Nucleotide Polymorphisms in 165 Human Genes Using Assembled Expressed Sequence Tags," *Genome Res.*, **9**, 1087–1092 (1999).

Goffeau, A., et al., "Life with 6000 Genes," *Science*, **274**, 563–567 (1996).

Gyapay, G., et al., "A Radiation Hybrid Map of the Human Genome," *Hum. Mol. Genet.*, **5**, 339–346 (1996).

Hadi, M., et al., "Functional Characterization of Ape1 Variants Identified in the Human Population," *Nucl. Acids Res.*, **28**, 3871–3879 (2000).

Halushka, M.K, et al., "Patterns of Single-Nucleotide Polymorphisms in Candidate Genes for Blood-Pressure Homeostasis," *Nat. Genet.*, **22**, 239–247 (1999).

Hayward, C., et al., "Molecular Genetic Analysis of the APEX Nuclease Gene in Amyotrophic Lateral Sclerosis," *Neurology*, **52**, 1899–1901 (1999).

Hedenfalk, I., et al., "Gene-Expression Profiles in Hereditary Breast Cancer," *N. Engl. J. Med.*, **344**, 539–548 (2001).

Hogenesch, J.B., et al., "A Comparison of the Celera and Ensembl Predicted Gene Sets Reveals Little Overlap in Novel Genes," *Cell*, **106**, 413–415 (2001).

Hustert, E., et al., "Natural Protein Variants of Pregnane X Receptor with Altered Transactivation Activity Toward CYP3A4," *Drug Metab. Dispos.*, **29**, 1454–1459 (2001).

Ingelman-Sundberg, M. and W.E. Evans, "Unraveling the Functional Genomics of the Human CYP2D6 Gene Locus," *Pharmacogenetics*, **11**, 553–554 (2001).

Kalow, W., "Pharmacogenetics, Pharmacogenomics, and Pharmacobiology," *Clin. Pharm. Ther.*, **70**, 1–4 (2001a).

Kalow, W., "Pharmacogenetics in Perspective," *Drug Metab. Dispos.*, **29**, 468–470 (2001b).

King, M-C, "Leaving Kansas . . . Finding Genes in 1997," *Nat. Genet.*, **15**, 8–10 (1997).

Lai, E., "Application of SNP Technologies in Medicine: Lessons Learned and Future Challenges," *Genome Res.*, **11**, 927–929 (2001).

Lander, E.S., et al., "Initial Sequencing and Analysis of the Human Genome," *Nature*, **409**, 860–921 (2001).

Liang, F., et al., "Gene Index Analysis of the Human Genome Estimates Approximately 120,000 Genes," *Nat. Genet.*, **25**, 239–240 (2000).

Liotta, L. and E. Petricoin, "Molecular Profiling of Human Cancer," *Nat. Rev. Genet.*, **1**, 48–56 (2000).

Lockhart, D.J. and E.A. Winzeler, "Genomics, Gene Expression and DNA Arrays," *Nature*, **405**, 827–836 (2000).

McLeod, H.L. and W.E. Evans, "Pharmacogenomics: Unlocking the Human Genome for Better Drug Therapy," *Annu. Rev. Pharmacol. Toxicol.*, **41**, 101–121 (2001).

Mewes, H.W., et al., "Overview of the Yeast Genome," *Nature*, **387**, 7–65 (1997).

Miki, Y., et al., "A Strong Candidate for the Breast and Ovarian Cancer Susceptibility Gene BRCA1," *Science*, **266**, 66–71 (1994).

Mohrenweiser, H.W. and I.M. Jones, "Variation in DNA Repair is a Factor in Cancer Susceptibility: A Paradigm for the Promises and Perils of Individual and Population Risk Estimation?" *Mutat. Res.*, **400**, 15–24 (1998).

Nickerson, D.A., et al., "DNA Sequence Diversity in a 9.7-kb Region of the Human Lipoprotein Lipase Gene," *Nat. Genet.*, **19**, 233–240 (1998).

Nwosu, V., et al., "Heterogeneity of Genetic Alterations in Prostate Cancer: Evidence of the Complex Nature of the Disease," *Hum. Mol. Genet.*, **10**, 2313–2318 (2001).

Ophoff, R.A., et al., "P/Q-type Ca²⁺ Channel Defects in Migraine, Ataxia and Epilepsy," *Trends Pharmacol Sci.*, **19**, 121–127 (1998).

Ophoff, R.A., et al., "Familial Hemiplegic Migraine and Episodic Ataxia type-2 are Caused by Mutations in the Ca²⁺ Channel Gene CACNL1A4," *Cell*, **87**, 543–552 (1996).

Ophoff, R.A., et al., "The Impact of Pharmacogenetics for Migraine," *Eur. J. Pharmacol.*, **413**, 1–10 (2001).

Pasinetti, G.M., "Use of cDNA Microarray in the Search for Molecular Markers Involved in the Onset of Alzheimer's Disease Dementia," *J. Neurosci. Res.*, **65**, 471–476 (2001).

Ponder, B.A.J., "Cancer Genetics," *Nature*, **411**, 336–341 (2001).

Riordan, J.R., et al., "Identification of the Cystic Fibrosis Gene: Cloning and Characterization of Complementary DNA," *Science*, **245**, 1066–1073 (1989).

Risch, N.J., "Searching for Genetic Determinants in the New Millennium," *Nature*, **405**, 847–856 (2000).

Roses, A.D., "Pharmacogenetics and the Practice of Medicine," *Nature*, **405**, 857–865 (2000).

Runnebaum, I.B., et al., "Progesterone Receptor Variant Increases Ovarian Cancer Risk in BRCA1 and BRCA2 Mutation Carriers Who Were Never Exposed to Oral Contraceptives," *Pharmacogenetics*, **11**, 635–638 (2001).

Sachidanandam, R., et al., "A Map of Human Genome Sequence Variation Containing 1. Million Single Nucleotide Polymorphisms," *Nature*, **409**, 928–933 (2001).

Salanoubat, M., et al., "Sequence and Analysis of Chromosome 3 of the Plant *Arabidopsis thaliana*," *Nature*, **408**, 820–822 (2000).

Saunders, A.M., "Gene Identification in Alzheimer's Disease," *Pharmacogenomics*, **2**, 239–249 (2001).

Sheffield, V.C., et al., "A Collection of Tri- and Tetranucleotide Repeat Markers Used to Generate High Quality, High Resolution Human Genome-Wide Linkage Maps," *Hum. Mol. Genet.*, **4**, 1837–1844 (1995).

Shen, M.-J., et al., "Nonconservative Amino Acid Substitutions Exist at Polymorphic Frequency in DNA Repair Genes in Healthy Humans," *Cancer Res.*, **58**, 604–608 (1998).

Shi, M.M., et al., "Pharmacogenetic Application in Drug Development and Clinical Trials," *Drug Metab. Dispos.*, **29**, 591–595 (2001).

Sorbi, S., et al., "Genetic Risk Factors in Familial Alzheimer's Disease," *Mech Ageing Dev.*, **122**, 1951–1960 (2001).

Strausberg, R.L., et al., "The Cancer Genome Anatomy Project: Building An Annotated Gene Index," *Trends Genet.*, **16**, 103–106 (2000).

Tavtigian, S.V., et al., "The Complete BRCA2 Gene and Mutations in Chromosome 13q-Linked Kindreds," *Nat. Genet.*, **12**, 333–337 (1996).

Venter, J.C., et al., "The Sequence of the Human Genome," *Science* **291**, 1304–1351 (2001).

Wang, D.G., et al., "Large-Scale Identification, Mapping, and Genotyping of Single-Nucleotide Polymorphisms in the Human Genome," *Science* **280**, 1077–1082 (1998).

Wang, W.W., et al., "A Single Nucleotide Polymorphism in the 5′Untranslated Region of RAD51 and Risk of Cancer Among BRCA1/2 Mutation Carriers," *Cancer Epid. Biomarkers Prev.*, **10**, 955–960 (2001).

Watson, J.D. and E. Jordan, "The Human Genome Program at the National Institutes of Health," *Genomics*, **5**, 654–656 (1989).

Weber, J.L. and K.W. Broman, "Genotyping for Human Whole-Genome Scans: Past, Present, and Future," *Adv. Genet.*, 77–96 (1989).

Weinshilboum, R., "Thiopurine Pharmacogenetics: Clinical and Molecular Studies of Thiopurine Methyltransferase," *Drug Metab. Dispos.*, **29**, 601–605 (2001).

Wilson, R.K., "How the Worm was Won. The *C. Elegans* Genome Sequencing Project," *Trends Genet.*, **15**, 51–58 (1999).

Wolf, C.R., et al., "Pharmacogenetics," *Br. Med. J.*, **320**, 987–990 (2000).

Xu, J., et al., "Genomewide Screen and Identification of Gene-Gene Interactions for Asthma-Susceptibility Loci in Three U.S. Populations: Collaborative Study of the Genetics of Asthma," *Am. J. Hum. Genet.*, **68**, 1437–1446 (2001).

Yu, A., et al., "Comparison of Human Genetic and Sequence-Based Physical Maps," *Nature*, **409**, 951–953 (2001).

Zanger, U.M., et al., "Comprehensive Analysis of the Genetic Factors Determining Expression and Function of Hepatic CYP2D6," *Pharmacogenetics*, **11**, 573–585 (2001).

THE IMPLICATIONS OF POPULATION GENETICS FOR PHARMACOGENOMICS

CRAIG L. HANIS, PH.D

I. INTRODUCTION

Few things are so readily apparent in the human population as the tremendous variation that exists among individuals. This variation is manifest in size, shape, appearance, and behavior. Underlying these differences is a complex interplay between genetic background and environmental exposures. At the DNA level alone, there is virtually an infinite number of possible genetic combinations. Indeed, any two randomly selected individuals will differ on average by approximately one DNA base change every 1000 base pairs. Given some 3 billion base pairs in the human DNA sequence, this results in 3 million differences between any two individuals. These differences may be so innocuous that they result in no detectable differences between the individuals other than the observed DNA sequence change. Other equally simple DNA changes may result in profound differences that compromise development, aging, disease processes, or response to environmental stimuli such as pharmacologic treatment. As the genetic underpinnings increasingly yield to investigation and elucidation, it will be possible to systematically "manipulate" these systems to alter the manifestation and course of disease.

Despite both the potential and realized variation, patterns emerge that are easily recognized. The recognition of these patterns can be exploited

Pharmacogenomics: Social, Ethical, and Clinical Dimensions, Edited by Mark A. Rothstein.
ISBN 0-471-22769-2   Copyright © 2003 Wiley-Liss, Inc.

to suggest, localize, and identify the underlying genes contributing to the pattern of interest. Pharmaceutical agents are often valued because of their broad applicability and efficacy. We are quickly moving to an era, however, where the genetic underpinnings will play an increasing role in guiding pharmaceutical discovery, clinical trials, and so-called personalized medicine (Stephens, 1999). To place these advances in the proper context, however, there must be a firm consideration of genetic variation and its frequency, structure, evolution, and phenotypic impact; that is, there must be a consideration of the population genetics of variation. This chapter briefly reviews the key considerations of population genetics that will most impact pharmacogenomics. In particular, it examines the properties of frequency, population structure, association/disequilibrium, and scale.

II. ALLELE FREQUENCIES AND THEIR EQUILIBRIUM PROPERTIES

Genetic changes at a specific site are generally termed polymorphic if they occur commonly in the population (i.e., two alternative forms with frequencies of 1% or more). Polymorphisms are classified into several major classes including single nucleotide polymorphisms (SNPs), short tandem repeats (STRs), and variable number of tandem repeats (VNTRs). STRs and VNTRs have proven extremely useful for the localization of disease genes through linkage analysis (Gyapay et al., 1994). Linkage, however, does not have fine enough resolution to lead to the identification of the disease-causing gene and allele. The latter require fine mapping, and it is in this area that the single nucleotide changes are most useful (Johnson et al., 2001). Furthermore, susceptibility alleles (whether changing disease susceptibility or response to treatment) are likely to fall into the class of single nucleotide differences. In Mendelian diseases, such single nucleotide changes result in truncated proteins, altered binding sites, or other amino acid changes that are responsible for disease. In the context of common complex diseases, such as cardiovascular disease, hypertension, diabetes, and obesity, it is likely that the same kinds of changes exist but with more subtle effects on disease. It is also likely in the context of common diseases that single nucleotide changes in noncoding regions of a gene may have subtle (or not so subtle) effects on regulation of a gene (Horikawa et al., 2000).

Ultimately, all polymorphisms arose as mutations. In a randomly mating, infinitely large population, gene (allele) and genotype frequencies are constant from generation to generation if no mutation or selection occurs. Hardy, an English mathematician, and Weinberg, a German physician, first demonstrated this in 1908. This notion of constancy of frequencies is the central pillar of population genetics. Genes that appear to behave in this fashion within a population are said to be in Hardy–Weinberg equilibrium.

This can be simply illustrated by considering a hypothetical locus having two alleles, A1 and A2, with frequencies p and q, respectively. Under random mating, the frequencies of the three possible genotypes involving A1 and A2, namely, A1A1, A1A2, and A2A2, will occur with frequencies p^2, 2pq, and q^2, respectively. The allele frequencies p and q and the genotype frequencies p^2, 2pq, and q^2 remain constant from generation to generation under random mating. In reality, the Hardy–Weinberg principle is so robust that most polymorphisms exist in equilibrium. One can test that a gene behaves in accordance with its Hardy–Weinberg expectations by estimating the frequencies of the alleles and then comparing the observed distribution of genotype counts to their expectations generated from the allele frequencies (i.e., are the observed numbers of each genotype in proportions equal to p^2, 2pq, and q^2?).

One of the obvious implications of genotypes being distributed in proportions p^2, 2pq, and q^2 is that it leads to a very uneven distribution of individuals in each genotypic group. Consider for a moment a gene having two alleles with the least frequent allele occurring 1% of the time. This implies that in a sample of 1000 individuals you may never observe an individual homozygous for the least frequent allele, whereas 20 are expected to be heterozygotes and 980 will be homozygotes for the common allele. In the context of treatment for common diseases, such a locus may not lead to opportunities for personalized medicine (although it may still be important in leading to an exploitable metabolic pathway). If the least frequent allele were to occur 10% of the time, then the distribution of genotypes in a sample of 1000 would be 10 homozygotes for the least frequent allele, 180 heterozygotes, and 810 homozygotes for the common allele. As the frequency increases, larger segments of the population will be grouped by genotype. If there are genotype-specific responses to treatment, then these frequencies will determine in large part the strategy for screening, classification, and treatment choice.

There are no circumstances for a gene with variation in Hardy–Weinberg equilibrium that will lead to equal numbers of individuals in each of the genotypic classes. Only in the case when allele frequencies are equal will the expected distribution be symmetrical. In such a case there will be equal numbers of the two homozygous classes (1/4 in each group), and 1/2 of the total sample will consist of heterozygotes.

A simple measure of the amount of variation possible in the population for a polymorphic locus is the expected frequency of heterozygotes, which is termed "heterozygosity" (Roychoudhury and Nei, 1988). Heterozygosity is also a reflection of the amount of linkage information available in a marker for mapping genes. Regardless of the number of alleles at a locus, genetic variation (heterozygosity) is always maximized when alleles are equally frequent. For two alleles, the maximal heterozygosity is 0.50. For systems with many alleles like STRs and VNTRs the heterozygosity is often above 0.80;

indeed, for mapping genes via linkage, sets of markers are chosen because of their high heterozygosity or other similar measures of their information content (Botstein et al., 1980).

Although genetic considerations lead to unbalanced numbers in each group, statistical considerations are often most advantageous with balanced numbers. Judicious statistical design is of paramount importance for controlling costs while maintaining adequate statistical power to test hypotheses. This is of particular concern in the context of clinical trials that require extensive study of individuals in the trial. By establishing a two-stage sampling for studies, both considerations can be reconciled. For example, an initial screen of individuals for a clinical trial may involve minimal examinations of a large number of potential participants followed by genotyping. Once candidates are genotyped, a balanced subset representing the genotypic groups can be selected for the full intervention. Statistical power is maintained (likely enhanced) while minimizing the required sample sizes. Such designs will be required to test the efficacy of treatments by genotype.

III. FACTORS AFFECTING THE EQUILIBRIUM PROPERTIES OF ALLELE FREQUENCIES

A number of "forces" can lead to departures from Hardy–Weinberg equilibrium. In defining Hardy–Weinberg equilibrium, three specific conditions were required, namely, random mating, no selection, and no mutation. In general, mutations are rare and have little effect on considerations regarding Hardy–Weinberg equilibrium. Similarly, although selection may be one of the driving forces leading to the distribution of alleles in the population, the time frame in which selection works results in a minimal impact in the scope of current studies. Before turning to a consideration of issues surrounding random mating, it is important to mention two other forces and one artifact that can lead to departures from equilibrium. The two forces are drift and migration. The artifact is the impact of laboratory error. Drift simply refers to the random changes in allele frequencies from generation to generation. Its impact is magnified in small populations. As human populations have undergone a variety of historical bottlenecks in their evolution, drift has played a significant role in establishing the contemporary distribution of alleles and their frequencies (Kimura, 1991). Even so, in modern studies sampling sufficient numbers to provide adequate statistical power for testing hypotheses will lead to generally stable allele and genotype frequencies.

Migration of individuals into a population can have a substantial impact on allele distribution stemming from two considerations. First, migration can act much like mutation except at a much higher frequency. That is, individuals coming into a population can change the allele distribution in the resulting mixed population. This is especially true if the allele frequency distributions are markedly different between the two mixing populations. The second consideration, although related, is that migration can lead to population substructuring in which the new population no longer undergoes random mating. This is discussed further below.

No laboratory is free from laboratory error. By their very nature, genetic polymorphisms pose a number of challenges for accurate genotyping of large numbers of individuals or polymorphisms (Sobel et al., 2002). Without enumerating all the difficulties, suffice it to say that laboratory error is likely the most common cause of departures from Hardy–Weinberg equilibrium and most often should be the first consideration if deviations from Hardy–Weinberg equilibrium are detected. As a consequence, considerations of Hardy–Weinberg equilibrium are central to genotyping quality control.

The last force to be considered is that of departures from random mating. Although human mating is decidedly a nonrandom process, little evidence exists that mating is other than random with regard to genetic variation at most loci that have been examined. However, if within a population there exists one or more subgroups (through such processes as migration) where mating is more likely to occur within or is restricted to subgroups, then there are a variety of consequences. For example, consider the case of a population consisting of two subgroups where there is no gene flow between the two groups. If the allele frequencies at a specific locus are sufficiently different between the two populations, then a sample drawn from the overall population without regard to structure may lead to spurious departures from Hardy–Weinberg equilibrium. Perhaps more importantly, such unaccounted differences may lead to spurious associations between a disease and a gene (Devlin et al., 2001). If, rather than disease, the trait of interest is response to pharmacologic treatment, then a spurious conclusion that there is an association may have profound effects on subsequent studies, trials, and intervention strategies.

In the simple case of two subgroups with no gene flow between them one could easily resolve the problems by explicitly incorporating population structure into any design and analyses. Unfortunately, most population substructure is not so readily apparent (Wilson et al., 2001). Because of the potential for compromising the inferential power of a study when there may be unrecognized population structure, a variety of specialized methods have been developed for testing associations between genes and disease. These generally rely on some type of family-based control where the transmission of genes can be followed (see Devlin et al., 2001 and Zhao, 2000 for recent reviews of such methods).

As populations amalgamate, opportunities for gene flow between the populations begin. This process of genetic admixture leads to considerations similar to those of population substructure. The difference is that with admixture the apparent substructure begins to break down. Over a number of generations of admixture, the characteristics of the parental populations are merged into a new population. With time, allele frequencies in the resulting population come to equilibrium at frequencies that represent the weighted average of the allele frequencies from the contributing parental populations. The weights represent the proportion of alleles derived from the ancestral populations and can be applied to populations or individuals (Hanis et al., 1986). An admixed population that has come to equilibrium poses no serious additional considerations for design and analysis. However, studies in an admixed population in the process of transition may result in erroneous conclusions regarding associations, either erroneously concluding that an association exists or erroneously concluding that one does not exist (Deng, 2001). Admixed populations need not be excluded from consideration for studies of association with disease. The types of family-based methods mentioned above for population substructure are also appropriate in admixed populations regardless of whether they have come to equilibrium.

Methods have also been developed for exploiting the differences in allele frequencies between ancestral populations for actual mapping of traits (Shriver et al., 1997). These latter methods, termed mapping by admixture linkage disequilibrium (MALD), demonstrate the widely varying frequencies of genetic polymorphisms across populations. The ideal genetic marker for these studies would be one in which one allele was virtually monomorphic in one ancestral population but the alternative allele was monomorphic in the other population. A large number of such markers exist and have been cataloged (e.g., Smith et al., 2001). Whereas this is important in the mapping context, it has perhaps greater implications for other studies and designs. Clearly, population groups differ with regard to frequency of disease and with regard to frequencies of genetic polymorphisms. This consideration alone can have a major impact on frequency of disease, strategies for screening to identify high-risk individuals, efficacy of drug response in the population, and the design of clinical trials. In none of these areas can frequency of alleles be ignored.

Acknowledging differences in allele frequencies among populations should lead to appropriate caution when interpreting the lack of consistency of findings from various population groups. It may be that the inability to demonstrate statistical consistency of results across populations reflects differential statistical power in the different groups owing to frequency differences among groups as opposed to biological differences of effects. This is not to say, however, that biological differences do not exist. In general, an allelic effect in one population should produce a similar effect in another population even though the frequencies may differ dramatically. It may be that differential responses to hypertension treatment by race (Prichard et al., 2000) may be partially explained by differing allele frequencies at genes involved in the metabolism of antihypertensive agents. This is illustrated in more detail in the consideration of scale effects below in this chapter.

IV. Simultaneous Consideration of Two or More Loci

So far, properties of a single genetic locus have been considered as though they were independent of all others. Although a complete consideration of multiple loci is clearly beyond the scope of this chapter, there are some underlying principles that, when considered, can have profound effects for disease and pharmacogenomic investigations. Biologically, genes are grouped by chromosome. Even so, if the physical distance between two genes on the same chromosome is sufficiently large, the two loci will behave independently of one another. For illustrative purposes, consider two loci A and B each having two alleles; namely A1 and A2 at the A locus and B1 and B2 at the B locus. Let p and q be the frequencies of A1 and A2, respectively, and r and s be the frequencies of B1 and B2. Assuming that these loci are in Hardy–Weinberg equilibrium, we expect the distribution of genotypes at the A locus (A1A1, A1A2, and A2A2) to occur with frequencies p^2, 2pq, and q^2. Similarly, B1B1, B1B2, and B2B2 should occur in proportions r^2, 2rs, and s^2. Independence of these loci simply means that the frequency of any joint combination such as A1A1B1B1 will be equal to the product of their individual frequencies; i.e., p^2r^2.

When the physical distance between two loci on a chromosome is sufficiently small, there are constraints placed upon the loci such that they may no longer behave in an independent fashion. This results from the low probability of genetic recombination between two physically close loci. Assume that the A and B loci are sufficiently close together that recombination is no longer free. As stated above, all polymorphisms ultimately have arisen as the result of mutations. Suppose that we could go back in time to a point where the A locus was polymorphic and the B locus was monomorphic for the B1 allele. A mutation in the B1 allele could give rise to the B2 allele. Because mutation is a rare process, we can ignore for discussion the possibility of recurrent mutations. At any point before the mutation in B1 occurs, allele A1 is always adjacent to B1 on a chromosome. Similarly, allele A2 is always adjacent to B1. These two locus enumerations of the alleles on a single chromosome (e.g., A1B1 or A2B1) are termed haplotypes. Assuming that the allelic form at the A locus has no impact on the chance of a mutation at B, then B2 can arise either on the A1 background or the A2 background. Suppose that it arises on the A1 background and leads to a new haplotype denoted A1B2.

The only way for B2 to occur on an A2 background is if there is another mutation, which already has been ruled out, or if a recombination event occurs in an individual who is heterozygous for both loci (A1B2/A2B1, where the forward slash separates the maternal and paternal haplotypes in an individual). With no recombination, A1B2 or A2B1 are inherited as units by an offspring. A recombination event would lead to gametes of the types A1B1 and A2B2. Note that the A1B1 type has been seen previously, but the A2B2 type is new. With sufficient time, recombination will break down the initial association between A1 and B2 to the point that in the population these two loci will appear to be independent despite their close physical proximity. The time required to break down these associations is a function of how close the two markers are. It is possible that the markers are so close that one expects there to be a continued association. So long as there remains an association, these markers are considered to be in disequilibrium. This is the basis for fine mapping and association studies.

Although this principle has been illustrated with two arbitrary genetic markers, there is no reason not to consider the situation in which one of these markers is a disease susceptibility locus. For example, let the B locus be a disease susceptibility locus. In this case we may not know the identity of the B locus (indeed, this is what we are trying to find), but the genetic disequilibrium between A and B will lead to an association between marker A and the disease. When such is found, one can then look in the region of marker A to see the array of genes that are present to find the functional gene and causative allele. At first glance, the strategy is straightforward. The paucity of demonstrated successes in moving from linkage to association to gene identification, however, indicates the complexity of the process (Altshuler et al., 2000). As the Human Genome Project and related data mature, it is anticipated that the pace of gene discovery in the context of common diseases will increase.

The most common paradigm for gene identification continues to be one involving moving from linkage to association and then gene identification. It has been suggested that disequilibrium and disease association could be exploited directly (Kruglyak, 1999). Although this has some obvious appeals related to much simpler sampling designs (e.g., association studies do not require more complex pedigree-based sampling), there remain considerable challenges because of the shear numbers of markers that must be typed to achieve adequate genome coverage (Kruglyak, 1999). At least for the foreseeable future, it is likely that the linkage-to-association paradigm will continue. This is not to say that large-scale association studies will not be done. Numerous groups are developing sets of SNPs and genotyping strategies based on pooled samples to move in this direction (Sham, 2001).

There may also be a compromise position in this regard in that, rather than systematic examination of the entire genome, it may be possible to systematically examine all genes. To this end, efforts are being made to enumerate the full array of genetic variation in all genes, establish their haplotypes (Stephens et al., 2001), and employ so-called "haplotype tags" to capture the information in a gene (Johnson et al., 2001). Among the advantages of this approach is that it will be possible to resolve the contribution of a specific gene to a trait or disease and not be left with the rather common conclusion that the observed association may be the result of a functional mutation yet to be identified that is in linkage disequilibrium with the polymorphism examined. Greatly aiding this process is the fact that although there are nearly enumerable haplotype possibilities at any given gene, the realized array of haplotype variation is a much reduced subset (Stephens et al., 2001). As with single polymorphisms, these haplotypes show marked variation in frequency among populations. There are some haplotypes that seem to be distributed ubiquitously across populations, but there are others that are characteristic of a single population (Stephens et al., 2001). Imbedded in these haplotype frequencies is a great amount of information regarding the evolution of these genes, potential selective constraints, indicators of population structure, and information that will permit the identification of disease and drug-response genes.

V. SCALE EFFECTS

In addition to the frequency considerations attendant to the examination of polymorphisms or haplotypes, one must also consider the impact of possible differences in the magnitude of effects of any putative loci. The magnitude of any effects is denoted as "scale" effects based on the notion from quantitative genetics that there will be a displacement from the overall population mean for a trait that is dependent on genotype. To illustrate the effects of scale and frequency, consider two well-known examples of genetic effects. These are the effect of the apolipoprotein E (apo E) polymorphism on cholesterol levels and the impact of the familial hypercholesterolemia polymorphism on cholesterol levels.

The apo E polymorphism is among the most widely studied genetic polymorphisms to date (see Eichner et al., 2002 for a recent review). Three common alleles occur; namely E2, E3, and E4. These alleles appear to be present in every population that has been examined. Furthermore, in all populations studied the E3 allele predominates although there is considerable variation in the frequencies of this and the other two alleles (Roychoudhury and Nei, 1988). These alleles differ with regard to their binding affinity to lipoprotein receptors (Cassel et al., 1984). The impact of the differential binding is observed in plasma cholesterol levels. Individuals having at least one E2 allele have on average a lower cholesterol level compared to the population mean, whereas those having an E4 allele tend to have higher cholesterol levels. These effects, however, are not large, with those carrying an E2 allele having an approximate 10 mg/dl lower mean cholesterol and those carrying an E4 allele an approximate 8 mg/dl higher average cholesterol.

These results have been shown to be consistent across populations (Hallman et al., 1991).

Whereas there are common variants for the apo E locus, familial hyper-cholesterolemia is an uncommon disease due to mutations in the LDL receptor locus. At least 421 mutations in the LDL receptor have been identified in subjects with familial hypercholesterolemia (Goldstein et al., 2001). Although there is a broad array of mutations in the LDL receptor gene that lead to similar disease, in the aggregate familial hypercholesterolemia is quite uncommon. Homozygous affected individuals occur at approximately 1 in 1 million individuals, whereas heterozygotes occur at approximately 1 in 500 individuals (Goldstein et al., 2001). For individuals carrying one or more such mutations, there is a dramatic impact on cholesterol level and risk for premature heart attacks. Heterozygotes have cholesterol levels from 350 to 550 mg/dl, whereas homozygotes have levels from 650 to 1000 mg/dl. The latter often experience a fatal heart attack before age 20 (Goldstein et al., 2001).

Clearly, there are dramatic differences on individuals in terms of the impact of genotypes at the apo E locus versus those at the LDL receptor locus. However, because the apo E polymorphism is much more common than familial hypercholesterolemia mutations, the overall impact on the population is larger for genetic variation at apo E than for that at the LDL receptor. Overall, genetic variation at the apo E locus explains some 4–10% of the variation in cholesterol levels among individuals in the population whereas variation at the LDL receptor explains only some 1% of the variation in the population. This demonstrates the importance of both scale and frequency effects. The implications for pharmacogenetics are primarily twofold. First, a locus that is highly polymorphic, but has a small-scale effect, can have a substantial impact on the overall population. It may be that an intervention exploiting these small differences may have a much more dramatic impact on public health than an intervention targeting only the extremes of the distribution. Second, the range of scale effects may define the boundaries of efficacy of various interventions.

Although we have illustrated the scale and frequency effects using the classic examples involving apo E and familial hypercholesterolemia, it is clear that a similar range of effects will be documented for response to treatment. Consider, for example, individuals with succinylcholine sensitivity (Weber, 1997). A low-frequency amino acid substitution polymorphism in serum cholinesterase gives rise to the succinylcholine sensitivity. Those homozygous for the polymorphism are rare in the population but have prolonged apnea after administration of succinylcholine, which is an effective muscle relaxant. Those who are normal metabolizers have no such adverse effect (Weber, 1997). This represents an example of a low-frequency polymorphism with a dramatic effect. On the other end of the spectrum are the modest differences in blood pressure reduction observed by Turner et al. (2001) among hypertensives differing in genotype at the G protein β3 subunit. In this case, the polymorphism is very common, but the effect is small albeit significant. Thus it has the potential for significant impact on public health although it may not have substantial effects on the individual.

VI. CONCLUSIONS

Approximately a century and a half ago, Mendel defined a series of simple rules regarding the transmission of discrete units from one generation to the next in peas based on careful observation of patterns and the application of simple statistics to those patterns. Even though the genome remained largely inaccessible until the last decade, many of the molecular characteristics regarding DNA, its storage, and transmission were inferred through careful experimentation. As the genome became more accessible, these properties were confirmed and genes began to yield to both investigation and manipulation. Coupling the unprecedented access to the genome that we enjoy today with careful observation holds the promise of identifying those genes and pathways that impact on the diseases that so dominate our health. These include cardiovascular disease, cancer, hypertension, diabetes, and obesity. The understanding that comes is accompanied by the prospects for identifying those at increased (and possibly early) risk and for developing intervention strategies to disrupt the natural history of disease. Central to identifying and elucidating these pathways is the exploitation of genetic variation.

Once identified, the impact in populations will largely be determined by the frequencies of alternative forms of genes. These frequencies will be critical for design of observational studies and clinical trials, but they must be balanced with considerations of the magnitude of any effects produced by the variation. As discussed above, because familial hypercholesterolemia occurs only rarely in the population, we can establish that the overall impact of the more common apo E variation in explaining differences among individuals in their cholesterol levels is greater. However, the intense study of the less common LDL receptor mutations has led to the elucidation of pathways that are now exploitable with pharmacologic agents, and not just in the context of extremes of the distribution. In all of these areas, the fundamental properties of genes and combinations of genes that have been the realm of population genetics will repeatedly come to bear. When appropriately considered, these population aspects should facilitate and speed the gene discovery and assessment processes.

REFERENCES

Altshuler, D., et al., "Guilt by Association," *Nat. Genet.*, **26**, 135–137 (2000).

Botstein, D., et al., "Construction of a Genetic Linkage Map in Man Using Restriction Fragment Length Polymorphisms," *Am. J. Hum. Genet.*, **32**, 314–331 (1980).

Cassel, D.L., et al., "The Conformation of Apolipoprotein E Isoforms in Phospholipid Complexes and Their Interaction with Human Hep G2 Cells," *Atherosclerosis*, **52**, 203–218 (1984).

Deng, H.W., "Population Admixture May Appear to Mask, Change or Reverse Genetic Effects of Genes Underlying Complex Traits," *Genetics*, **159**, 1319–1323 (2001).

Devlin, B., et al., "Unbiased Methods for Population-Based Association Studies," *Genet. Epidemiol.*, **21**, 273–284 (2001).

Eichner, J.E., et al., "Apolipoprotein E Polymorphism and Cardiovascular Disease: A HuGE Review," *Am. J. Epidemiol.*, **155**, 487–495 (2002).

Goldstein, J.L., et al., "Familial Hypercholesterolemia. In: Scriver CR, et al. (eds), *The Metabolic & Molecular Bases of Inherited Disease*, 8th ed., New York: McGraw-Hill, pp. 2863–2913 (2001).

Gyapay, G., et al., "The 1993–94 Genethon Human Genetic Linkage Map," *Nat. Genet.*, **7**, 246–339 (1994).

Hallman, D.M., et al., "The Apolipoprotein E Polymorphism: A Comparison of Allele Frequencies and Effects in Nine Populations," *Am. J. Hum. Genet.*, **49**, 338–349 (1991).

Hanis, C.L., et al., "Individual Admixture Estimates: Disease Associations and Individual Risk of Diabetes and Gallbladder Disease among Mexican Americans in Starr County, Texas," *Am. J. Phys. Anthropol.*, **70**, 433–441 (1986).

Hardy, G.H., "Mendelian Proportions in a Mixed Population," *Science*, **28**, 49–50 (1908).

Horikawa, Y., et al., "Genetic Variation in the Gene Encoding Calpain-10 is Associated with Type 2 Diabetes Mellitus," *Nat. Genet.*, **26**, 163–175 (2000).

Johnson, G.C., et al., "Haplotype Tagging for the Identification of Common Disease Genes," *Nat. Genet.*, **29**, 233–237 (2001).

Kimura, M., "Recent Development of the Neutral Theory Viewed from the Wrightian Tradition of Theoretical Population Genetics," *Proc. Natl. Acad. Sci. USA*, **88**, 5969–5973 (1991).

Kruglyak, L., "Prospects for Whole-Genome Linkage Disequilibrium Mapping of Common Disease Genes," *Nat. Genet.*, **22**, 139–144 (1999).

Prichard, B.N., et al., "New Approaches to the Uses of Beta Blocking Drugs in Hypertension," *J. H. Hypertens.*, **14**(Suppl 1), S63–S68 (2000).

Roychoudhury, A.K., and M. Nei, *Human Polymorphic Genes: World Wide Distribution*, Oxford University Press, New York, pp. 3–18 (1988).

Sham, P., "Shifting Paradigms in Gene-Mapping Methodology for Complex Traits," *Pharmacogenomics*, **2**, 195–202 (2001).

Shriver, M.D., et al., "Ethnic-Affiliation Estimation by Use of Population-Specific DNA Markers," *Am. J. Hum. Genet.*, **60**, 957–964 (1997).

Smith, M.W., et al., "Markers for Mapping by Admixture Linkage Disequilibrium in African American and Hispanic Populations," *Am. J. Hum. Genet.*, **69**, 1080–1094 (2001).

Sobel, E., et al., "Detection and Integration of Genotyping Errors in Statistical Genetics," *Am. J. Hum. Genet.*, **70**, 496–508 (2002).

Stephens, J.C., "Single-Nucleotide Polymorphisms, Haplotypes, and Their Relevance to Pharmacogenetics," *Mol. Diagn.*, **4**, 309–317 (1999).

Stephens, J.C., et al., "Haplotype Variation and Linkage Disequilibrium in 313 Human Genes," *Science*, **293**, 489–493 (2001).

Turner, S.T., et al., "C825T Polymorphism of the G Protein Beta(3)-Subunit and Anti-hypertensive Response to a Thiazide Diuretic," *Hypertension*, **37**, 739–743 (2001).

Weber, W.W., *Pharmacogenetics*, New York, Oxford University Press (1997).

Weinberg, W., "Uber den nachweis der vererbung beim menschen. Jahresh. verein f. varerl," *Naturk in Wurttemberg*, **64**, 368–382 (1908).

Wilson, J.F., et al., "Population Genetic Structure of Variable Drug Response," *Nat. Genet.*, **29**, 265–269 (2001).

Zhao, H., "Family-Based Association Studies," *Stat. Methods Med. Res.*, **9**, 563–587 (2000).

RESEARCH AND DEVELOPMENT CHALLENGES AND CONSIDERATIONS

GENOME RESEARCH AND MINORITIES

HENRY T. GREELY, J.D.

I. INTRODUCTION

"All these perplexities develop more and more the dreadful fruitfulness of the original sin of the African trade."

In 1820, when James Madison wrote these words to his friend the Marquis de Lafayette, genetics—let alone pharmacogenomics—was inconceivable. But it was already apparent that the problems stemming from America's enslavement of Africans would be complex and long-lived. Over 180 years later, this original sin has real consequences for genomics research, adding new issues to the factors that any person or any group should consider when deciding to participate in such research.

The descendants of slaves, along with many other minorities in the United States, have had an historical, and to some extent continuing, experience of oppression and discrimination. The scientific and medical professions participated in that experience. The "scientific racism" of the nineteenth century predated the (re)discovery of genetics in 1900; one could date it at least to Linnaeus' eighteenth-century classification of humans into four subspecies with its assignment of superiority to his own European subspecies. When the science of genetics arrived with the twentieth century, genetic explanations were quickly advanced for the perceived inferiority of minorities.

Pharmacogenomics: Social, Ethical, and Clinical Dimensions, Edited by Mark A. Rothstein. ISBN 0-471-22769-2 Copyright © 2003 Wiley-Liss, Inc.

At the same time, European-American physicians and scientists undertook medical research with minority groups, sometimes with little regard for the consequences of the research on their minority subjects. The Tuskegee study of untreated syphilis was perhaps the ironic highpoint of this disregard—ironic in that a study that appears genuinely to have been motivated to help African Americans as a population so ignored its own devastating consequences on some individual African Americans. The high incidence of medical research at inner city teaching hospitals has, in recent decades, preserved the feeling among some ethnic groups, particularly African Americans, that they are being used as "guinea pigs." These experiences have led some ethnic minorities to view scientific and medical research with understandable concern.

Human genetics research, of course, has raised concerns for many people, inside and outside minority groups. But members of minority groups who are asked to take part in such research should have special concerns, both about human genetics research in general and about pharmacogenomics research in particular. This chapter discusses those concerns in three categories: general concerns with special implications for minority groups, special concerns about the effects of the research on racism, and fears about the effects of successful pharmacogenomic research on members of minority groups. Although ultimately research into human genomics offers great potential benefit to all peoples, it needs to be undertaken in ways that protect the peoples, and individuals, who take part in it. This chapter makes specific recommendations for protections that researchers should provide to minority groups participating in genomics research—and that individuals participating should demand.

II. General Concerns with Special Implications for Minority Groups

Current rules for human subject research are inadequate to deal with several concerns raised by modern human genomics research (Greely, 2001). Research rules designed mainly to protect individuals from the risks of physical harm do not protect either individuals or the groups to which they belong from risks that are less tangible, but no less important. Genetic research for both medical and anthropological purposes raises these concerns. The risks can be magnified for minority groups because of their relative lack of power and the possibility of discrimination against them. Of course, the term "minority groups" encompasses vastly different populations that confront very different issues. African Americans may well have different concerns, and different ways to address them, than Hmong refugees from Laos or Hutterite religious colonists. Nonetheless, genetic research with minority groups raises distinctive questions in five areas:

informed consent, control over research materials and data, confidentiality, return of relevant information, and commercialization.

A. INFORMED CONSENT

In the United States, as in most other countries, individual informed consent is generally required before research may be done with a living, identifiable human being. Informed consent for research generally has two different justifications. First, it allows the potential subject to weigh for himself or herself the balance of risks and benefits from his or her participation in the proposed research. Second, it embodies the individual's autonomy and control over interventions in or about his or her own body.

When genetic research is about *groups* of people, these justifications can appear in a different, collective perspective. Research that is "about" a group of people—the Old Order Amish, the Irish, the Ashkenazim—has some potential risks and benefits that affect the entire group. Under existing rules, the decision of a few individual members of the group to participate in the research could lead to the publication of research results about the group that could have negative consequences for the group. Its reputation might be stigmatized, its members' insurability could possibly be hampered, and its culture could be changed—all without any consent from the group itself.

The realization that at least some of the risks of genetic research on groups fall on the entire group, not just on those group members who participate, led the North American Regional Committee of the Human Genome Diversity Project (HGDP) to call for researchers to obtain group consent to genetics research whenever feasible (North American Regional Committee, 1997; Greely, 1997a; Greely, 1997b). The North American Committee's "Model Ethical Protocol" declared that the HGDP would only collect samples in North America when it had both the consent of the individual and, where it was possible, the consent of the "culturally appropriate authorities" of the group involved. If the group had not consented, the Project would not collect samples even from an enthusiastically willing individual subject. The Model Ethical Protocol recognized that both defining the group and ascertaining its "culturally appropriate authorities" would often be difficult; it pointed out, though, that in some cases it would be easy. For example, a federally recognized Native American tribe would, presumptively at least, be a "group" and would have, in its tribal government, at least one culturally appropriate authority. It conceded, though, that some other ethnic groups, such as Irish Americans or Ashkenazi Jews, would not have any "culturally appropriate authority." In those cases, it mandated broad discussion in the local community where collection was taking place.

This suggestion of group consent kicked off a lively debate that, five years later, persists in the literature of research ethics. The North American Committee's position has found many opponents (National Research

Council, 1998; Juengst, 1998a, 1998b; Reilly, 1998; Reilly and Page, 1998) and few supporters (Freeman, 1998). It has, however, spawned substantial discussion of an intermediate position, sometimes called "community consultation," which would require researchers to discuss their proposed research broadly with the affected group, although not to seek or obtain their formal consent (Foster et al., 1999; Foster and Sharp, 2000; Foster et al., 1998).

The issue of group consent or consultation is not an easy one. There are difficulties in implementing such a requirement, and it would restrict an individual's choice to participate in research. As Juengst has argued, it might give a false impression that culturally defined groups had a genetic identity. And, for many researchers, it would be one more barrier, consuming time and money, between them and their research. Yet, for minority groups, group consent, *de jure*, or, through a strong consultation requirement, *de facto*, appears quite appealing. Only with group consent or consultation requirements can the entire group learn about, discuss, and make either a decision or a recommendation on participation. Without such a requirement, genetic research about the group might be undertaken and its results published without almost any group knowledge and with no group control.

If minority groups have reason to be particularly worried about the use of genetic research about them, some requirement of group consent or consultation becomes an essential procedural step toward addressing those fears. This is particularly true because some of the other protections, discussed below, would require negotiations and choices that could only be done on a consistent basis by the group and not by its individual members. Although in almost every case except that of federally recognized Native American tribes and Native Alaskan villages, minority groups in the United States do not have "governments" that can play a role in group consent or consultation, usually there will be organizations, local or national, that aim to represent the group's interests. Those organizations, perhaps with some training or federal financial support, could consider the consequences of particular research plans and negotiate protections for their groups. The difficulty in selecting representative organizations that will fairly represent the groups, including the fear that the organization will largely protect its own interests, makes this solution less than perfect. It remains, I believe, better than both the status quo and the suggestion that potential research subjects be warned of possible group effects during the individual informed consent (Juengst, 1998a).

B. CONTROL OVER RESEARCH MATERIALS AND DATA

In an older and simpler model of research, samples are given by research participants to individual scientists, for use for a single purpose, by that scientist's own laboratory. This model was never entirely accurate. Among other things, researchers have long had an obligation to provide materials to others who are seeking to replicate their published results (Kennedy, 2002). But in the genomics era, the variations from the simple model may be great. The same set of samples, or data derived from those samples, may be shared with hundreds of laboratories, which might use it for purposes far beyond those expected by the people whose DNA and data are being shared. Although a DNA sample is a finite resource, if blood from an individual has been preserved as a transformed cell line, a set of indefinitely reproducing white blood cells from the patient, a potentially infinite amount of DNA can be derived from that sample. And, of course, once the samples have been transformed into data, electronic file sharing makes them easily and broadly accessible without any limits.

The mobility of samples and data means that samples or data provided by research subjects could be used by people they never met—and hence with whom they never developed a relationship of trust. They may also be used for purposes different from those for which the samples were collected and possibly purposes that the research subject does not like. Research subjects will rarely learn of these secondary users and uses of their samples and data. When they do, the collecting researcher may be able to point to broad language in a consent form authorizing their sharing of samples or broader research aims. Nonetheless, participants may feel misled and harmed. For example, at least one band of Canadian native peoples alleged publicly that it had been mistreated when samples it claimed it had given for purely medical research were put to anthropological uses (Kleiner, 2000; They Need Your DNA, 2000).

Research subjects may want to limit the research use of their materials and data to scientists they trust and ends of which they approve. It currently may take a lawyer's reading of the consent form to determine whether the materials and data may be shared or used for other purposes. Even if the form does limit who can use the research materials and data and for what purposes, the extent to which those forms can and will be enforced is limited. Minority groups with special concerns about the use of samples and data from their groups need to reach clear and enforceable understandings with researchers. Groups may want to limit the research use of their samples and data to the questions initially posed or further questions expressly approved by the group. Similarly, they may wish to limit to whom the samples and data can be transferred.

C. CONFIDENTIALITY

Human genetics research has long recognized the importance of confidentiality to many of the people who agree to participate in the work. Researchers typically agree to keep an individual subject's identity confidential as far as possible. Of course, perfect confidentiality is difficult to achieve. A researcher with the best intentions in the world may end up breaching confidentiality for reasons ranging from computer hackers to court orders. Close-to-perfect confidentiality could be achieved by making all samples and data completely anonymous, so that neither the researchers nor anyone else can connect a particular sample or information with a particular person. Such anonymity has its costs, though. The researchers cannot tie any new information about the subject to the existing file, nor can they return any information to the subject.

Confidentiality can also be gained at the group level through qualified anonymity. Some have suggested that instead of identifying a particular ethnic group as the subject of the research, publications could add some intentional vagueness (Foster and Freeman, 1998). Publications could talk of research with "subjects from a Southwestern Native American tribe" or "U.S. residents born in Southeast Asia." This kind of anonymity also has its costs. The research loses some value if readers do not know whether the group studied was Navajo or Pueblan, Khmer or Thai. And the research is made difficult or impossible to replicate. Furthermore, the chances that the research will lead to some medical benefit for the research population will inevitably be diminished if no one other than the original researchers knows the studied group's identity. Still, although there are costs to concealment, there are also costs to openness for participating minority groups. The group itself should consider how much anonymity, if any, it wants and then negotiate on the subject with the researchers.

D. Return of Relevant Information

The question of returning relevant information derived from research to individual research subjects is a vexing one. On the one hand, the information might have substantial medical or other importance to the research participant. On the other hand, research findings are rarely definitive, research laboratories are rarely certified for clinical testing, and researchers are often not clinicians. The risk of returning misleading information can be substantial, whereas the cost of not returning clearly relevant information could be a preventable death.

Once again, an issue for individual research subjects gives rise to a slightly different issue for minority groups participating in research. In addition to the return of individually meaningful information, groups participating in genetic research may want the return of information learned about their group. To be useful, the information may have to be translated, perhaps into languages other than English and certainly into languages other than "Science." Just sending reprints of scholarly articles to research participants will rarely provide meaningful information. If asked, researchers should plan, and budget, for postresearch trips to explain their findings and answer questions. The caveat is important. Some groups may not want to know the research findings, considering them irrelevant or disruptive. The researchers should have an obligation to offer to explain their results to the participating group, not to force unwanted information on them.

E. COMMERCIALISM

In the idealized human genetics research of the past, individual professors worked with families that suffered from a high incidence of a particular disease, searching for genetic links that would allow testing for, and, it was hoped, ultimately prevention or treatment for, that family's genetic disease. The professors were motivated by altruism, the families hoped, ultimately, for prevention of or treatment for that family's genetic disease. This picture was never entirely true. Pharmaceutical companies, and then biotechnology firms, have long played some role in genetic research. Professors, even the most idealistic, no doubt were interested in research not just for altruism but for tenure, grants, and the Nobel Prize. And research subjects often would receive small cash payments for travel or other expenses.

However accurate the idealistic picture may once have been, it clearly is much less common today. The rise of biotechnology firms, often with academic researchers as holders of stock or stock options, has added new commercial interest to genomic research. So have universities, hungry for the royalty payments encouraged by the Bayh–Dole Act of 1980 and demonstrated by the Cohen–Boyer patent on recombinant DNA, which, during its 17-year life, earned Stanford and the University of California about $200 million. Outside of governments, human genomic research is now largely done by corporations, by academic scientists working with corporations, or by academic scientists whose universities are eager to find commercial opportunities in faculty research. Exceptions continue to exist, particularly in areas of human genetics not focused on disease, but they are not the rule. At the same time, at least some of the research has moved away from a focus on "family diseases." So-called associative research looks broadly for correlations between genomic data and medical information; pharmacogenomic research tries to find links between genetic variations and responsiveness to drugs. In both cases, the health of the research subject and his or her family is much less directly implicated.

Under these circumstances, it may be unrealistic to expect research subjects to give scientists their DNA, family histories, and medical records for free. If the scientists are looking for millions, prospective research subjects may well wonder why their contribution should be uncompensated. On the other hand, it is extremely rare that any one individual research subject leads to a genetic breakthrough. Locating and cloning disease genes usually requires contributions of medical histories and DNA from thousands of subjects. Any one subject who held out for payment could be bypassed.

Group research, on the other hand, may provide a group with realistic bargaining power. Researchers who want to study a particular group or a condition found at a high rate in that group may not have feasible alternatives. Compensation to the entire group, for the benefit of the group, also seems fairer than compensation only to those group members who are asked to participate. Of course, if the group is being researched because its members suffer from a high rate of a particular disease, the members may have a personal incentive to participate that surpasses money. On the other hand, those incentives also are affected by commercialization. Canavan disease is a deadly childhood genetic disease found at unusually high levels among Ashkenazic Jews. Parents of children with the disease several years ago provided assistance, including biological samples and funding, to a young scientist, Dr. Reuben Matalon, to look for genes responsible for the disorder. Their efforts were successful; the scientist cloned the gene in 1993 while working at Miami Children's Hospital, which patented his discovery. In October 2000, some of the parents, and the Canavan Foundation, sued, arguing that Miami Children's Hospital's licensing policy made the prenatal test for Canavan disease too expensive and too hard to obtain. The Canavan families and the Canavan Foundation, having contributed to the research, want some control over its fruits.

Minority groups asked to participate in genetic research may well seek some kind of share of any financial benefits from the research. The idea of sharing financial benefits has been endorsed, in one form, by the Ethics Committee of the Human Genome Organization, an international scientific group that seeks to facilitate and coordinate human genomics research. In April 2000, that Committee issued a "Statement on Benefit Sharing." In general, the statement rejected the idea of direct sharing of benefits with research participants in favor of a broader obligation on "profit-making entities [to] dedicate a percentage (e.g., 1%–3%) of their annual net profit to healthcare infrastructure and/or to humanitarian efforts" (HUGO Ethics Committee, 2000). After noting that any promised benefits should not serve as an undue inducement to individuals or communities to participate in research, the Committee went on to argue that:

> In the very rare case where the extended family or a small group/tribe harbours an unusual gene, yet the research eventually benefits those with another disorder, justice may require that the original group deserve recognition. In this situation, benefits could be provided to all members of the group regardless of their participation in the research. Limiting the returns to only those who participated could create divisiveness within a group and is inconsistent with solidarity.

Minority groups, especially those disproportionately poor or excluded from good health care, may want to negotiate some share of the financial benefits derived from research with their group. It is important that those involved realize that most research does not lead to pharmaceutical products and most products do not lead to large revenues. A small royalty, however, to be used for the overall benefit of the group and not just of the research participants, might well be appropriate. So might direct provision of health benefits to the community (North American Regional Committee, 1997). When minority groups, particularly disadvantaged groups, and their members play indispensable roles in research leading to successful products, some sharing of the financial benefits seems just.

III. HUMAN GENOMICS RESEARCH AND RACISM

The problems discussed above apply to a wide range of groups, from disease organizations to extended families. Research with minority groups in the United States runs into another risk, not encountered by other kinds of groups—racism. The connections between race and genetic research are complicated and unclear (Lee et al., 2001; Foster, 2002), but research into human genomics *should* undercut the idea that human "races" have any significant genetic basis.

Although not doubting their cultural reality, human population geneticists deny that human genetic races exist (Cavalli-Sforza, 2000; Templeton, 1998). Humans are too similar and their variations too widespread to fit into separate slots. Humans are much more genetically homogeneous than chimpanzees, our nearest relatives, and the differences that exist on a continental scale are, quite literally, superficial—differences in our surfaces that interface with the outside world and its different conditions, such as our skin, noses, and eyes. Almost all genetic variations are found in almost all human populations. Thus, for the ABO blood types, every (or almost every) population has some members who are type A, B, AB, and O. The percentages of each type vary from group to group, but each group contains all four variants. Finally, each human group contains much more genetic variation within its own group than there is between groups. The genetically "average" Irish person is much more similar to the genetically "average" Japanese person than either is to most members of his or her own group. Human population geneticists firmly believe that this genetic meaninglessness of race is established by existing data and will only be further confirmed by any new data. But although such research should help further bury "scientific racism," when warped by the nation's original sin of slavery and its concomitant racism, genetic data could end up furthering racism through stigmatization, through intentionally misused results, or through a process that mistakenly focuses on ethnic differences.

A. STIGMATIZATION

So how could this research abet racism? One possible route would be through disease-based stigmatization. Many ethnic groups have higher, and lower, rates of some genetic diseases than other groups. Phenylketonuria is found at unusually high rates among the Irish. Sickle-cell anemia is particularly common among sub-Saharan Africans and their descendants (along with some Mediterranean populations, including those from one region of Greece), whereas the related β-thalassemia occurs at high rates in both the Mediterranean and Southeast Asia. Tay–Sachs disease has a relatively high incidence among Ashkenazic Jews and French-Canadians. Huntington disease is found at astoundingly high levels in some villages in Venezuela along the shores of Lake Maracaibo. In most cases, the explanation is thought to be the bad luck called the "founder effect"—a rapid expansion of a population whose founders, by chance, included people with the genes that cause those diseases. In some cases, as with sickle-cell anemia and β-thalassemia, the high levels of disease are actually tied to an evolutionary advantage termed "heterozygote advantage." People who carry one allele for those two hemoglobin disorders do not have the diseases but are much less susceptible to malaria. Their improved survival in malarial regions (like sub-Saharan Africa, the Mediterranean basin, and Southeast Asia) compensate for the increased death rates of those who carry two copies of the allele and have the disease.

Whether the high rates of particular diseases stem from the founder effect, heterozygote advantage, or other unknown causes, it is possible that populations with a higher rate of such genetic diseases would be stigmatized by others as particularly unhealthy. The consequences could, in theory, include employment and insurance discrimination, marriage disadvantages, and the development of a public impression of genetic inferiority. At least some minority groups have faced this; African Americans confronted irrational discrimination when testing for sickle-cell trait was widely introduced in the 1970s (Duster, 1990). Other groups have more recently expressed concern about possible stigmatization. One newspaper article reported the following:

> "While Jews understand the cool rationale as to why their genes are so coveted by scientists, they worry the reasoning will be lost on the public who may be misled into believing Jews have more genetic disorders than other groups.
>
> 'Society doesn't do well with subtleties,' said Judith Tsipis, a biology professor at Brandeis University. 'We live in an era of one-liners.'" (Wen, 2000).

This concern about stigmatization is, at its heart, an odd one. Public health statistics on varying levels of disease exist, both for nations and, in some cases, for ethnic groups within nations. Those statistics already show that some groups have higher rates of certain diseases than other groups do—but lower rates of other diseases. Even if observers focus only on the diseases with increased incidence, genetic research into the precise alleles responsible for the disease should not add information about the relative incidence of the disorder in different ethnic groups. Even if the stigma lies not in the higher rate of the disease but in the disease's genetic nature, in many cases, particularly with the classic "simple Mendelian" genetic disorders, the genetic nature of the disease will already be well known. The stigma, if any, must come from the greater publicity given the genetic findings as a result of research. Even there, it is not clear whether believing that a disease is the result of a person's genes is more or less stigmatizing than believing it is the result of diet, environment, habits, or bad luck.

B. INTENTIONALLY MISUSED RESULTS

The population geneticists' argument that human genetics undercuts racism presupposes that people will engage in their science honestly and fairly. Demagogues or racists are both creative and not bound to the standards of fair argument. Given that some genetic variations will be more common in some populations than others, it would be simple to claim that a variation found in a favored group had good consequences, but one found in a disfavored group was negative. One can even imagine, once the chimpanzee genome is sequenced, a careful search by racists to find regions of genomic sequence where a higher than average percentage of a disfavored group happens to have the same base pair as at least one chimpanzee.

The risk of this kind of intentional misuse of population genetics against minority groups—or, perhaps, by some minority groups in their own favor—cannot be dismissed. The population is largely ignorant of population genetics but is convinced that genes are magic (Nelkin and Lindee, 1995). In athletics, for example, popular discussion continues about whether certain ethnic groups are genetically better at particular sports (Anderson et al., 2000). Kenyans, for example, have been acclaimed as genetically prepared to be excellent distance runners, disregarding not only the effects of culture and environment, but the rotation over the last century of the leading distance runners from country to country, with Finns, Moroccans, and now East Africans dominating at different times. Our cultural predispositions make it possible for racists to twist the results of any genomic research with minority groups, no matter how accurate and unbiased, to try to make their points. This danger should be considered as one of the risks of any minority group participation in genomics research.

On the other hand, the risk can be used to obtain a commitment by the researchers. Researchers should feel compelled to protect their research subjects from harm, including the protection of minority groups from the racist misuse of their genetic data. Scientists who wish to engage in genetic research with minority groups should commit themselves to combating the misuse of their results. The founding document of the HGDP expressed it this way:

Human history—and the human present—is full of racism, xenophobia, hypernational-
ism, and other tragedies stemming from beliefs about human populations. In the past,
some of those tragedies have been perpetrated by, or aided by, the misuse of scientific
information. All those involved in the HGD Project must accept a responsibility
to strive, in every way possible, to avoid misuse of the project data. *[emphasis in*
original] (HGDP, 1994).

All genetic researchers whose might be similarly misused should accept
a similar responsibility—and the groups with which they work should
demand such a commitment.

C. Misfocused Process

Some genetics research with minority groups seems, by its nature, to require
classification by ethnic status. Thus a researcher seeking to find genetic connections between, for example, a Native American population and a native
Siberian population would need to label people as being members of those
groups. But not all anthropological research needs to use ethnic identities.
In the early discussions about the HGDP, two of its prime movers, Luca
Cavalli-Sforza and Allan C. Wilson, disagreed strongly on the Project's proposed sampling strategy. Cavalli-Sforza wanted to sample people from a
broad range of ethnic groups, defined largely on the basis of language.
Wilson wanted to do a random survey, sampling a certain number of people
found in arbitrarily defined squares of equal area on the Earth's land surface.
If one of the people sampled in the Parisian square was a Japanese tourist,
so be it. Cavalli-Sforza's view won out, in part for logistical reasons and at
least in part because of Wilson's untimely death. But Wilson's position
showed that even a large scale anthropological research project did not need
to characterize its participants by ethnicity.

Genetic research for medical purposes can also often dispense with
ethnic or racial identifiers, particularly in the area of pharmacogenomics. The
goal of pharmacogenomics is to find the underlying genetic variations that
affect, or control, individuals' responses to drugs. In some cases, those
responses are currently characterized in racial or ethnic terms, but, if successful, pharmacogenomics could dispense with saying, for example, "most
Asians respond poorly to this drug" and replace it with "humans with the
following genetic variation, of whatever ethnicity, respond poorly to this
drug."

The problem may arise if the search for those variations is conducted in
racial or ethnic terms. Studying an African American population to find the
genetic variations correlated with certain drug responses may tend to solid-
ify, in the minds of the public as well, perhaps, as in the minds of the
researchers, the idea that there are significant genetic differences that underlie racial categories. Even though the goal of the research may be to dispense

with these ethnic tags, if it proceeds by using them, it may reinforce their significance.

The answer to this is straightforward: avoid, whenever possible, the use of ethnic descriptions or ethnically defined populations in genetic research. Particular genetic variations linked to hypertension will no doubt found in some African Americans with high blood pressure—and in some other non-African Americans. Other African Americans with hypertension will have other genetic variations, which they will share with other non-African Americans. Members of minority groups should take part in human genetics research to build information that will improve human health. In many cases, though, there will be no need for them to participate as part of an ethnically defined population and good reasons to avoid it.

IV. PHARMACOGENOMICS AND MINORITY GROUPS

Finally, participating in pharmacogenomic research raises even more complicated questions for minority groups and their members. The basic idea of pharmacogenomics is to divide patient populations for application of particular drugs by allowing prediction of the safety and efficacy of the drug for particular individuals. The consequences of this market segmentation for minority groups are deeply unclear.

Note first that, despite its popularity as an industry slogan, it remains uncertain whether pharmacogenomics will ever play a significant role in medicine. The scientific basis for the assumption that different drug reactions are usually (or even often) caused by different genotypes remains unproven. Even if the scientific basis is solid, substantial business, regulatory, and liability issues—explored in other chapters of this book—remain. It is sobering to think that the consequences of carrying different alleles of various genes in the cytochrome P450 gene family for drug metabolism have been known for over a decade. These genes affect the effectiveness and safety of numerous common and important pharmaceuticals. But testing of patients for their pharmaceutically relevant variants of different cytochrome P450 genes scarcely exists. The consequences of research into pharmacogenomics for minority research may not prove to be worth worrying about because pharmacogenomics may never significantly develop.

But assume that those barriers are overcome and a pharmaceutical company markets an important drug based on pharmacogenomics. Assume, for example, that pharmacogenomics research allows the firm to know which 80% of the population will benefit from the drug, which 10% will receive absolutely no benefit, and which 10% will be seriously harmed by it. The short-term consequence would be to improve health outcomes of everyone, not just the 80% who would benefit but the other 20% who should not take the drug. The dynamic consequences are less clear. If such pharmacogenomic research were done, would pharmaceutical companies pour research funds into the search for a drug, or drugs, for the other 20% of the affected population or would they be content with 80% of the market, leaving the 20% as "orphan genotypes"? If pharmacogenomic research were not used, would the pharmaceutical companies successfully develop a drug that would treat all 100% or would they abandon the whole market, including the 80% who could benefit from the existing drug? How that question is answered may have major implications for whether the research has good or bad consequences, but, at this point, it is unanswerable. It depends on too many unknown facts—and may well vary from drug to drug, disease to disease.

Finally, whatever the general consequences of such a pharmacogenomic drug, its effects on minority groups are unknowable at this point. It is tempting to think that members of American minority groups will be found disproportionately among those for whom developed drugs are either not safe or not effective, but this assumption may well be wrong. It is based, at heart, on the myth that consistent genetic differences separate racial and ethnic groups. We know they do not, in general—but that does not mean that the frequency of some genetic variations of significance for pharmacogenomics may not vary by ethnicity.

So, what is to be done? This dilemma argues strongly for incentives to encourage firms to find drugs for "orphan genotypes." One route might be a revision of the Orphan Drug Act, which provides added market exclusivity to the makers of drugs with a small market, to cover drugs whose pharmacogenomic market is small. Direct grants, loans, and NIH funding could also encourage development of effective medications for as many people as possible. This may have a disproportionately useful effect on minority groups, or it may not. But, with pharmacogenomics, we all find ourselves as potential carriers of such orphan genotypes. Minority, plurality, or majority, we each live behind a Rawlsian veil of ignorance, not knowing whether this research will help us or hurt us (Rawls, 1971). As John Rawls argues, confronted by that uncertainty, we should hedge our bets and try to ensure that we will be well treated, no matter what our characteristics turn out to be. Protection of those with orphan genotypes is an insurance policy for all of us that seems likely to be worth the price.

V. CONCLUSION

Slavery, and racism, its inextricably intertwined sibling, are America's original sin. They continue, nearly five generations after emancipation, to influence many aspects of our society. Genomics research is no exception. This research might help alleviate human suffering, inside and outside minority groups, and it would be a shame if members of minority groups shunned participation in it. But at the same time, the burdens of history impose special risks on, and raise special questions for, minorities considering such research. Both researchers and potential research subjects need to consider those questions carefully in the ambitious hope of erasing that sin, or at least limiting its ongoing damage.

REFERENCES

Anderson, J.L, et al., "Muscles, Genes, and Athletic Performance," *Sci. Am.* (Sept. 2000).

Cavalli-Sforza, L.L., *Genes, Peoples, and Languages.* New York: Northpoint Press (2000).

Duster, T., *Backdoor to Eugenics.* New York: Rutledge (1990).

Foster, M.W. et al., "The Role of Community Review in Evaluating the Collective Risks of Human Genetic Variation Research, *Am. J. Hum. Genet.*, **64**, 1719–1727 (1999).

Foster, M.W. and R.R. Sharp, "Involving Study Populations in the Review of Genetic Research," *J. L. Med. Ethics*, **28**, 41–51 (2000).

Foster, M.W., et al., "A Model Agreement for Genetic Research in Socially Identifiable Populations," *Am. J. Hum. Genet*, **63**, 696–702 (1998).

Foster, M.W. and W.L. Freeman, "Naming Names in Human Genetic Variation Research," *Genet. Res.*, **8**, 755–757 (1998).

Freeman, W.L., In: Weir, R.F. (ed), *Stored Tissue Samples: Ethical, Legal, and Policy Implications.* Iowa City: Univ. Iowa Press, pp. 267–301 (1998).

Greely, H.T., "The Control of Genetic Research: Involving the Groups Between," *Hous. L. Rev.*, **33**, 1397–1430 (1997a).

Greely, H.T., "The Ethics of the Human Genome Diversity Project: The North American Regional Committee's Proposal Model Ethical Protocol," in *Human DNA Sampling: Law and Policy—International and Comparative Perspectives.* Kluwer Law Intl, pp. 239–256 (1997b).

Human Genome Diversity Project, Summary Document (1994), available at *http://www.stanford.edu/group/morrinst/hgdp/summary93.html.*

HUGO Ethics Committee, Statement on Benefit Sharing (April 9, 2000), available at *http://www.gene.ucl.ac.uk/hugo/benefit.html.*

Jones, J., *Bad Blood: The Tuskegee Syphilis Experiment*, 2d ed., New York: Maxwell McMillian Intl (1993).

Juengst, E.T., "Group Identity and Human Diversity: Keeping Biology Straight from Culture," *Am. J. Hum. Genet.*, **63**, 673–677 (1998a).

Juengst, E.T., "Groups as Gatekeepers to Genomic Research: Conceptually Confusing, Morally Hazardous, and Practically Useless," *Kennedy Inst. Ethics J.*, **8**, 183–200 (1998b).

Kleiner, K., "Blood Feud," *New Scientist*, 7 (Sept. 30, 2000).

Lee, S.S.-J., et al., "The Meanings of 'Race" in the New Genomics: Implications for Health Disparities Research," *Yale J. Hlth. Policy, Law, Ethics*, **1**, 33–75 (2001).

Madison, J., "Letter to Marquis de Lafayette, Nov. 25, 1820," in *Letters and Other Writings of James Madison*, Vol. 3, 190 (Philadelphia, 1865).

National Research Council, *Evaluating Human Genetic Diversity*, Washington, DC: National Academy Press (1997).

Nelkin D. and M.S. Lindee, *The DNA Mystique: The Gene as Cultural Icon* New York: Freeman (1995).

North American Regional Committee, Human Genome Diversity Project, "Proposed Model Ethical Protocol for Collecting DNA Samples," *Hous. L. Rev.*, **33**, 1431–1473 (1997).

Rawls, J., *A Theory of Justice*. Cambridge, MA: Belknap Press of Harvard University Press (1971).

Reilly, P.R., "Rethinking Risks to Human Subjects in Genetic Research," *Am. J. Hum. Genet.*, **63**, 682–685 (1998).

Reilly, P.R. and D.C. Page, "We're Off to See the Genome," *Nat. Genet.*, **20**, 15–17 (1998).

"They Need Your DNA," *New Scientist*, 3 (Sept. 30, 2000).

Wen, P., "Jews Fear Stigma of Genetic Studies," *Boston Globe*, p. F1 (Aug. 15, 2000).

5

DRUG DEVELOPMENT STRATEGIES

PENELOPE K. MANASCO, M.D. AND TERESA E. ARLEDGE, D.V.M.

I. INTRODUCTION

Genetics will fundamentally change the practice of medicine and the process of drug development. The timing of this massive change is not certain; however, it is likely to occur in the next 5–10 years. The reasons for the change include the evolution of the science and technology in the fields of the human genome, single-nucleotide polymorphism (SNP) maps, genotyping, and bioinformatics. We are living through a momentous time when the convergence of scientific and technological developments has resulted in the possibility of new approaches to unraveling the mysteries of disease and health.

DNA was described less than 50 years ago. In 2000, the draft sequence of the entire human genome was released by the Human Genome Project (HGP), a publicly funded effort, and Celera, a privately held company. Through the discoveries enabled by the HGP, the SNP Consortium (a group of 11 pharmaceutical companies, 5 academic centers, 2 information technology companies, and the Wellcome Trust) expanded the map of the genome that had been used for the past 10 years from 400 markers to 1.7 million markers. The impact of this new map can be imagined in the following way. If you think of the distance from New York to California as the genome, the road signs would change from one every 7.5 miles to one every 9 feet. The markers (or road signs) are SNPs. The SNP Consortium released the map into the public domain so that anyone doing gene mapping experiments could use this new SNP map. The original 400 markers were difficult to

Pharmacogenomics: Social, Ethical, and Clinical Dimensions, Edited by Mark A. Rothstein.
ISBN 0-471-22769-2   Copyright © 2003 Wiley-Liss, Inc.

measure and required significant laboratory personnel time. In contrast, SNPs are much easier to measure and high-throughput assays can be developed, decreasing the cost and time required to do whole genome scans.

Advances in measurement of gene expression have also been phenomenal. In five years, the numbers of genes, sensitivity of the assays, and reproducibility of results have also increased significantly as has the ability to analyze the data. The measurement of gene expression has been used in several ways in cancer genetics and cancer pharmacogenomics. Gene expression has led to better ways to classify cancers and to select the appropriate therapy (Miyazato et al., 2001; Birner et al., 2001).

Through a different avenue of technological development, the field of bioinformatics has also developed into a specialty in its own right. Computational experts take data from multiple sources (genetic data from different species, different tissues, and different types of studies, including the scientific literature) and make sense of it (Searls, 2001). Data exploration techniques that were developed through the study of vast amounts of biological data, data from space, and even data from the banking industry are now being focused on the problems of understanding the copious complex data from the genome.

The growth of the biotech industry has helped the field of genotyping technology development to bloom. Many companies with competing technologies are trying to meet the challenge of taking genotyping from a cost of $10.00 per genotype down to $.10 in a period of five years—with throughput increasing exponentially with the advent of high-throughput genotyping using SNPs.

The stage is now set to use all of these discoveries to improve the way we diagnose, treat, and even prevent disease. The changes in understanding disease will lead to new targets and better therapeutics, as well as changes in drug development. Although this chapter is designed to discuss the changes that the genetic and genomic revolution will make in drug discovery and development, the changes to the rest of the practice of medicine will be similarly astounding.

II. THE PHARMACEUTICAL INDUSTRY

The pharmaceutical industry faces many challenges today. Despite an increase in research and development (R&D) spending of more than $30 billion per year; there has actually been a decline in new drugs approved by the FDA on a yearly basis, with fewer than 30 new drugs approved in 2001. The attrition rate is still exceedingly high, and only one in 1000 compounds that are developed actually makes it to the market and only one in 10 of those compounds is a commercial success (as defined by sales of over $500 million per year). Each compound today costs approximately $800 million and over 10–15 years to develop and bring to the marketplace. When each compound fails, it is often not clear whether the failure can be attributed to the characteristics of the specific compound or the target. Most companies have taken the approach of bringing several compounds in different chemical classes forward for each biological target to try to minimize the risk that the toxicity of a single compound will derail the evaluation of a molecular target. Thus if the lead compound has unacceptable safety issues associated with early testing, a compound from another class of drugs is less likely to have the same safety concerns.

The costs of adverse events can be measured in many ways. Since 1997, 13 drugs were taken off the market because of unacceptable side effects (*http://www.fda.gov/fdac/features/2002/chrtWithdrawals.html*). The costs to the patients are significant, both to those who suffer the adverse events and to those who responded to the medicines and were unable to take them once they had been removed from the market. Lazarou et al. (1998) estimated that the deaths from adverse events from drugs was between the fourth and sixth leading cause of death in the United States. Johnson and Bootman (1995) estimated that in 1995, the cost of morbidity and mortality of drugs was approximately $76 billion. In 2001, Ernst and Grizzle (2001) updated the outputs from the Johnson and Bootman model and estimated the total annual cost of drug-related problems among ambulatory Americans at $177.4 billion. The FDA has made the issue of drug safety such a high priority that it has started a new Office of Postmarketing Surveillance (http://www.fda.gov/cder/present/dia-699/opdra2-dia/). Recent publications by the FDA have stressed that individualized therapy through pharmacogenetics and pharmacogenomics offers the hope of maximizing benefit and minimizing risk to patients (Lesko, 2002).

Not only is safety a key concern, better defining the responder population is also critical. Table 5.1 presents a review of the data from the Physicians Desk Reference (PDR) showing the variability in efficacy rates (as defined by the percent of responders) for every class of drugs. The percent of responders range from a low of 25% (oncology products) to a high of 80% (Cox2 inhibitors), with the majority of drugs having a responder rate of 50–60%. There are costs associated with lack of efficacy, including direct costs such as additional visits to the health care provider and loss of productivity for the patient as well as the indirect costs of continuing to suffer from the signs and symptoms of the illness while trying to find an efficacious drug.

III. CURRENT APPROACHES TO DRUG DEVELOPMENT

Drug discovery and development is a long and arduous process taking an average of 15 years from identification of a potential target for a medicine (e.g., a receptor at which a medicine will work) to marketing a product. The

TABLE 5.1. The Efficacy Rate of Medicines Varies

Condition	Ave. Annual Rx Cost ($)	Efficacy Rate (%)
Alzheimer's	1,500	30
Analgesics (Cox2)	1,350	80
Asthma	330	60
Cardiac arrythmias	650	60
Depression (SSRI)	700	62
Diabetes	1,300	57
HCV	5,000	47
Incontinence	1,000	40
Migraine (acute)	240	52
Migraine (prophylaxis)	600	50
Oncology	3,500	25

Source: Physicians' Desk Reference (2002).

basic processes are illustrated in Figure 5.1. Genetics can impact each phase of the process. It is important to know that the impact of genetics and genomics on society will be felt soonest in genetics predicting the safety and efficacy of medicines in the marketplace. In contrast to popular belief, the impact will be greater in the traditional area of small molecules than in gene therapy.

The process of drug discovery encompasses identifying a biological target that appears to play a role in the disease of interest or a role in the symptom expression of the disease. From that point, small molecules are usually designed to increase the level of the target or block its activity. These molecules must also be designed so that the chemical synthesis is not too expensive and is amenable to making large quantities. Additional requirements for the chemical include properties that make it stable in heat and light and easily absorbed by the human gastrointestinal tract. Because people generally do not do well taking medicines more than twice a day, the chemical must also last in the body long enough to enable patients to take only two pills a day. Last but not least, the medicine must be nontoxic.

Genetics has been touted as playing a role in the very first part of this process: identifying new targets for the design of medicines. To date, this approach has been disappointing. Even though many susceptibility genes have been identified (e.g., NIDDM 1, low-density lipoprotein receptor-related protein 5, ApoE), few, if any, have resulted in new drug candidates. While the target work continues, there is value in using genetics for drug discovery for better disease understanding and the identification of new disease pathways that were not identified using traditional approaches.

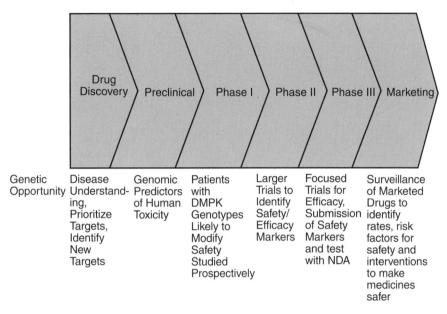

Genetic Opportunity	Disease Understanding, Prioritize Targets, Identify New Targets	Genomic Predictors of Human Toxicity	Patients with DMPK Genotypes Likely to Modify Safety Studied Prospectively	Larger Trials to Identify Safety/ Efficacy Markers	Focused Trials for Efficacy, Submission of Safety Markers and test with NDA	Surveillance of Marketed Drugs to identify rates, risk factors for safety and interventions to make medicines safer
Time frame For Impact	>15 years	10-15 years	5-10 years	5-7 years	5-7 years	3-5 years

FIGURE 5.1. Genetic Opportunities Occur Throughout the Drug Development Process.

Identification of new pathways will provide new insights into the pathogenesis of disease and new opportunities to develop preventative therapies. Clearly, discoveries such as the finding that ApoE4 allele confers susceptibility to Alzheimer disease has provided significant new avenues of research into the etiology of Alzheimer disease, stroke, and head trauma—insights that would not have been possible without the field of genetics.

The field of pharmacogenetics (the study of the genetic causes for variability of response to medicines) has provided additional disease insights. For instance, by studying the patients that responded to L-dopa, Martin et al. were able to identify a subset of patients with Parkinson disease who carried a susceptibility gene that had only been reported in a rare form of familial Parkinson disease in Japan (Martin et al., 2001). By studying the response to leukotriene modifiers, Drazen et al. (1999) and Anderson et al. (2000) were able to identify a group of polymorphisms along the leukotriene pathway that explained a significant proportion of patients who do not respond to leukotriene modifiers. The pathway and enzymes shown is illustrated in Figure 5.2. Not only may these results eventually be used to produce guidance to health care providers as to which patients should use leukotriene modifiers, these data may provide an additional clue to a subset

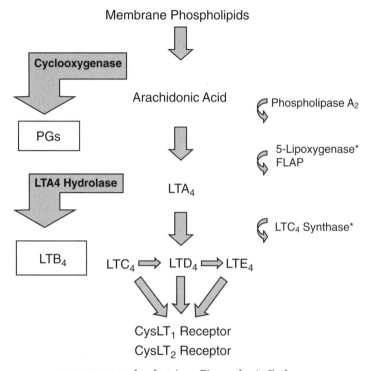

FIGURE 5.2. Leukotriene Biosynthesis Pathway.

of patients with a different etiology for their asthma. Furthermore, some of the polymorphisms identified are much more common in African Americans, thus providing insights into differences in asthma in different ethnic groups.

Most, if not all, pharmaceutical companies are now identifying the genetic variability around the targets for their molecules. This is absolutely critical to be sure that the chemical molecule binds to the most common form of the biological target and not a rare form or a form found only in animals.

The costs of drug development increase as a drug progresses through the development pipeline. Preclinical development is the time when chemical compounds are tested in the laboratory to learn as much as possible about how medicines work "in a test tube." These types of experiments can be done with many compounds in a relatively short time and with relatively low cost. This is also the time that animal testing begins to see whether the chemical compounds are safe. These studies help scientists to determine how medicines will be dosed in humans. They are also important to understand whether any toxicity is related to the medicines. Toxicological tests are time consuming and labor intensive. They are often the rate-limiting step in the preclinical development of compounds, and a winnowing of the number of compounds is necessary before toxicological testing begins. If high-throughput tests could be developed to predict which compounds would be toxic in animals or humans, this would significantly improve the chances of safe and effective medicines getting to market, because many more compounds could be tested. The field of genomics offers real hope in this area, a field sometimes called toxicogenomics. Through this approach, scientists will study the gene expression profile of various tissues from various animals in response to drugs that are known to cause toxicity. It is hoped that a specific profile can be identified that is predictive of toxicity. This approach is in the development phase but offers tremendous opportunity to streamline the drug development process.

IV. Clinical Development

A. Phase 1 Trials

Phase 1 trials are designed to identify the early tolerability of new medicines, usually in healthy volunteers. At this time, the pharmacokinetics and pharmacodynamics of a medicine are determined in humans, thus forming the base of knowledge on how medicines should be dosed.

A genetic basis for drug response is not a new concept. As early as 1902, Archibald Garrod hypothesized that genetic variance in a biochemical pathway for the detoxification of a foreign substance was the cause of alcaptonuria (Garrod, 1902). During World War II, it was noted that hemolysis related to antimalarial treatment was much more common among African American soldiers, leading to the identification of inherited variants of glucose-6-phosphate dehydrogenase (G-6-PD). It was during this time that scientists discovered that the prolonged muscle relaxation and apnea after suxamethonium in some patients was due to an inherited deficiency of a plasma cholinesterase. Peripheral neuropathy was observed in a significant number of patients treated with the antituberculosis drug isoniazid, leading to the identification of genetic differences in acetylation pathways.

The current approach to drug development has been to avoid putting compounds into development that have as their primary route of metabolism one of the polymorphic 450 enzymes, particularly P4502D6. Nevertheless, it may be better to have a drug metabolized through a pathway that is well understood and for which appropriate dosing recommendations can be made than through a pathway where the polymorphic drug-metabolizing enzymes are not as well understood. It is likely that in the future, study of subjects who are known to have polymorphisms in the drug-metabolizing enzymes will lead to accurate dosing recommendations for a wider range of patients and perhaps better drug tolerance. Furthermore, studies of drug-drug interactions should include subjects with polymorphisms of the involved drug-metabolizing enzymes.

We are likely to see this type of approach for new medicines, but we should not lose sight of the need to do similar research on existing medicines. At present, there are many potentially important developments in pharmacogenetic research that have not made their way into clinical practice because translational studies have not been included. Although funding for the National Institutes of Health (NIH) has increased significantly over the past 10 years, the research needed to take basic science discoveries into clinical applications has not kept up. It is this research, funded by the Health Resources and Services Administration (HRSA) and the Agency for Health-care Research and Quality (AHRQ), that is critical to implementation of pharmacogenetics into clinical practice.

Most of the work involving pharmacogenetics of drug-metabolizing enzymes is in the investigational stage, but one test has made it into clinical practice. The thiopurine methyltransferase (TMPT) test identifies polymorphisms in the gene for the enzyme responsible for metabolism of thiopurine and other immunosuppressive drugs used to treat leukemia and other diseases (Weinshilboum, 2001). If patients have two copies of one of the polymorphisms of the gene that decrease activity, they are at risk for serious dose-related side effects, such as bone marrow suppression and even death. Carrying even one copy of the polymorphism can put patients at risk for increased side effects. At the Mayo Clinic, Memorial Sloan-Kettering, and other medical centers, genotyping of patients before treatment ensures that the appropriate dose is prescribed and patients recover from their cancer and are not harmed by their treatment.

B. Phase 2 Trials

Phase 2 trials are usually conducted in several hundred to a thousand subjects with the disease to be treated by the new medicine. Through this approach, preliminary evidence of efficacy can be obtained without exposing an excessive number of patients. The clinical trial population is very closely defined, excluding patients with concomitant diseases, the young, and the elderly. Phase 2 trial design will likely change significantly when pharmacogenetic approaches are incorporated into mainstream drug development. During this phase, evidence of a correlation between an efficacy and/or safety parameter and a pharmacogenetic marker could be identified. To enable this type of finding, collection of DNA will need to be part of the design of the study. Because genetic studies must be replicated to ensure that the findings are not spurious, two trials with commonly defined end points (safety or efficacy) will be needed to enable analysis across studies. Furthermore, the size of the studies may need to be larger to ensure adequate sample size to identify the pharmacogenetic marker.

The analysis of safety data may require larger data sets, so care in ensuring that a consistent phenotype is collected will be important. Furthermore, the benefit to the drug development program could be that additional data to predict adverse events could be available at the time a drug is marketed.

Collecting and banking DNA according to good laboratory practice (GLP) standards will be a critical activity during drug development. Tests used to identify genetic or genomic markers for safety and efficacy may be different from the tests used in clinical practice once a drug is on the market. Having banked samples with accompanying phenotypic data, companies will be able to transition easily from a test used during the research of drug development to a commercially available test. Furthermore, they can continue to develop better medicine response tests for the marketplace with these critical reference data and samples.

C. PHASE 3 STUDIES

Smaller, more efficient Phase 3 trials for markers of efficacy should be possible with pharmacogenomics. Figure 5.3 extrapolates data from the study of Kuivenhoven et al. on the effect of polymorphisms in the cholesterol ester transfer protein on differential response in prevention of vessel narrowing by Pravastatin to calculate new numbers for a Phase 3 trial (Kuivenhoven et al., 1998). The size of the study could be 156 if only the responders were studied, in contrast to 786 if all subjects were sampled. It must be recognized that although the treatment population is much smaller in a clinical trial in the responder population, three times as many subjects would need to be screened to identify the appropriate subpopulation. Nevertheless, it is likely that patient recruitment time would be reduced, the cost of the study would be significantly lowered, and the magnitude of effect would make the pharmacoeconomic justification much stronger than that of evaluating the entire population.

This is not the only example. Recently, polymorphisms in the G protein β3-subunit gene have been shown to predict response to hydrochlorothiazide (Turner et al., 2001). Psaty et al. identified a subset of patients with a variant of the α-adducin gene that were associated with a lower risk of myocardial infarction and cerebral hemorrhage with diuretic therapy (Psaty et al., 2002). In addition, Genaissance has indicated that it has identified markers that predict the response to the statins, a class of drugs used to lower cholesterol.

Drug safety is one of the first places where we are likely to see the benefits of pharmacogenetics. Pharmacogenetic approaches differ according to the rate of adverse events. Common adverse events with rates of 1% or more can be evaluated during a traditional drug development program. Com-

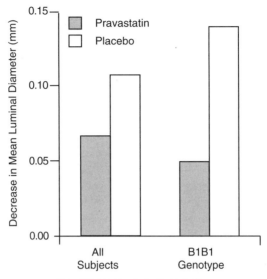

FIGURE 5.3. Smaller Phase III Trials are Possible when Basing Sample Size on Efficacy.

monly, 3000 patients or more are included in a regulatory submission; thus one can likely identify major genetic effects. The concern over the safety of marketed medicines has been the focus of the FDA's Center for Drug Evaluation and Research (CDER). Recently, CDER developed a new division, OPDRA, that will take on monitoring adverse events as well as more active collection of safety data.

There are numerous examples where genetic markers have been studied to identify predictors of a safety response. The difficulty has been that many cannot be replicated because of small sample sizes or different study populations. Recently, two separate groups reported an association between the development of hypersensitivity syndrome after treatment with abacavir and a polymorphism in HLA B57 (Mallal et al., 2002; Hetherington et al., 2002). Figure 5.2 illustrates that a single marker (HLA B57) predicts 50% of the cases. Mallal's data showed an even higher rate of predictability for the marker. The effects of TNF-α could largely be explained by the effects of HLA B 57.

In contrast, rare adverse events, with rates of less than one per thousand, will require different approaches. A case-control approach will be needed, and it will likely require release of the drug into the general population to

TABLE 5.2. Single Polymorphism can Predict
50–70% HSS with Abacavir

	TNF-α	HLA B57
Cases (HSS)	25/58	39/84
	43%	46%
Controls	7/99	4/113
	7%	3.5%

identify adverse events and to collect sufficient numbers of patients and controls to evaluate the genetic basis of these events. Thus a new approach to postmarketing surveillance will likely evolve. This approach has been used for a number of rare events. For instance, Pirmohamed et al. (2001) moved beyond the traditional approach of studying only the drug-metabolizing enzymes to study downstream immunologic events that would act as the final common pathway for numerous drugs that cause Stevens–Johnson syndrome (SJS), including carbamazepine. They showed that a genetic variation in tumor necrosis factor was associated with patients that developed SJS while on carbamazepine.

The Centers for Education and Research in Therapeutics (CERTs), funded by AHRQ, have funded an effort in SJS, and both AHRQ and the National Institute of General Medical Sciences (NIGMS) have funded efforts in prolonged QT. Several high-profile drugs have been withdrawn due to prolongation of QT, including cisapride.

One additional strategic issue must be recognized in drug development. Submission of data to support the use of a medicine response test must be coordinated with the submission of the application for marketing of the medicine. In the U.S., both centers are under the umbrella of the FDA, but this is not necessarily true in all countries. The timing of these submissions is critical to ensure that the test will be available at the time of the approval of the drug. Furthermore, as mentioned above, for maximum efficiency in the approval of subsequent test improvements, DNA samples of patients whose data support the claims for the medicine response test need to be maintained according to GLP.

The strategy for identifying genetic markers is evolving. Initially, it was thought that measuring only one polymorphism in a gene was sufficient to see whether there was an association between a gene and an outcome. It was also thought that polymorphisms were only important if they were in the coding region of the gene. Both of these assumptions have proven to be false. Many important polymorphisms have been identified in the regulatory control regions and in other noncoding regions. Furthermore, there are now many reports that show that in two SNPs within a few thousand base pairs of each other, one may be associated with a phenotype and another may not.

Furthermore, the associated SNP may also be associated with SNPs much farther away. The concept of haplotype analysis as an approach to identifying important gene:phenotype associations is now becoming the method of choice. This approach has been used by Drysdale et al. (2001) in their work on β-receptor polymorphisms and response to β-adrenergic drugs for asthma. The tools are now available to support a whole genome scan approach to identifying genetic markers to predict safety or efficacy. The whole genome SNP map was developed by the SNP Consortium (http://snp.cshl.org/) and released to the public domain. Several companies are developing marker sets to provide whole genome genotyping. The price is decreasing, and over the next 2–3 years this approach will be truly feasible. The benefit of this approach is that scientists can identify important new pathways that have not been previously recognized as important. The opportunity for new insights is truly exciting.

There will be an ethical advantage to this approach as well. Instead of using markers such as ApoE4 to predict safety and efficacy, which may carry significant collateral (and sensitive) information, the use of SNPs should minimize the risk of release of additional information.

V. Postmarketing Surveillance

Postmarketing surveillance is critical to ensuring the safety of medications because of the limitations of Phase 2 and 3 studies to identify rare, serious adverse events. Typically, these studies are too short and the study population is fairly healthy. Once a drug is marketed, the population of patients using the drug expands to include sicker patients, older patients, and other populations.

Currently, the OPDRA is using several techniques in postmarketing surveillance, which goes beyond just reviewing reported ADRs. As we know, not all serious ADRs are reported to the FDA; therefore, the OPDRA is also reviewing extramural drug data to better understand prescribing practices, reporting rate estimates, risk factor identification, off-label uses, and other critical factors in postmarketing surveillance. Additionally, OPDRA is involved in compliance studies, incidence studies, and active surveillance of specific events, such as acute, severe hepatotoxicity.

Pharmacogenetics can add significantly to the tools available for postmarketing surveillance. The current drug development effort does not include numbers of research subjects adequate to detect and understand serious, rare adverse events. Whereas some have advocated having a slower launch for new drugs or testing the drug products in more patients (both of which will add to the development time and cost of the new drug products), it is also possible to envision a pharmacogenetic postmarketing surveillance program for new drug products.

Pharmacogenetic postmarketing surveillance could be implemented through a network of trained sites in the U.S. and Europe where consumers would agree to have a blood sample stored for future testing if needed. With the use of the Internet, systems are being developed to enable education of both patients and physicians about medicines. Patients or their physicians could then report the adverse events, and accurate rates could be ascertained for the adverse events. Tests can be developed and evaluated in a prospective manner much more quickly, thus enabling patients who are responding to medicines to remain on treatment.

Furthermore, instead of just identifying a signal, the additional data and DNA samples enable scientists to rapidly study groups of patients that develop the adverse event and compare them with groups of patients who do not develop the event. Once a rare, serious adverse event has been identified, a genome-wide SNP scan could be performed on samples from patients with the serious, rare adverse event and compared with patients without the event, thus identifying the SNP markers associated with the rare adverse event.

The informative SNPs could be tested prospectively through the surveillance network, and decreases in rate could be confirmed. If confirmed, the tests would be rapidly approved and implemented as part of the risk-reduction strategy for the marketed drug.

This new type of pharmacogenetic surveillance will obviously require some discussion among regulatory authorities, bioethicists, and the pharmaceutical industry before implementation. It is critical that health care providers and patients understand how the blood or DNA sample may be used in future research (this is especially critical for patients chosen as controls). It is probable that, until this method of surveillance becomes the accepted norm in medical practice, institutional review board (IRB)/ethics committee (EC) approval will be required. Additionally, patients will need to provide informed consent before participating as part of the surveillance network. Other issues that will need to be addressed include any potential conflict with the European Union Data Privacy Act if data are transferred from Europe to the U.S. and any potential problems with health insurance in the U.S. if a patient participates in a pharmacogenetic postmarketing surveillance study. Impact on health insurance would be a significant concern because, although SNP markers are much less likely than candidate gene analysis to yield results that might be predictive of future health concerns, the informational risk, while smaller, is still there.

VI. ARE WE READY FOR PHARMACOGENETICS?

There will be a number of people who will be concerned about whether patients and health care providers are ready for medicine response tests and pharmacogenetics. Although the practice of medicine will undergo a radical change once medical response tests are available for a number of different drug products, it is likely that the shift will occur much more rapidly and easily than most people believe. For instance, the medical community and patients have rapidly accepted the need to test for viral mutations as a normal part of HIV therapy. Additionally, although Herceptin was only marketed in 1998, testing breast cancer biopsy specimens for overexpression of the HER-2 gene is now considered the standard of care.

Not only has the medical community been responsive to new approaches to prescribing medication, but the regulatory community has also started preparing for the influx of new submissions with pharmacogenetic or pharmacogenomic information. For example, FDA and other regulatory agencies have organized expert seminars and educational series to address issues in pharmacogenetics/pharmacogenomics. FDA is meeting with pharmacogenetic working groups to identify issues that will need to be resolved; it also has begun the process of developing long-range policies and discussing these issues with other regulatory agencies. Although there are still many issues to overcome, such as how to present these data in regulatory submissions, design of clinical trials, ad hoc and post hoc analyses, and development of a medicine response test, regulatory agencies are taking a proactive stance to prepare for the new information and technology that will soon be forthcoming (Lesko, 2002).

VII. SUMMARY

As Janet Woodcock and Lawrence Lesko (FDA) stated in a recent publication (Lesko and Woodcock, in press), "the central issue is not whether Pharmacogenetic or Pharmacogenomic-guided drug prescriptions will happen, but when and how." The practice of medicine and the way in which pharmaceutical companies develop drugs will undergo a radical shift in the upcoming years. It is critical that everyone involved in drug development and patient health care is ready for this paradigm shift so that implementation can be made smoothly and we can reap the benefits of the exciting new tool, pharmacogenomics.

REFERENCES

Anderson, W., et al., "A Single Nucleotide Polymorphism in the 5-Lipoxygenase Gene is Associated with a Reduced Response to Leukotriene Antagonists," *Am. Rev. Respir. Crit. Care Med.*, **163**(5), A198 (2000).

Birner, P., et al., "Evaluation of the United States Food and Drug Administration-Approved Scoring and Test System of HER-2 Protein Expression in Breast Cancer," *Clin. Cancer Res.*, **7**(6), 1669–1675 (2001).

Drazen , J., et al., "Pharmacogenetic Association Between ALOX5 Promoter Genotype and the Response to Anti-Asthma Treatment," *Nat. Genet.*, **22**, 168–170 (1999).

Drysdale, C.M., et al., "Complex Promoter and Coding Region Beta 2-Adrenergic Receptor Haplotypes Alter Receptor Expression and Predict in Vivo Responsiveness," *Proc. Natl. Acad. Sci. U.S.A.*, **97**, 10483–10488 (2000).

Ernst, F., and A. Grizzle, "Drug-Related Morbidity and Mortality: Updating the Cost-of-Illness Model," *J. Am. Pharm. Assoc.*, **41**, 192–199 (2001).

Garrod, A.E., "The Incidence of Alcaptonuria: A Study in Chemical Individuality," *Lancet*, **2**, 1616–1620 (1902).

Hetherington, S., et al., "Genetic Variations in HLA-B Region and Hypersensitivity Reactions to Abacavir," *Lancet*, **359**, 1121–1122 (2002).

Johnson, J., and J. Bootman, "Drug-Related Morbidity And Mortality. A Cost-Of-Illness Model," *Arch. Intern. Med.*, **155**, 1949–1956 (1995).

Kuivenhoven, J.A., et al., "The Role of a Common Variant of The Cholestereryl Ester Transfer Protein Gene In The Progression Of Coronary Atherosclerosis," *N. Engl. J. Med.*, **338**, 86–93 (1998).

Lazarou, J., et al., "Incidence of Adverse Drug Reactions in Hospitalized Patients, A Meta-Analysis Of Prospective Studies," *JAMA*, **279**, 1200–1205 (1998).

Lesko, L., and J. Woodcock, "Pharmacogenomic-Guided Drug Development: Regulatory Perspective Pharmacogenomics" (in press).

Mallal, S., et al., "Association Between Presence of HLA-B *5701, HLA-DR7, and HLA-DQ3 and Hypersensitivity to HIV-1 Reverse-Transcriptase Inhibitor Abacavir," *Lancet*, **359**, 727–732 (2002).

Martin, E.R., et al., "Association of Single-Nucleotide Polymorphisms of the Tau Gene with Late-Onset Parkinson Disease," *JAMA*, **286**, 2245–2250 (2001).

Miyazato, A., et al., "Identification of Myelodysplastic Syndrome-Specific Genes by DNA Microarray Analysis with Purified Hematopoietic Stem Cell Fraction," *Blood*, **98**, 422–427 (2001).

Physicians' Desk Reference, 55th ed. Montvale, NJ: Medical Economics Company (2001).

Pirmohamed, M., et al., "TNFα Promoter Region Gene Polymorphisms in Carbamazepine-Hypersensitive Patients," *Neurology*, **56**, 890–896 (2001).

Psaty, B.M., et al., "Diuretic Therapy, the _-Adducin Gene Variant, and the Risk of Myocardial Infarction or Stroke in Persons with Treated Hypertension," *JAMA*, **287**, 1680–1689 (2002).

Searls, D.B., "Mining the Bibliome," *Pharmacogenomics J.*, **1**, 88–89 (2001).

Turner, S.T., et al., "C825T Polymorphism of the G Protein Beta (3)-subunit and Anti-hypertensive Response to a Thiazide Diuretic," *Hypertension*, **37**, 739–743 (2001).

Weinshilboum, R., "Thiopurine Pharmacogenetics: Clinical and Molecular Studies of Thiopurine Methyltransferase," *Drug. Metab. Dispos.*, **29**, 601–605 (2001).

DRUG DEVELOPMENT, REGULATION, AND GENETICALLY GUIDED THERAPY

DAVID W. FEIGAL, JR., M.D., M.P.H. AND STEVEN I. GUTMAN, M.D.

I. INTRODUCTION

In 1906, the first consumer protections for medical products were created in the new U.S. Food and Drug Administration (FDA) and the Public Health Service (Young, 1992). Although focused initially on drugs and biological products, the law was expanded in 1976 to include medical devices, including in vitro diagnostic (IVD) tests and reagents (Food and Drug Law Institute, 1997). Across all medical products, the FDA's mission is to ensure safe use of investigational products, ensure a high level of manufacturing quality, approve marketing entry and claims by assessing the evidence that products are safe and effective, and monitor products in clinical use through the Medical Device Reporting System for adverse events to ensure that serious problems are identified and appropriately remedied.

Scientific advances in clinical pharmacology have assumed an increasingly important role in drug discovery and development (Parascandola and Keeney, 1983). The ability to develop assays to quantitatively identify and measure drugs and metabolites in biological fluids allowed pharmacology to provide a "tool kit" to guide new drug development. Animal pharmacology, both pharmacokinetics (PK) and pharmacodynamics (PD), provides

Pharmacogenomics: Social, Ethical, and Clinical Dimensions, Edited by Mark A. Rothstein.
ISBN 0-471-22769-2 Copyright © 2003 Wiley-Liss, Inc.

the rationale for selecting both a safe starting dose for first human use and, often, the rationale for the therapeutic effect. Drug absorption, distribution, metabolism, and excretion (ADME) studies in animals are expected to be performed preliminary to human use of a new candidate drug. The correlation of drug and metabolite concentrations with toxicity in animals and in human trials improves the safety of human use of the drug. The therapeutic window between the minimal dose for the intended effect and the dose at which toxicity becomes unacceptable is better understood when pharmacokinetic relationships to effectiveness and toxicity are defined. The evaluation of the route of administration and the interactions of drugs with each other has used pharmacokinetics to identify potential problems and to optimize intended effects.*

Given the established role of pharmacology in the planning and conduct of clinical trials, particularly early trials, what can knowledge about a patient's genome or genetic expression potentially add?

Much of the potential of this genomic information lies in its potential to contribute to clinical pharmacology by helping health care providers select drugs that will confer the greatest benefit with the least likelihood of harm. Generally, this will be done by identifying genotypic subgroups, not otherwise identifiable, that are responsible for differences in drug absorption, rate of drug metabolism, activity of drug metabolites (good and bad), drug induction of metabolism, metabolically mediated drug-drug interactions, and drug-disease interactions.

The ability of scientists involved in drug development to identify genotypic subgroups with variable metabolic behavior leading to differences in drug exposure and clinical response to drugs is expected to have a profound impact on clinical trials. This information, if used prudently and proactively, will allow for better selection of patients in terms of optimizing efficacy or minimizing risk, for better understanding of variability in patient response, and also potentially for better demonstration of efficacy through use of more refined and homogeneous target populations. FDA is prepared to interact actively with the developers of new therapies to ensure that the genotypic information being used in clinical studies is scientifically sound and that it is applied using good scientific principles in the conduct of genotypic testing and the design of efficacy and safety studies. FDA, sponsors, health care providers, and patients all stand to benefit through the use of this new type of information.

Although pharmacology usually plays a critical role in drug development, the principles of pharmacology are less often followed in the practice of medical therapeutics, which is often performed by using predetermined or empiric prescribing patterns. Clinically there is often a broader set of rela-

---

* The FDA website with the Agency guidance on these topics can be found at: *http://www.fda.gov/opacom/7pubs.html*

tionships between diagnostic information, in many instances without a strong pharmacological basis, and clinical therapeutics, which comes from a variety of sources, including patient histories and physicals, patient demographic information, and in some cases routine or special laboratory testing. Genomic information has an impact on both—predicting the pharmacology and revealing important information about the patient. It therefore offers the unique possibility of providing insights that will help predict drug behavior or toxicity in patients with particular precision and reliability. Use of this information in providing a better mechanism for selection of drug treatment for individual patients will revolutionize modern medical practice and create powerful new forms of quality care. FDA expects genotyping to become a potential key feature in providing evidence of effectiveness, estimating risk, and individualizing drug dosages in future drug approvals.

II. REGULATION OF DIAGNOSTIC DEVICES

Diagnostic tests are regulated by an overlapping set of consumer protection laws administered by two federal agencies. Medical device manufacturers are regulated by the FDA at the Center for Devices and Radiological Health under the authority of the Medical Device Amendments of 1976.* This law put into place a series of general controls, special controls, and requirements for premarket review of testing devices being sold as kits or systems in interstate commerce. All clinical laboratories either using these devices or performing testing developed by themselves as in-house procedures are subject to ongoing accreditation by the Centers for Medicare and Medicaid Services following the Clinical Laboratory Improvement Amendments of 1988 (CLIA). CLIA instituted a comprehensive program of regulatory oversight for laboratories, including requirements for appropriately trained personnel, quality control and quality assurance systems, and proficiency testing.

The setting in which an IVD device is used and the intended use of the device determine how it is regulated (See Tables 6.1 and 6.2).

Research use tests are subject only to labeling controls. As long as properly labeled "for research use only," these devices are exempt from requirements for premarket review, good manufacturing practices (the quality system regulations), and requirements for postmarket surveillance. A common example in drug development is the use of assays to measure drug levels in early-phase clinical trials. The results of these assays are used to assess drug effects but not to make individual diagnostic decisions.

---

* The Center for Biologics Evaluation and Research (CBER) regulates medical devices including IVDs used for blood and blood components under the Public Health Service Act and the Food, Drug and Cosmetic Act. CBER also regulates all IVDs to diagnose and monitor retroviral infections, such as human immunodeficiency virus.

TABLE 6.1. DIAGNOSTIC INFORMATION AND THE THERAPEUTIC RESPONSE

Diagnostic Information	Therapeutic Response	Examples	
		Information	Response
Risk	Prophylaxis	CD4 count	PCP prophylaxis
Preclinical disease detection	Early treatment	Cervical dysplasia	Cryotherapy
Diagnosis	Treatment	TSH level	Replacement therapy
Disease subset	Subset-specific treatment	Her2Neu	Her2Neu monoclonal antibody treatment
Disease response or resurgence	Treatment modification	Rising PSA	Orchiectomy or hormone therapy
Adverse experience	Treatment modification	G6PD	Malaria drug choice modification
Treatment monitoring	Treatment individualization	Cyclosporine levels	Cyclosporine dosing

Although there will be FDA interest in the quality of the assays, such tests are evaluated indirectly as part of the drug submission rather than independently as a medical device.

Investigational use tests are also subject to labeling controls. These are expected to be used in formal clinical studies for diagnostic purposes under the supervision of an institutional review board (IRB) and with proper informed consent when appropriate. Because most laboratory tests are considered exempt from the requirements for the investigational device exemption (IDE) submission used for most device studies, these are generally not reviewed by FDA during the investigational phase of study. However, they are subject to design controls, and the FDA expects good investigational design and laboratory quality controls in their conduct. The FDA encourages manufacturers voluntarily to submit protocols for review to ensure that appropriate scientific premarket issues are addressed in the course of new test studies.

IVDs developed for use at a single site and offered commercially only at that site are considered laboratory testing services, in-house tests, or so-called "home brew" tests. These have historically been a widely used practice for test development, and a broad menu of tests is available in this mode. These tests are subject to CLIA regulation. If the test is performed by using commercially prepared and purchased active ingredients (so-called analytic specific reagents or ASRs), FDA does impose requirements on both the

TABLE 6.2. Regulation of In Vitro Diagnostic Tests

Setting	In Vitro Diagnostic	Regulation
Assay used for basic research	Nonclinical research for nondiagnostic purposes	Labeling requirements only
Assay used for clinical research to determine diagnostic effect	Clinical research for diagnostic purposes	Labeling requirements; Investigational Device Exemption submissions only in unusual cases; Subject to CLIA if results are reported out
Single-site IVD	"Home brew" or "in-house" IVD made with an analyte specific reagent	FDA requires labeling disclosing the "in-house" nature of the test but has no premarket review requirement; CLIA requires analytical validation and quality control systems
Single-site IVD	"Home brew" or "in-house" IVD made with "in-house" reagents	FDA requests labeling disclosing the "in-house" nature of the test, but no regulatory requirements are in place; CLIA requires analytical validation and quality control systems
Commercial reagents for use in "home brew" tests	Analyte specific reagents	FDA: most 510(k) exempt; blood bank and high-public health-risk reagents subject to premarket reviews; CLIA oversight of laboratories using these reagents
IVD kits or systems for sale at multiple sites	Regulated IVDs	FDA: 510(k) or PMA premarket submissions CLIA: Oversight dependent on categorization of test (high or moderate complexity or waived)
Point of care	Claims vary; may be moderate or high complexity or CLIA waived tests	FDA: 510(k) or PMA CLIA: Oversight depending on categorization of test
Self-care	Over-the-counter IVDs	FDA: 510(k) or PMA; not subject to CLIA; tests are automatically considered waived.

manufacturers and users but does not require premarket applications. Manufacturers of ASRs are required to register and list with FDA, to make the ASRs using quality system regulations, to label these devices as building blocks for "home brew" tests, and to restrict sales to laboratories approved to perform "high-complexity" tests under CLIA. Laboratories using ASRs must be accredited as high complexity, must follow CLIA requirements, and must label these tests to clarify their status as "home brew" tests and to include a disclaimer that the FDA has not reviewed the test. It is the responsibility of the testing laboratory to meet the requirements for IRB oversight and informed consent when such tests are investigational.

If the test offered at a single site is performed using active ingredients also made at that same site, the in-house or "home brew" test is considered to be outside of the requirements of the ASR rule and is not subject to any FDA regulations. FDA requests that laboratories label their tests using appropriate disclaimers, but this is voluntary.

When an IVD is developed as a kit or system to be used with specific equipment and is sold to multiple laboratories, it is considered a device in interstate commerce and is subject to premarket review. When the IVD is a novel test, premarket approval (PMA) will be based on the analytical and clinical validation that will determine whether the test is safe and effective for clinical use. When there is evidence that the IVD is substantially equivalent to a legally marketed device, FDA clears such tests under section 510(k) of the Device Amendments to the Food, Drug and Cosmetic Act. The necessary level of evidence and requirements are described in more detail below.

Finally, some diagnostic tests are designed to be used at the point of care, such as in an emergency room, a primary care provider's office, or clinic without the supervision of a clinical laboratory. These tests vary considerably in complexity and performance characteristics and are, depending on the complexity category assigned through the CLIA program, considered high or moderate complexity or may be waived. The latter designation allows for point-of-care testing with no CLIA regulatory oversight. There are also diagnostic devices designed to be used by patients themselves, the most familiar being blood glucose meters. To be approved for such use, these must be approved by FDA through either the PMA or section 510(k) procedures. By statute, these tests automatically qualify for waived status under CLIA.

III. FDA Regulation

There are five fundamental elements of FDA regulation of medical devices: (1) facility registration and product listing; (2) safe use of investigational devices; (3) premarket notification (unless exempt) and product labeling;

(4) good manufacturing practices (quality systems requirements); and (5) corrective and preventive actions for device failures.

Medical devices are regulated on the basis of a three-level risk classification. The highest risk products are the class 3 products that require premarket applications, almost always with clinical data that demonstrate that the product is safe and effective for the intended use. By default, a novel product is a class 3 product unless there is an approval application for initial approval as a class 2 device (the de novo process). Clinical trials for class 3 products before they are approved usually require an IDE, which is similar to the IND required for investigational drugs.

IVDs may be exempt from IDE requirements if the following four conditions are met: The testing is (1) noninvasive; (2) does not require invasive sampling presenting significant risk; (3) does not introduce energy into a subject; and (4) is not used as a diagnostic procedure without confirmation of the diagnosis by another, medically established diagnostic device or procedure. Being exempt from an IDE does not relieve an investigation from IRB or informed consent requirements, only from filing an IDE with FDA.

Approved IVD tests are required to have standard product labeling. The current labeling requirements* include provisions dealing with intended use(s), summary and explanation of test, information on specimen collection and preparation, procedures, results, limitations of the procedure, expected values, and specific performance characteristics.

Class 2 medical devices are intermediate-risk products, most of which require premarket notification with a section 510(k) application.† In 1976, when the device amendments to the Food, Drug and Cosmetic Act became law, the existing class 2 medical devices were "grandfathered" and allowed to continue being marketed without FDA review. New class 2 products were allowed to be cleared for marketing if they were "substantially equivalent" to an existing marketed product (a predicate product) based on "special controls" tailored to each device group. About 10% of the 800 different groupings of class 2 products require clinical trial data as the special control, but most special controls require performance testing by other methods.

Class 1 medical devices, which include some simple IVDs, are usually exempt from the premarket section 510(k) application process, and their manufacturers are only required to register and list, follow good manufacturing practices, and report device failures. As experience develops with a given medical device it can be reclassified downward from class 3 to class 2 and eventually to class 1.

---

* see 21 CFR § 809.10(b).
† A useful reference on the 510(k) process for IVDs can found at:
http://www.fda.gov/cdrh/manual/ivdmanul.html

IV. FDA Compared with CLIA

FDA device regulation is focused on the device and the device manufacturer. CLIA, on the other hand, focuses on laboratory quality, including the quality of the laboratory test results provided by the devices used, whether developed in-house or as a test kit in commercial distribution to multiple laboratories. The programs differ substantially in approaches and in data requirements. FDA requires unique submissions for each test under its purview, evaluates both performance and labeling, and requires demonstration of analytical validity and clinical validity as appropriate. CLIA inspects laboratories using a system approach based on key probes of the operating system. CLIA requires a demonstration of analytical performance and quality control but does not require a showing of either clinical validity or clinical utility.

FDA routinely requires analytical data on tests under review to demonstrate that they measure what they claim to measure. Analytical performance is usually directed at evaluating a test's accuracy or bias compared with a predicate or reference test, precision or repeatability, and analytical specificity and analytical sensitivity.

If the test is well known and clinical performance can be imputed from the analytical performance, then further studies of the device are not generally required. However, if clinical performance cannot be reasonably determined on the basis of the analytical information used to validate the test, then additional clinical validation is usually required. This validation is most frequently done against established laboratory or clinical surrogate end points and used to characterize clinical sensitivity (the ability of a test to correctly identify disease) and clinical specificity (the ability of a test to correctly identify the absence of disease). FDA prefers that testing be performed by using receiver operating curve analysis and using a population reflective of that in which the test is expected to be used. FDA will sometimes request information on the predictive value of a positive and/or negative test either in populations with predetermined prevalences of the disease state or by the use of modeling to extrapolate performance across populations that might have different prevalences of disease. FDA is cautious in the use of predictive values in submissions and labeling because prevalences can vary markedly from setting to setting and because basic performance characteristics themselves (i.e., clinical sensitivity and specificity) may exhibit spectrum bias.

Clinical utility takes the concept a step further and shows that the patient's medical outcome is favorably influenced by knowledge of the diagnostic test. An example would be that the tight insulin control that is only possible with frequent home blood glucose monitoring has a positive impact on delay of complications of diabetes. FDA does not require that a demonstration of clinical utility be performed in support of submissions but does expect a test to have biologically or medically plausible uses to justify pre-market approval.

V. PHARMACOGENOMIC IN VITRO DIAGNOSTICS

The initial applications of genomics to drug development are likely to involve determining the genetic information and studying the drug effects with respect to genotype in conjunction with traditional pharmacology. If trials are designed to show better clinical outcomes in subsets predicted by pharmacogenomic information, then the foundation is in place to determine analytical validity, clinical validity, and clinical utility of the genotyping test.

What's left is to have a coordinated product development plan. There have been examples of drugs developed with therapeutic drug monitoring where the diagnostic was not commercially ready at the same time as the drug. In some cases, this resulted in the need to link the approved drug to an experimental diagnostic. To prevent such an occurrence the strategy for introduction of the diagnostic to the market should be discussed and worked out early in the drug development plan. FDA can make binding agreements about the evidence needed for approval of the device. During product development the Device and Drug or Biologics reviewers may be able to work from a single combined IND/IDE application, if that would be helpful to the sponsor. Issues such as transition from use of a central laboratory to developing a commercial diagnostic should be discussed with the Center for Devices and Radiological Health.

VI. NEW REGULATORY CHALLENGES IN ASSESSING THE USE OF PHARMACOGENOMIC DIAGNOSTICS AND THE USE OF TARGETED THERAPEUTICS

FDA has comprehensive postmarket requirements in place for surveillance of both drugs and devices. The options and requirements for reporting adverse events are outlined on the MedWatch section of the FDA web page (http://www.fda.gov/medwatch/). In addition, both drugs and devices may be subject to postmarket studies (phase IV commitments) in conjunction with the premarket review process. No matter how well constructed premarket studies may be, unexpected changes in performance, or in terms of efficacy or safety, may be observed in real-world use. FDA will take advantage of its full armamentarium of postmarket risk management tools to track these new products and to develop new strategies (e.g., guidances) as appropriate. Although the agency is hopeful that existing regulations will be sufficient, if needed, new regulations, guidances, and authorities will be developed to ensure proper use and performance of these important new medical products. A particularly challenging area is the common and legal practice of off-label use. FDA has both limited authorities and resources in this area and hopes to partner and leverage with professional groups to assist in proper dissemination of knowledge about new medical products so that this information can be available to inform physicians about good medical practice.

VII. SUMMARY

An overlapping set of consumer protections is in place to ensure the quality of clinical laboratories, that investigational diagnostic devices are safely used, that the devices are manufactured with a satisfactory and consistent quality, that product failures are detected and corrected, and that diagnostic devices will result in safe and effective use of the clinical information. Device by device, claim by claim, it is FDA's responsibility to evaluate the evidence that it takes to substantiate these claims, to check the source data, and to periodically inspect the manufacturer's facilities and quality systems. Confidence is more easily lost than regained. CLIA and FDA oversight are important processes to ensure the timely realization of the public health benefits of medical progress.

FDA recognizes that its mission to promote and protect public health provides it with a dual challenge: to keep bad products off the market while ensuring rapid technology transfer for safe and effective new products. Nowhere is this more challenging than in the area of pharmacogenomics, where new technologies and analytical platforms, patterns of information generation, and interpretation of results may be complex and demanding. FDA seeks to apply good science to its regulatory processes to ensure that an appropriate balance is maintained in meeting the challenge of this work.

REFERENCES

Food and Drug Law Institute (U.S.), "Basic outlines on medical device law and regulation: a collective work by top legal and regulatory experts in the food and drug field." Washington, DC: Food and Drug Law Institute, ii, 490 (1997).

Parascandola, J. and Keeney, E., *Sources in the History of American Pharmacology.* Madison, WI: American Institute of the History of Pharmacy, p. 5 (1983).

Young, J.H., *The Medical Messiahs: A Social History of Health Quackery in Twentieth-Century America.* Princeton, NJ: Princeton University Press, pp. xiii, 498 (1992).

Intellectual Property and Commercial Aspects of Pharmacogenomics

Allen C. Nunnally, J.D., Scott A. Brown, J.D., and Gary A. Cohen, J.D.*

I. Introduction

Pharmacogenomics is defined as the use of genetic information to optimize drug discovery and development by studying how various genetic traits affect drug responsiveness (Rudolph, 2001). Used as a tool in the pharmaceutical industry, pharmacogenomics allows for the creation of personalized medicine responsive to the particular biochemistries of individuals, increasing efficacy and reducing or eliminating undesirable side effects by tailoring drugs in accord with genetic traits shared by those in subgroups of disease populations. The genetic information utilized in pharmacogenomics, as well as the technology and know-how surrounding it, are valuable commodities for companies engaged in the drug development process, in which the costs to develop a pharmaceutical product stand in the hundreds of millions of dollars. Intellectual property protection for pharmacogenomics technologies and resulting products is thus critically important to the entities working in this area to develop optimized drug therapies. These proprietary assets are necessarily licensed and exchanged, in turn, among

Pharmacogenomics: Social, Ethical, and Clinical Dimensions, Edited by Mark A. Rothstein.
ISBN 0-471-22769-2   Copyright © 2003 Wiley-Liss, Inc.
*The views expressed in this chapter are solely those of the authors and do not necessarily represent the positions of Hale and Dorr LLP, Millennium Pharmaceuticals, or any related entities.

pharmacogenomics entities requiring access to outside resources to effectuate the best use of their own shares of the field's collective intellectual property.

This chapter examines the forms of legal protection available to those in the business of pharmacogenomics to protect their intellectual property and how such protection is applied to discover more effective drugs, advance the state of the medical art, and create value for shareholders. Furthermore, this chapter explores the types of business models and commercial arrangements—including licensing and collaborative partnerships—used by entities in the field to optimize pharmacogenomic end points and therapies. Only by understanding the methods of protection for such important intellectual property as genes, expressed sequence tags (ESTs), single-nucleotide polymorphisms (SNPs), genomic databases, and technology platforms, as well as the spectrum of relationships built around such property to capture its value, can the commercial aspects of pharmacogenomics be fully understood.

Although analysis of the intellectual property protection and interplay among the commercial entities controlling those protected assets is indispensable in gaining an appreciation of the driving forces in pharmacogenomics business, the scope of this chapter does not reach many other important aspects of the field. Notably, the multitude of regulatory, social, and ethical considerations that must be addressed by the industry are not analyzed in this discussion. Such important topics, however, are treated elsewhere in this book. This chapter provides a comprehensive overview of the most important intellectual property and commercial aspects of pharmacogenomics, providing the reader with a framework for the industry from a business standpoint.

II. Intellectual Property Protection

Two types of intellectual property protection are critically important to pharmacogenomics: patents and trade secrets. Patent protection is undoubtedly foremost in the industry as a vehicle providing exclusivity in the use of genes, ESTs, SNPs, and diagnostic and therapeutic processes using these important genetic components and markers. Also relevant in the field is trade secret protection, which proscribes misappropriation of proprietary information such as scientific know-how, genomic database content, and combinations of software and other computer and robotic technologies comprising important technology platforms. Employing various combinations of protection for the intellectual capital these companies painstakingly develop, acquire, and nurture creates value by enabling valuable licensing arrangements of such property and the ultimate development of marketable pharmacogenomic products.

A. PATENTS

The patent system is an essential foundation of the biotechnology industry. Through patent protection, those spending extraordinary amounts of money on research and development are able to recoup costs and earn appropriate returns for shareholders by benefiting from limited periods of exclusivity over their discoveries. The importance of this protection becomes obvious as one considers that the average research and development costs for a single new drug has traditionally been estimated at $500 million (Robbins-Roth, 2000) and may actually now stand at as much as $800 million according to a recent study by the Tufts Center for the Study of Drug Development (Pear, 2001). Where the marginal cost in pharmaceutical production may be nil, the recoupment mechanism that the patent system provides guards against the threat of free riders who would otherwise interlope and undersell innovators by producing a drug at prices approaching cost during the patent holder's period of exclusivity. The patent system thus allows the developer to recover its tremendous research and development investment during the exclusivity period.

1. THE PATENT STATUTE. A patent represents a "contract" between an inventor and the public. The inventor offers full disclosure of his or her invention in sufficient detail to instruct the public about the invention and its use. In exchange for public disclosure and the societal benefits derived therefrom, a patent holder is granted exclusive rights to make, use, and sell the invention and the right to exclude all others from the same for a period of 20 years. Thus inventions that would otherwise remain preciously held secrets are made known for greater advancement of the public good.

The U.S. patent statute provides that "any new and useful process, machine, manufacture, or composition of matter" is protectable subject matter (35 U.S.C. § 101). Under the Patent and Trademark Office's (PTO) current interpretation of the U.S. patent laws, individuals or entities may patent specific genes and gene markers, as well as diagnostic and therapeutic methods surrounding such genetic components, discovered in the course of their work (Campbell, 2001). The issue of gene patenting, however, remains highly controversial (Murray, 1999; Kieff, 2001; Rai, 2001; Rai, 1999).

As with any patent application, to obtain a patent on a gene, genetic component (e.g., EST or SNP), or diagnostic or therapeutic method, the applicant must demonstrate the utility, novelty, and nonobviousness of the discovery. The applicant must also supply a disclosure and written description of the subject of the proposed patent that is sufficient in detail to show that the inventor had possession of the claimed subject matter at the time of application and to enable one skilled in the field to make use of the claimed subject matter. Although the fundamental elements of patentability have been in place for decades for protection of such innovations as new mechanical devices, chemicals, and various scientific processes, only recently have such elements been applied to genetic innovations in biotechnology.

2. Interpretation of the Patent Statute by the Courts. The seminal case that paved the way for the patenting of genetic material is *Diamond v. Chakrabarty* (1980). In *Chakrabarty*, the Supreme Court ruled that claims did not fall outside the purview of patent protection merely because they pertained to living organisms. The Court found that a genetically engineered microorganism of sufficient utility brought into existence as the product of human ingenuity was plainly patentable. The Supreme Court's approval of patents on living organisms under the patent statute signaled a call for the lesser-included pieces of life—DNA sequences—to be patented as well. Since *Chakrabarty*, the Federal Circuit (the specialized federal appellate court in charge of patent appeals and subordinate only to the United States Supreme Court) has ruled that, under the patent statute, DNA is a chemical composition to be held to the same nonobviousness standard as other standard chemicals in patent cases (*Amgen v. Chugai Pharmaceutical Co.*, 1991). As with other chemical compositions, to satisfy the written description requirement, an applicant must disclose a sufficiently detailed chemical structure of the specific DNA molecule; describing functional utility and methods for isolating a DNA molecule alone are not sufficient. Thus, when a previously undiscovered gene (satisfying the novelty requirement) encoding a given protein with a given function (satisfying the utility requirement) is discovered, its discoverers may satisfy the written description requirement by providing a written description of the gene sequence (sufficiently detailed chemical structure of the molecule) in the patent application. Assuming the discovery was not obvious based on prior art and that the disclosure enables those skilled in the art to use the gene, the patent will be issued, granting the discoverer a limited period of exclusivity over the use of his or her innovation.

Gene fragments such as ESTs, genetic markers such as SNPs, and the diagnostic and therapeutic processes surrounding each are other major subjects of interest in the area of pharmacogenomics for which commercial enterprises are seeking patent protection. The issue of patentability of such genetic elements, however, has not yet come squarely before the courts and is the subject of ongoing debate in the scientific and legal communities. Applying current judicial interpretation to the patenting of genes and their mutant variations, however, may lend predictive perspective. When a gene's sequence varies from that of the normal or wild-type gene, the protein product encoded by that variant, mutant gene may give rise to a disease state (Snustad et al., 1997). Even small variations in the genetic sequence may have an enormous effect on the function of the protein that the gene encodes (Purves et al., 1995). Although a mutant gene form that differs from a previously patented wild-type form by only a few nucleotides might be considered prima facie obvious under some circumstances, a showing of unexpected properties of the mutant gene form not present in the previously patented form, such as correlation to disease, will satisfy the nonobviousness standard (Murray, 1999). Case law in the Federal Circuit clearly allows for more than one sequence to be patented on variations of the same gene (different mutations or mutant vs. wild type) in such instances (Murray, 1999). Therefore, particular SNPs, representing variations in the human genome, might be held patentable by the courts where a relationship between the presence of an SNP in individuals' genomes and the onset of or susceptibility to a disease or responsiveness to a drug is demonstrated (Murray, 1999).

Notwithstanding speculation as to likely judicial interpretation in the event the issue is raised in an actual case, the PTO serves as the gatekeeper of patents, determining what innovations qualify under its understanding of the statute and judicial precedent. An understanding of PTO practice provides perhaps the best current instruction on the issue of genetic patenting.

3. THE PTO AND THE UTILITY AND WRITTEN DESCRIPTION GUIDELINES. Application of the statutory patent requirements to genetic material in the new age of biotechnological discovery continues to engender substantial debate over the propriety of granting exclusivity, however limited, on genetic material that chemically maps out living organisms. As the issuer of patents and first arbiter of patentability, the PTO is forced to determine difficult issues regarding patentability that have not yet come before the courts and to establish policy to be applied to all applications. One such issue before the PTO is that of genetic patenting and specifically the application of the utility and written description requirements to genetic patents. To clarify its position on these elements, the PTO published revised versions of its guidelines concerning utility and written description in 2001 that clearly reflect the PTO's support of the patenting of human genetic material consistent with, and elaborating upon, judicial interpretation (Utility Examination Guidelines, 2001; Written Description Guidelines, 2001).

In its announcement of the revised guidelines, the PTO explicitly rejected public comments suggesting that genes and other genetic components should be unpatentable because they are naturally occurring phenomena and not new compositions of matter and that DNA should be viewed as the constitutionally protected and fundamental core of humanity, not a marketable invention. In response to the assertion that DNA has little utility, the PTO unequivocally stated that a "purified DNA molecule may meet the statutory utility requirement if, e.g., it can be used to produce a useful protein or if it hybridizes near and serves as a marker for a disease gene" (Utility Examination Guidelines, 2001). Likewise, in response to the suggestion that only genes with known, functional protein products ought to be patentable, the PTO maintains, "the utility of a claimed DNA does not necessarily depend on the function of the encoded gene product. A claimed DNA may have a specific and substantial utility because, e.g., it hybridizes near a disease-associated gene. . . ." (Utility Examination Guidelines, 2001).

The PTO's statements with respect to patenting of whole genes are consistent with established judicial precedent.

Although the issue of patenting genes that code for important proteins of known function is controversial but well settled from a legal standpoint, attempts to patent mere fragments of genes or smaller pieces of genetic material give rise to even more acrimonious debate and less legal certainty (Holman and Munzer, 2000). Most of this discussion has centered on ESTs and SNPs. An EST, or expressed sequence tag, is a genetic sequence fragment that may be used as a probe to positively identify or "tag" a whole gene. ESTs were used initially by Craig Venter and Mark Adams to identify many human genes (Davies, 2001). Their methods sought to "shortcut" the sequencing of those portions of the human genome that are most relevant to drug discovery—those portions encoding proteins—by "tagging" expressed regions for focus in further research. With the tag, or EST sequence, in hand, elucidation of the entire coding gene would be facilitated.

An SNP, or single-nucleotide polymorphism, is a mere single base pair variation from the reference genetic sequence found in most members of a population. SNPs are the most common points of departure between the genomes of individuals, which are, for all apparent differences among people, virtually identical (Davies, 2001). SNPs located within the coding regions of genes themselves may have a direct effect on the protein production that is biologically benign, or profoundly negative, such as in causing disease (Davies, 2001). SNPs may thus give rise to disease, increased susceptibility to disease, sensitivity or insensitivity to certain drug treatments, or adverse side effects from those treatments. Alternatively, SNPs may mark the presence of disease-causing genes or indicate disease susceptibilities or drug sensitivities. Scrutiny of inheritance patterns of SNPs in families afflicted with genetic disorders has revealed links to previously unknown disease genes in close proximity to SNPs within afflicted individuals' genomes (Davies, 2001). In this way, SNPs can demarcate errant disease genes, thereby facilitating the tracking and identification of the disease genes themselves. The study of these small genetic differences is thus critically important in understanding disease and its treatment. A full understanding of the SNPs of various disease subpopulations provides the potential for individualization of drug treatment on the basis of genetics.

The heated debate that has arisen over attempts to patent ESTs and SNPs stems from several main sources. One argument against such patenting is that protection for ESTs and SNPs as research tools or probes provides exclusive use to materials critical in discovering and fully characterizing genes of interest and their corresponding proteins. It is argued that these patents would prematurely grant exclusive means of access to full-length genes and, in effect, allow for clear pathways to eventual patenting of the whole genes for the fragment patent holder (Holman and Munzer, 2000; Rai, 2001). Thus, some opine, the patenting of an EST or an SNP, before the patentee or anyone else has been able to make these important full characterizations, would make it impossible for other researchers to work toward the goal of fully characterizing the gene without incurring licensing fees. Such a situation, it is argued, critically inhibits essential research.

Proponents of patenting of ESTs and SNPs that meet the criteria for patentability counter that the patent system exists to ensure that inventions are disclosed to fuel further innovations. They argue that all industries use patented inventions to build on and advance technology. There is no credible evidence, they suggest, that patenting of ESTs and SNPs would cause any more of a chilling effect on identification of whole genes than it has on other technologies.

A somewhat more subtle but related point in opposition to the patenting of ESTs and SNPs is the potential for such patents to block the use of subsequently discovered full-length genes that include the lesser patented fragments (Rai, 2001). The issue of such blocking is not clear, but John Doll, Director of Biotechnology Examination at the PTO, suggests that some patents on gene fragments may, under certain circumstances, indeed block use of full-length genes that are the subject of independent patents (Rai, 2001). Analogizing to the picture tube as a patented innovation that could block the use, manufacture, or sale of a television incorporating that technology, Doll hints at a mechanical comparison that may or may not overcome scrutiny under judicial construction of DNA as a subset of chemical technology (Rai, 2001). This issue has not yet been before the courts (Rai, 2001). Moreover, the possibility of patent coverage over an earlier-stage invention is not a new phenomenon. Such circumstances exist in all areas of developing technology and have not historically resulted in stunting of further innovation and technology growth.

A third concern of gene patenting opponents is the attempt to patent ESTs or SNPs, for which a biological function has not been identified. Because there is inherent uncertainty as to the identity of the genes that may be associated with ESTs and SNPs, some guesswork is involved in making descriptions of the utility of the genetic material. Critics voice concern about the potential to make vague or inaccurate claims as to the utility of ESTs and SNPs and the consequences of winning protection on useful research tools to which the patentees do not deserve exclusive rights.

Again, in the face of these criticisms about patents on ESTs and SNPs, proponents maintain that such genetic materials legitimately qualify under the patent statute and its current interpretation by the courts as useful inventions (Koneru, 1998). These commentators contend that to disallow patents on such "research intermediates" useful as tools in genetic discovery is to ignore that research itself is an industry that we should seek to promote by providing patent protection for innovation (Koneru, 1998). Proponents of EST and SNP patents note that the licensing that would be required to use such patented inventions for research is no different than what occurs in other industries.

The PTO has responded to some of the challenges to patenting of ESTs and SNPs by promulgating its new Utility and Written Description Guidelines. Under the new Utility Guidelines, the PTO provides its interpretation of the patent laws allowing patenting of not only whole genes, but also ESTs, SNPs, and other useful and adequately described genetic components and markers as long as all of the traditional requirements for patentability are met. In response to public comments suggesting that ESTs should not be patentable, the PTO explicitly states "ESTs which meet the criteria for utility, novelty, and non-obviousness are eligible for patenting when the application teaches those of skill in the art how to make and use the invention" (Utility Examination Guidelines, 2001). This statement concisely posits that an application for an EST or an SNP must express sufficient utility and an adequate written description for a patent to issue, and the Guidelines flesh out each requirement.

Section 112 of the patent statute requires that the "specification shall contain a written description of the invention. . . ." (35 U.S.C. § 112). According to the Guidelines, "a patent specification must describe the claimed invention in sufficient detail that one skilled in the art can reasonably conclude that the inventor had possession of the claimed invention" (Utility Examination Guidelines, 2001). Possession of an invention may be demonstrated in a number of ways such as through words, structures, or formulas, and under the Federal Circuit's current treatment of DNA as a chemical composition no different from any other nonbiological chemical, a sufficiently detailed disclosure of its chemical structure (i.e., the gene sequence) is a necessary component of written description. Although there is a strong presumption that the written description of an originally claimed invention is adequate, the Guidelines state that written description may be inadequate where "the claims require an essential or critical feature which is not adequately described in the specification and which is not conventional in the art or known to one of ordinary skill in the art" (Utility Examination Guidelines, 2001). This caveat is directed largely toward those who seek to patent biomolecules and directs patent applicants to comply with the aforementioned conventions established by the Federal Circuit. To procure a patent on an EST or SNP, the patent applicant must be sure to describe the chemical structure (sequence) of the biomolecule in addition to its function.

A thornier issue than written description for ESTs and SNPs, however, is that of utility. Section 101 of the patent statute requires that the subject of a patent must be a "useful invention." Under the Guidelines, an application satisfies the utility requirement if the invention has "specific, substantial, and credible" utility such that "a person of ordinary skill in the art" would recognize the claim as credible "in view of disclosure [contained in the patent application] and any other evidence of record . . . that is probative of the applicant's assertions"(Utility Examination Guidelines, 2001). The Guidelines warn against attempts at advancing theories of "throw-away" utility (e.g., suggesting that an invention would be useful as fodder for a landfill) and "nonspecific" utility (e.g., a utility that would apply generically to many unrelated DNA molecules, such as use as a chromosome marker), which will not be accepted. The requirement that the assertion of utility have specificity is aimed at protecting against patents on ESTs and SNPs whose functional utility is not apparent to applicants at the time of filing. In addition, the "substantial" language protects against attempts to advance utility theories that are largely abstract or not of significant real-world use.

Thus, applicants attempting to file thousands of EST and SNP patents in the hope of procuring protection at an early stage and later discovering the potential locked in some subset of their many biomolecules face a high hurdle in obtaining patents. John Doll has, in fact, stated that consequent to the recently issued Guidelines, many outstanding applications on ESTs will "have a difficult time" meeting the utility requirement (Rai, 2001). This represents a strong indication that bulk applications claiming utility based on categories such as tissue typing, homology, and conditions under which sequences were obtained (Holman and Munzer, 2000) are likely to experience a high level of scrutiny that will result in the rejection of many applications.

The PTO's statements clearly evince its conceptual support for the patenting of human genes and gene markers, such as ESTs and SNPs, that meet utility and other statutory requirements. As the PTO asserts, genes encoding both normal proteins (encoded by wild-type genes) and defective, disease-causing proteins (encoded by mutant genes) are indubitably useful. Sequences marking disease genes or drug responsiveness also carry the requisite utility. The PTO's affirmative indication that useful ESTs and SNPs, like genes of known function, with sufficiently demonstrated utility are patentable subject matter sets an important industry and legal standard.

4. PATENT POLICY DEBATE.   Current standards of patentability for genetic subject matter have led to substantial debate over patent policy in the scientific and legal communities. Critics of genetic patents often bemoan the perceived high costs associated with licensing fees for therapeutic and diagnostic products. Some researchers cite what they claim are prohibitive licensing and royalty costs, and some physicians bristle at the thought of having to pay licensing fees to conduct genetic tests on their patients for diseases implicating known gene or marker sequences (Rafinski, 1999).

Massachusetts Institute of Technology professor Jonathan King decries the human gene patenting underlying pharmacogenomics and the biotechnology industry as per se wrongful, stating that "Genes derive from millions of years of evolution and are, in the deepest sense, products of nature . . . not the inventions of individuals, corporations, or institutions" (Schmidt, 2001). Likewise, both the American College of Medical Genetics (ACMG, 1999) and the College of American Pathologists (CAP, 2000) have issued position statements condemning human genetic patents. Dr. Aubrey Milunsky, Director of the Center for Human Genetics at the Boston University School of Medicine, expresses concern that high licensing fees for use of patented genetic material could "ultimately exclude [everyone but the patent holder] from working on [a particular] gene," inhibiting potentially invaluable research contributions from the medical research community (Aoki, 2000). Critics in the medical community contend that protecting the use of such genetic material only inhibits discovery of patient maladies while increasing patient costs.

The issue of human gene patenting is not simple; indeed, it may be wrongful to derail medical innovation by denying patent protection to highly creative biotechnology companies focused on discovering new medicines. Industry insiders point out that patents are essential for biotech companies, noting that without patent protection such companies could not recoup the vast capital resources invested in research and development needed for identification and clinical testing of new treatments for human disease (Aoki, 2000). Biotech industrialists also contend that patents provide a necessary incentive to innovate in exchange for a period of exclusivity, without which companies would not make such enormous investments in the innovations. Those in the industry maintain that competition remains healthy and that, in the final analysis, biotech companies are tremendous resources in curing diseases for the benefit of all (Aoki, 2000).

Patents also help promote the exchange of knowledge because they publicly explain an invention or discovery in sufficient detail to be understood and used by others. Although the patent holder enjoys time-limited, exclusive rights over the innovation, researchers have a judicially created limited defense to infringement for purely scientific, noncommercial experimentation or inquiry. Thus a patent does not create an absolute bar to scientific advancement, even during the patent term. Rather, it is commercial free ridership at which the statute takes aim. The danger for companies engaging in costly research and development is that, without patent protection, competitors would be free to sell products or services derived from another's labor without having incurred the high costs of making the initial discovery and developing technology to use it. Although prohibiting human gene patents may appeal to those who see genetic information as the untouchable, shared property of all humankind, the practical effects on genetic medical advancement would be severe and far-reaching.

The reliance of biotechnology on patent protection is perhaps best evidenced by an incident in which the mere public perception of a departure from the current status quo sent the U.S. biotechnology market sector into a tailspin (Davies, 2001). In March 2000, President Bill Clinton and British Prime Minister Tony Blair made a brief joint announcement regarding the Human Genome Project, stating that "the human DNA sequence and its variations, should be made freely available to scientists everywhere" (Davies, 2001). Although the two leaders noted that "[i]ntellectual property protection for gene-based inventions will also play an important role in stimulating the development of important new health care products,"

(Davies, 2001) the public biotechnology markets reacted mercurially. The announcement seemed to the uninitiated to foreshadow the death of gene patents, and the resultant sell-off saw biotechnology stocks lose a tremendous percentage of their value. For example, Celera's stock dropped from a previous high of $290 per share to $100 per, and $50 billion in value was drained from the leading American biotechnology companies within two weeks (Davies, 2001). The public terribly misperceived the message of the announcement; proprietary innovations of biotechnology companies in the area of genetics were, in fact, safe. Nonetheless, the market felt staggering effects as investors withdrew on the perception that such companies would flounder without the protection of patents for their genetic advances.

Market realities necessitate the incentive the patent regime supplies to drive the biotechnology industry, making an outright ban on gene patents a remote possibility. Rather, a statutory amendment could potentially narrow the scope of protection to address specific concerns by some critics of the system without eliminating gene patenting wholesale. Such a measure was taken in 1995, when Representative Greg Gansky, a medical doctor, led Congress to pass an amendment to the patent statute exempting medical practitioners from patent infringement for use of patent-protected medical procedures (35 U.S.C. § 287 (c)(1)). An exemption similar to that afforded to medical practitioners in the context of patented medical procedures could be implemented in the area of human genetic patenting to provide relief to clinical medical researchers, and perhaps some academics, should policy makers determine that such a limitation is desirable.

Given the exclusive rights that gene patents confer, coupled with the enormous predictive medical value inherent in their subject matter, it comes as no surprise that these patents are hotly contested and that their protection profoundly influences pharmacogenomics and the biotechnology industry. In light of the utility of these genetic components as predictors of disease, disease susceptibility, and drug sensitivity, protection for gene patents is critical in maintaining the impetus for innovation and in driving the industry. The proprietary value of important genes and genetic components is enormous to holders of patents of such subject matter, and under the current protective regime biotechnology companies continue to make tremendous capital and research investments in the area of genetic research (Davies, 2001). The patent system is thus at the crux of the pharmacogenomic infrastructure, providing a legally protective, incentive-based framework to ensure the development of personalized medical therapies. Although current policy is subject to judicial review and legislative mandate, the strongest available indications support continuation of the status quo.

B. TRADE SECRETS

Another form of intellectual property protection important to biotechnology and pharmacogenomics is trade secrets. Trade secret law serves to protect proprietary information against theft and misappropriation. Unlike patent protection, which is codified as a body of federal statutory law, trade secret protection is provided under various state statutes and bodies of state common law. Most states have adopted some form of the Uniform Trade Secrets Act (UTSA), which was developed by legal scholars in 1979 and amended in 1985 to serve as a model for state trade secret law. Under the UTSA, acquisition of valuable secret information such as formulas, patterns, compilations, programs, devices, methods, techniques, or processes by improper means (e.g., theft, bribery, misrepresentation, espionage, or breach of a duty to maintain secrecy) constitutes an illegal misappropriation if the acquirer of such information knew or had reason to know that the information was secret and the owner of such information made reasonable efforts to protect the information.

Unlike the more stringent subject matter requirements of patent law, the only limitations on the subject matter of a trade secret are that the information actually be secret (not generally known or readily ascertainable by proper means) and valuable (having actual or potential independent economic value). Trade secret law is thus able to fill in gaps in the protection of an entity's intellectual property by complementing what is protected by patent. Information and innovations surrounding patented subject matter or subject matter that is otherwise valuable but not patentable, from which competitors might benefit from a taking—free from research and development costs or purchase price, is protected under trade secret law. Evidence of the importance of trade secret protection to various manufacturing firms, including those in biotechnology and pharmacogenomics, is demonstrated by a 43% incidence of trade secret components in intellectual property disputes from a 1994 study of 530 Massachusetts companies (Lerner, 1994). An examination of how trade secret law operates to protect know-how, such information as genomic database content, and technology platforms of combined software, computer, and robotic technologies adds essential breadth and depth to the understanding of the intellectual property regime that supports pharmacogenomic development.

1. Nondisclosure, Noncompetition, and Assignment Agreements and the Protection of Know-How. Preemployment confidentiality, noncompetition, and assignment agreements are an important industry standard in scientific research-oriented fields. When a new employee is hired, he or she is frequently required to sign a contract in which the employee assigns rights to all inventions created during and in the scope of employment and promises to keep secret all proprietary company information he or she gains access to while working for the company. Such contracts often also involve noncompetition provisions that prohibit departing employees from working for direct competitors for a given interval of time after their departure, although the contractual limitations will not have legal

effect where they are overbroad (e.g., foreclosing the possibility of any reasonable subsequent employment) or extend unreasonably into the future. Nondisclosure, noncompetition, and assignment contracts are evidence of a company's diligence in maintaining its trade secrets. In disputes in which the defense challenges the diligence of the plaintiff in maintaining information as secret in an attempt to pierce the trade secret protective shield, plaintiffs may present such agreements as persuasive rebuttal.

Under noncompetition provisions, former employees are prevented from using trade secret information in starting their own companies or using the information to work for competitors. Ex-employees are not, however, proscribed from using "general skill, knowledge, training ,and experience" acquired through work for their former employer in the industry (Restatement (Third) of Unfair Competition, 1995). Finding the balance between such general information and know-how and trade secrets can be a complex process, but the standard of reasonableness governs all such controversies.

Assignment provisions of employment contracts provide that all rights to any invention invented by an employee during the course of employment (whether protectable under patent, trade secret, or other intellectual property mode) are assigned to the employer. Although the actual inventor must always be named on a patent application, the exclusive rights to the invention remain with the employer. Even in the absence of a contract that specifically provides for such assignment of intellectual property rights, the employer will generally take all rights to an invention made in the course of employment as a "work made for hire" where the act of invention constitutes part of the normal course of employment for the employee in question. In the absence of an agreement, if an invention is invented with employer resources but outside the scope of employment, the employer will have a nonexclusive "shop right" to practice the invention. Only where an employee invents on his or her own time using his or her own resources will that employee take full and exclusive rights in the invention.

Confidentiality, noncompete, and assignment agreements are used to protect the information, innovations, and inventions obtained through expenditure of an entity's time and resources from misappropriation and use by other entities. In addition to use in employment contracts, nondisclosure provisions are used in licensing arrangements in which a company sells the use of proprietary information to a limited class of licensees, that is, a class that is sufficiently small in size to reasonably ensure that the disclosure to such a class does not effectively undermine claims protected against free riders' unauthorized use of information and methods. Databases and platform technology are two types of trade secret-protected subject matter germane to pharmacogenomics.

2. DATABASE AND PLATFORM PROTECTION. The genomic database is an extremely important tool in pharmacogenomics research. The information

compiled in such databases is protected under trade secret law. Companies engaging in pharmacogenomics research utilize databases of their own creation or purchase licenses or subscriptions for access to database information services of genomics companies. Through subscriber agreements, under which information accessed is kept confidential and used only for agreed-upon research purposes, companies' trade secrets in proprietary databases are effectively maintained. Although a significant portion of the information contained in a genomics database may not be appropriate subject matter for patent, the collection, compilation, and organization of database content and tools useful in using it are valuable for biotechnology and pharmacogenomic research and protectable as trade secrets.

Comprehensive genomics databases and the proprietary software tools used for viewing, browsing, and analyzing data mined from them greatly hasten pharmacogenomic research and development. Genomics companies use and sell information services that provide raw genomic sequence information, gene expression data, protein composition and structure, and data on genetic variation and gene function, as well as proprietary software and computer technology to make efficient research use of the information. Database information services are indispensable in facilitating and expediting development of pharmacogenomic therapies and products by providing some of the raw starting materials and reference points necessary for efficient development.

Another key element in pharmacogenomic development often protected under trade secret is the "technology platform," a core of integrated technologies used to drive innovation. Various integrated platforms of biochips, microarrays, imaging and robotics equipment, software, and high-powered computers are used for high-throughput screening, sequencing, genotyping, haplotyping, expression and transcription profiling, and a host of other analyses. Such sophisticated combinations of technology constitute valuable methodologies for pharmacogenomic development, and protection against their misappropriation is important to the platform owners vying for rights to patentable developments stemming from their use. Companies may use proprietary technology platforms for their own pharmacogenomic development or license use of the platforms to other developers through various commercial arrangements. License agreements for the use of proprietary platforms provide for confidentiality, shielding integrated systems comprised of component technologies not otherwise protected under patent and derived from and utilized with trade secret know-how from disclosure that would pierce trade secret protection.

Understanding that there is more to the protective intellectual property regime than patents is critical to a full appreciation of the nature and effect of commercial dealings in pharmacogenomics. Licensing of both trade secret databases and technology platforms, in addition to licensing of patented technology, forms the cornerstone of the commercial arrangements that drive this nascent industry.

III. COMMERCIAL ARRANGEMENTS AND DEVELOPMENT

Pharmacogenomics may be thought of as a cutting edge discovery and development tool used to optimize pharmaceutical drug innovation by personalizing medicine through genetic profiling and identification of appropriate disease subpopulations. The measured success of pharmacogenomics depends on the success of commercial relationships between various pharmacogenomics entities and pharmaceutical companies with whom they collaborate in improving the quality of therapeutic products. Licensing of pharmacogenomic platforms and technologies and collaborative partnerships built around such pharmacogenomic resources pair pharmacogenomics entities with pharmaceutical development entities that have the development and marketing resources to effectively bring drugs to consumers. Thus, cooperative arrangements between pharmacogenomic innovators and the traditional pharmaceutical industry ("Big Pharma") promise to enhance the quality of diagnosis and drug treatment for patients.

Cooperation between pharmacogenomics companies and pharmacogenomics developers, however, is not the only mode by which pharmacogenomics is integrated into drug development. Examination of the current climate indicates that pharmaceutical companies are increasingly developing their own pharmacogenomic analysis capabilities, while pharmacogenomics companies are beginning to capitalize on market fragmentation stemming from pharmacogenomic technology in order to develop their own niche drugs for smaller subpopulations. Analysis of prevailing attitudes and beliefs in the marketplace and existing commercial relationships in pharmacogenomics at this early stage lays a foundation for predicting the role of pharmacogenomics in the dawning era of personalized, genetics-based medicine.

1. THE COMMERCIAL STATUS OF PHARMACOGENOMICS. With its potential to provide for analysis of disease risk and customization of drug therapies, pharmacogenomics has been acclaimed as an exciting, emerging focus of biotechnology and pharmaceutical development (Davies, 2001; Robbins-Roth, 2001). Although there can be little dispute as to the role pharmacogenomics will play in the future of medicine, and the eventual commercial value it will realize, future payoffs must be distinguished from present attitudes and realities that shape the commercial status of pharmacogenomics today. The commercial arrangements formed around pharmacogenomics are, of course, dictated by these market attitudes and conditions and should thus be analyzed from a market perspective.

For all its promise, Big Pharma has embraced pharmacogenomics slowly and for limited purposes to this point. Though its advocates extol its potential to improve the likelihood and speed of regulatory approval as well as marketability for late-stage and market drugs, the pharmaceutical industry has not historically embraced implementation of pharmacogenomic analysis in these areas, perhaps because of skepticism about the ability of pharmacogenomics to increase business, and some claim, from fear of revealing otherwise unknown side effects and risks that would actually hamstring regulatory approval efforts (Longman, 2001). Some believe that large drug companies may shy away from risks associated with fragmentation of the market, whereby target populations for their drugs are winnowed by identification and characterization of disease subpopulations, dampening the blockbuster market impact of their drugs (Longman, 2001; Ma, 2001). Although most of what the pharmaceutical industry fears lies at the heart of the intended human benefits of pharmacogenomics, these economic uncertainties and risks have thwarted efforts of pharmacogenomics advocates to inject pharmacogenomic analysis into the near-term agendas of Big Pharma (Longman, 2001).

The importance and potential of pharmacogenomics, however, are not lost on the broader pharmaceutical industry. Despite declining to make use of pharmacogenomic technology in present clinical trials, some Big Pharma players are banking clinical samples for retrospective analysis (Longman, 2001). The preferred role that pharmaceutical companies envision for pharmacogenomics lies as part of an integrated discovery program that utilizes pharmacogenomic analysis to head off tremendous development costs for drugs that would later die anyway and to focus on target populations for increased efficacy (Longman, 2001). Integration of pharmacogenomic techniques in discovery and at earlier stages in the development process does 'not inhere risks to development investments already made for late-stage compounds, and the use of pharmacogenomics in discovery and early stages in which speculation is an accepted part of the process is a much easier sell for unproven pharmacogenomic technology (Longman, 2001).

Still, pharmacogenomics holds massive potential for application in late-stage development, as pharmacogenomic tests on patients in expanded clinical trials will greatly facilitate identification of disease subpopulations for better targeting with the drug in development. Limiting pharmacogenomic application to discovery and early-stage development blocks valuable clinical research opportunities. From the traditional pharmaceutical industry's perspective, however, inserting pharmacogenomics into the pharmaceutical development phase, even where the technology is more proven, does not eliminate the issues of added cost and time associated with its use, the expanded clinical trials that would be required for meaningful pharmacogenomic analysis in development, or the undesired effects of market segmentation that cut against blockbuster application (Longman, 2001).

Thus, even as the technology begins to yield its analytic potential, questions linger as to how broadly pharmaceutical companies will seek to apply pharmacogenomic analysis to development. Because of the medical benefits pharmacogenomic analysis brings to bear in development, consumer groups are likely to demand and force broad-based application, especially for drugs whose targeted conditions do not have clear progress markers (Longman, 2001). The utility of pharmacogenomic analysis is high for diseases, such as cancer, in which toxicity is an issue, switching from an ineffective drug regimen wastes precious time, and indications of progress are not immediately obvious. On the other hand, pharmacogenomics may have lower utility for conditions with a clear treatment marker, such as high cholesterol. With the high potential for pharmacogenomic preselection of patients in areas such as oncology, depression, Alzheimer disease, schizophrenia, and obesity, it is reasonable to expect early pharmacogenomic development in these fields (Ma, 2001). Many companies have already demonstrated a significant commitment to pharmacogenomics, and some experts predict that within 10 years, 20–30% of new drugs will be marketed in conjunction with diagnostic genetic tests and that the FDA will require submission of genetic testing data for even more drugs (Ma, 2001).

Ultimately, to become widely accepted by the broader pharmaceutical industry, pharmacogenomics will need to deliver on its promise of providing the right drugs for the right patients at the right times. If it does, the value equation for promoting investment will be maintained by more effective treatments that head off expensive, unsuccessful drug therapy. Thus, innovators will still be able to recoup their increased development costs while simultaneously lowering high marketing costs where drugs are backed by stronger scientific evidence of efficacy and eliminating the use of wasteful, ineffective therapies from the system. Pharmacogenomics is a technology of tremendous import to the furtherance of drug therapies that pharmaceutical companies will increasingly and ultimately embrace for the mutual benefit of the pharmaceutical industry and the patients it seeks to treat.

Regardless of the role pharmacogenomics plays as an analytic technology for the pharmaceutical industry, however, major questions arise, including who is likely to supply such analysis and under what circumstances. Attitudes the pharmaceutical industry holds about pharmacogenomics are critical determinants in answering these questions. First, pharmaceutical companies are quick to recognize that, as valuable as pharmacogenomic services may be, pharmacogenomic analysis may be available from a variety of external and internal sources and that it is the pharmaceutical companies themselves that hold the most valuable assets—drug compounds and clinical patient samples (Longman, 2001). The vision that the industry holds for pharmacogenomics as a piece of a fully integrated discovery platform suggests that pharmacogenomics entities must pull together programs for target discovery and validation that incorporate multiple complementary technologies and resources along with pharmacogenomics. For pharmacogenomics providers, horizontal expansion to multifaceted platforms may be the key to providing the broad-based discovery services that the pharmaceutical industry desires and winning financially sustaining deals of higher value with Big Pharma.

An alternative mode for companies involved in pharmacogenomics is to expand vertically by in-licensing technologies and clinical candidates for their own internal product development. In the absence of high-value deals with pharmaceutical companies, marketing smaller niche drug products or off-patent follow-on drugs tailored to individual genetic profiles could provide revenue streams necessary for financial success (Longman, 2001). In an industry historically dominated by a handful of large pharmaceutical developers, pharmacogenomics could well herald an era in which smaller companies will be able to enjoy success through development of individualized drug products for smaller markets.

Collaboration between pharmacogenomics entities and the larger pharmaceutical industry, assimilation of pharmacogenomic analysis into internal pharmaceutical discovery and development, and the emergence of smaller pharmacogenomics entities as niche drug developers are each likely to be important in the commercial development of pharmacogenomics. By analyzing the nature and structure of existing pharmacogenomics collaboration agreements and examining how pharmaceutical companies and pharmacogenomic entities are positioning themselves for noncollaborative development, a commercial status and trajectory for the industry may be established.

2. COLLABORATION WITH BIG PHARMA.   One way in which pharmaceutical companies have shown commitment to pharmacogenomics at this stage in its commercial development is through collaboration and alliance agreements with pharmacogenomics entities. In fact, many major pharmaceutical companies already participate in the SNP Consortium, the nonprofit venture aimed at creating a public, human genome-wide SNP map. Beyond this base-level involvement, however, many pharmaceutical companies have also entered into private collaboration arrangements involving or centering around pharmacogenomics. Such agreements engage a variety of pharmacogenomics entities for a range of services.

Many entities have been active in collaborating with pharmaceutical companies through pharmacogenomics agreements. An early step in the history of such collaboration took place in 1997 when the Genome Center at MIT's Whitehead Institute for Biomedical Research and a group of leading pharmaceutical companies (Bristol-Myers Squibb (BMS), Affymetrix, and Millennium Pharmaceuticals) formed the Functional Genomics Consortium. The alliance provides the Whitehead Institute with $8 million in annual research funding and access to genomics and microarray technologies for five years and is aimed at developing new tools and methods to uncover the genetic basis of human disease, study human genetic variability, and match gene expression patterns with healthy and pathological cellular processes (Sigal, 2001). The Consortium's programs have focused on uncovering genetic markers for diseases and isolating disease genes in areas such as cardiovascular disease, diabetes, inflammatory conditions, and asthma as well as predicting clinical phenotypes in oncology through expression profiling. In 1999, the Consortium granted a royalty-free license for its important reduced representation SNP discovery technology to the SNP Consortium. In contrast with the broader scope of the SNP Consortium, the Whitehead group keeps a narrower focus on discovering SNPs in coding regions of genes, which are believed to be more relevant in indicating disease predisposition and variability in drug response (Sigal, 2001).

A $90 million 1999 alliance between Becton Dickinson and Millennium Pharmaceuticals for gene fingerprinting in cancer diagnostics stands as another important example of pharmaceutical company collaboration in this field. Also, after their joint pledge of contributions to the Functional Genomics Consortium, Millennium's predictive medicine division and BMS again joined forces with a $32 million 1999 pharmacogenomics deal. The companies are studying pharmacogenomics-related issues in oncology for tubulin polymerizing targets (e.g., Taxol) and Ras farnesyl transferase inhibitors in the hopes of developing new anticancer therapies for specific tumor types and disease subpopulations. The deal is structured to provide Millennium certain royalties on drug sales contingent on successful development of useful pharmacogenomic tests as well as a percentage of sales for Taxol dependent on increasing demand for or slowing sales erosion of the now off-patent drug (Longman, 2001).

Continuing to push the envelope in the field, BMS later signed a three-year collaboration in 2000 with the Karolinska Institute in Stockholm, Sweden for clinical trials for cancer treatment based in pharmacogenomics. The first clinical partnership for BMS in pharmacogenomics is aimed at furthering personalized oncology treatments by identifying and validating clinically relevant molecular markers (Sigal, 2001). Later that year, BMS signed an agreement with Orchid BioSciences for industrial-level genotyping for SNPs discovered by the SNP Consortium and the Functional Genomics Consortium. Another alliance in 2000 with PPGx provides BMS access to bioinformatics software and pharmacogenomic tools and services that will enhance the company's efforts. The early and aggressive approach BMS has taken with pharmacogenomic dealings is probably not typical for this early stage of the industry, but the oncology giant may have felt that as a leader in the field it was incumbent upon it to know as much as possible for use now and moving forward with the best drug development and probably also wanted to test the pharmacogenomics waters (Longman, 2001).

A number of other companies have made important pharmacogenomics collaborations as well. In 1997, Genset entered into a $42.5 million agreement with Abbott to develop pharmacogenomic tests for gauging drug response, one of the earliest and largest deals centering around the technology (Ma, 2001). Also notable is a $25 million 1997 deal between Incyte Genomics and GlaxoSmithKline (then SmithKline Beecham) to form diaDEXUS, a joint venture in genetic diagnostics representing another important early pharmacogenomic commercial arrangement (Ma, 2001). Soon afterward, in early 1998, deCODE entered into a $200 million genomics and pharmacogenomics agreement with Roche to identify disease genes through genetic analysis of the uniquely homogeneous Icelandic population. Another diagnostic deal was reached between deCODE and Roche in 2001 (Ma, 2001).

Still other pharmacogenomic commercial arrangements include agreements between Ciphergen and Pfizer to predict toxicity of Pfizer drug compounds, Lynx and AstraZeneca to discover SNPs that can be used to develop pharmacogenomic tests and discover drug targets, and Orchid and AstraZeneca to investigate SNPs related to areas of Astra's therapeutic work (Longman, 2001). CuraGen has also had substantial commercial dealings in pharmacogenomics. Under its 2001 alliance with Bayer, CuraGen will supply the pharmaceutical giant with 80 drug targets, provide access to CuraGen's functional genomics and bioinformatics tools, and determine whether drugs arising from the targets are suitable for individual patients by using pharmacogenomic analysis. Under a separate $124 million agreement, CuraGen will use these analytic capabilities to advise Bayer of toxigenicity and side effects in drug screening, and the partners will create a database of toxicity response genes for mutual use (Longman, 2001). Database subscription deals include the licensing of Genaissance's haplotype database to Johnson & Johnson, Pfizer, and AstraZeneca for use in disease and drug response correlations, and companies such as Celera Genomics, Oxford GylycoSciences, and Lexicon Genetics are including pharmacogenomic elements in their agreement proposals as well (Longman, 2001).

The deal making between large pharmaceutical companies and smaller companies offering pharmacogenomic tools and services represents a significant investment on the part of Big Pharma in pharmacogenomics. Despite reluctance with respect to how the technology is to be used at this stage, it seems clear that pharmacogenomics will be firmly entrenched in the future of drug development. Given the significant role the technology will play, it is reasonable to expect that Big Pharma and smaller companies already holding pharmacogenomic assets will explore alternate, independent business structures for realizing the potential of pharmacogenomics.

3. ASSIMILATION AND NICHE MARKETS.   Although many of the types of collaborations and alliances made up to this point should fairly predict future dealings, large pharmaceutical companies are also developing their own pharmacogenomic programs, and smaller pharmacogenomic asset-holders are building complete drug programs for smaller market drugs around strong pharmacogenomic discovery and analytic tools. In fact, the predictive and personalizing potential pharmacogenomics holds stands to drive major changes in the structure of the drug industry. As Big Pharma takes advantage of pharmacogenomics via internal and external mechanisms

to focus more efficient efforts at developing and marketing successful block-buster drugs, the consequent market fragmentation implicated in the use of pharmacogenomics creates potential for smaller pharmacogenomics entities to capitalize on niche markets. Both effects are important to examine.

Increasingly, large pharmaceutical companies are establishing their own pharmacogenomics programs and divisions in-house. Outside pharmacoge-nomic service providers seek cash, equity, or royalties in products that reach the market, and large pharmaceutical companies that control the drug com-pounds and clinical patient samples that truly drive drug development may not be willing to pay the asking price for pharmacogenomic technology that could be cultivated internally. Although alliances with biotechnology com-panies offering integrated discovery platforms that include pharmacoge-nomic components should continue to be a staple in pharmaceutical business, large pharmaceutical companies will be motivated to use phar-macogenomics in their own homegrown discovery work and eventually further along in development as the specifics of the drug and the market demand.

Already, Big Pharma research and development programs are expand-ing to incorporate pharmacogenomics technologies. Pfizer's President of Global Research and Development, John Niblack, has cited the advantages of larger R&D programs that use new understanding about the genome to predict medical applications for drug candidates (Borchardt, 2001). For now, Pfizer is using pharmacogenomics early in its discovery process to aid in pri-oritizing drug targets. Philip Vickers, Head of Genomic and Proteomic Sci-ences at the company, notes that pharmacogenomic information is still only one piece of the process, but Pfizer's overall approach to research has been aggressive, as the company seeks to realize the full potential of the genomic analysis (Longman, 2001; Borchardt, 2001). Pharmacogenomics will help companies such as Pfizer make better predictions about therapeutic respon-siveness and toxicity for narrower focus on appropriate target patients. Where a drug candidate might only be 50% effective across a broad disease category, the compound might have 85% efficacy in a subpopulation com-prised of 60% of those in the broad disease category. Similarly, a drug that is highly toxic in 25% of patients may only be toxic in 2% of patients of an appropriately identified subgroup. Although there are costs in making such determinations, the ability to establish these type of predictive guidelines will help pharmaceutical companies drastically improve efficiencies for patients and reduce marketing costs, because evidence-based, highly effec-tive products are inherently easier to sell.

The same pharmacogenomics-based subdivision of disease populations that stands to help Big Pharma narrow its focus will also lead to market seg-mentation that opens the door for smaller entities to move into the pharma-ceutical business. Traditionally, the costs of research, development, clinical trials, and marketing for drugs have been so staggering and the risks of failure at various points long into the process so great, that small entities cannot sustain themselves in producing drug products. Rather, gigantic pharmaceutical companies with vast resources have shouldered the burden of producing drugs. Even where a small company overcomes long odds to find and develop the one candidate in a thousand that will graduate from target to approved product, replicating such a performance for a second product is a trying task where the same adverse odds must be faced. Venture capital for small start-ups simply is not adequate to break into the pharmaceutical world outright, and companies successful in getting the support to develop a first product are often bought by the bigger fish in the Big Pharma pond.

An exception to this paradigm, of course, is the small niche market. Biotech companies such as Amgen, Genzyme, and Transkaryotic Therapies have been successful in capitalizing on markets for rare diseases that afflict a relatively small percentage of the population, relying on genetics-based information to make strategic discovery and development decisions and avoid pouring money into dead-end compounds. Still, such entities must face constant pressure to produce new products and continue their growth while competing with one another and ever-present Big Pharma for scarce resources. Even where these companies try to forge a path into the pharmaceutical business, the enormous cost of clinical trials leads smaller companies to sell joint development rights to larger pharmaceutical entities. Advances in pharmacogenomics, however, are likely to help make smaller companies independently engaging in the pharmaceutical business less exceptional.

The fragmentation resulting from pharmacogenomic advances should not be overstated: Big Pharma will continue to develop blockbuster drugs with broad application for large disease populations. The difference is that many of these blockbuster drugs, while still being indicated for high percentages of patients in broad disease categories, will have a somewhat narrower focus that results in fewer failed drug regimens. Simultaneously, identification of smaller subpopulations creates an opportunity for the growth of small pharmaceutical entities with strong pharmacogenomic predictive and development technologies. Thus, pharmacogenomics stands to more clearly bifurcate the pharmaceutical industry along the lines of market size. Whereas Big Pharma will continue to dominate high-stakes blockbuster drug markets, many smaller companies using pharmacogenomics-based technology, like biotechnology entities that have used genetics-based technologies to thrive in rare disease markets, may seek to capitalize on innovation in smaller disease subpopulation markets.

V. Conclusion

Intellectual property protection is a critical foundation for the successful commercial arrangements and development that drive the pharmaceutical and biotechnology world. The legal framework that protects innovations springing from costly research and development creates essential incentives to discover and invent. It also positions various entities to work collaboratively and productively with one another, free from fears of unchecked free ridership, theft, and misappropriation of proprietary resources. The workings of commercial pharmacogenomics neatly illustrate this interplay. Biotechnology and pharmaceutical companies are working together and independently with their patent-protected genes, genetic components, and pharmacogenomic methods and processes, as well as with other proprietary pharmacogenomic information, databases, and know-how, to develop better, personalized disease therapies.

By forging alliances involving licensing of these pharmacogenomic technologies and information, the varied resources of different entities are making possible discovery and development of successful new drugs and therapeutics more narrowly tailored for patients with specific, genetic-based disease profiles. To date, most commercial pharmacogenomic dealings have involved large pharmaceutical companies collaborating with smaller biotechnology companies that use pharmacogenomic assets and capabilities for discovery and prioritization of drug targets. However, as pharmacogenomic technologies become more advanced and add increased predictive capacity, large pharmaceutical companies are working to develop their own proprietary pharmacogenomic resources, and biotechnology entities with strong proprietary pharmacogenomic resources stand poised to capitalize on market segmentation resulting from pharmacogenomic identification of disease subpopulations. In each case, the ability of these entities to create value through collaboration or thrive independently rests on their ability to control the intellectual property that underlies their dealings and production.

The pace at which biomedical innovation and advancement is occurring in the new Genetic Age is remarkable and exciting but also challenging to both the traditional intellectual property regime and familiar business dealings and structures in the pharmaceutical field. As the medical world aspires to leave behind an era of unfocused trial and error with drugs for a new period of personalized, evidence-based medicine rooted in pharmacogenomic diagnostics and therapeutics, there could be no more important time for thoughtful and forward-looking jurisprudence that balances preservation of important incentives for research and development against needs for access to new discoveries. Although the issue of intellectual property protection for genetic discoveries remains ethically controversial, adaptation of the intellectual property regime by the courts and the PTO to accommodate new innovations in the genetics area seems to be a success. Fruitful commercial dealings and development in the biotechnology and pharmaceutical fields expeditiously and consistently produce new medical therapeutics. These innovations are beneficial and accessible to the public and indicate a highly functional and appropriately balanced intellectual property system. Recent progress suggests a continuation of the system's high functionality as applied to commercial pharmacogenomics moving forward.

REFERENCES

American College of Medical Genetics (ACMG), "Position Statement on Gene Patents and Accessibility of Gene Testing," *Genet Aw. Med.*, **1**, 237 (1999).

Amgen v. Chugai Pharmaceutical Co., 927 F.2d 1200 (Fed. Cir. 1991).

Aoki, N., "Patent Applications Booming in Biotech: Studies in Human Genetic Code," *Boston Globe*, p. D1 (Aug. 30, 2000).

Borchardt, J.K., "The Business of Pharmacogenomics," *Modern Drug Discovery*, **4**, 35–36, 38–39 (2001).

Campbell, C., "Legal Issues Associated with Pharmacogenomics," In: Tadmor, B. and M.H. Tulloch (eds.), *The Business Case for Pharmacogenomics*, 2nd ed., Woburn, MA: AdvanceTech Monitor, pp. 8–23 (2001).

College of American Pathologists (CAP), Gene Patents Detrimental to Care, Training, Research (July 5, 2000), *www.cap.org/html/advocacy/issues/genetalk.html* (last visited June 21, 2002).

Davies, K., *Cracking the Genome: Inside the Race to Unlock Human DNA*, New York: Free Press (2001).

Diamond v. Chakrobarty, 447 U.S. 303 (1980).

Guidelines for Examination of Patent Applications Under the 35 U.S.C. 112, Paragraph 1, "Written Description" Requirement (Written Description Guidelines), U.S. Patent and Trademark Office, 66 Fed. Reg. 1099–1111 (2001).

Holman, M.A. and S. Munzer, "Intellectual Property Rights in Genes and Gene Fragments: A Registration Solution for Expressed Sequence Tags," *Iowa L. Rev.*, **85**, 735–848 (2000).

Kieff, F.S., "Facilitating Scientific Research: Intellectual Property Rights and the Norms of Science—A Response to Rai and Eisenberg," *Nw. U. L. Rev.*, **95**, 691–705 (2001).

Koneru, P., "To Promote the Progress of Useful Articles?: An Analysis of the Current Utility Standards of Pharmaceutical Products and Biotechnological Research Tools," *IDEA: J. L. Technol.*, **38**, 625–671 (1998).

Lerner, J., "The Importance of Trade Secrecy: Evidence from Civil Litigation," *Harv. Bus. Sch.*, working paper #95-043 (Dec. 1994).

Longman, R., "Why Don't Big Pharmas Buy Pharmacogenomics?," *In Vivo*, pp. 18–44 (Dec. 2001).

Ma, P., "The Pharmacogenomics Opportunity: How and When to Become a Player," In: Tadmor, B. and Tulloch, M.H. (eds.)., *The Business Case for Pharmacogenomics*, 2nd ed. Woburn, MA: AdvanceTech Monitor pp. 64–85 (2001).

Merges, R.P., et al., *Intellectual Property in the New Technological Age*, 2nd ed.. New York: Aspen Law & Business (2000).

Murray, J., "Owning Genes: Disputes Involving DNA Sequence Patents," *Chi.-Kent L. Rev.*, **75**, 231–257 (1999).

Pear, R., "Research Cost For New Drugs Said to Soar," *New York Times*, C1 (Dec. 1, 2001).

Purves, W.K., et al., *Life: The Science of Biology*, 4th ed. Sunderland, MA: Sinauer Associates, Inc. (1995).

Rafinski, K., "Hospital's Patent Stokes Debate on Human Genes," *Seattle Times*, p. 1A (Nov. 14, 1999).

Rai, A.K., "Regulating Scientific Research: Intellectual Property Rights and the Norms of Science," *Nw. U. L. Rev.*, **94**, 77 (1999).

Rai, A.K., "Evolving Scientific Norms and Intellectual Property Rights: A Reply to Kieff," *Nw. U. L. Rev.*, **95**, 707–713 (2001).

Restatement (Third) of Unfair Competition § 42, Comment C (1995).

Robbins-Roth, C., *From Alchemy to IPO: The Business of Biotechnology*, Cambridge, MA: Perseus Publishing (2001).

Rudolph, N.S., "Introduction," In: Tadmor, B. and Tulloch, M.H. (eds.), *The Business Case for Pharmacogenomics*, 2nd ed. Woburn, MA: AdvanceTech Monitor, pp. 8–23 (2001).

Schmidt, C.W., "Cashing in on Gene Sequences," *Modern Drug Discovery*, **4**, 73–74 (May 2001).

Sigal, E., "The Impact of Pharmacogenomics in Oncology," In: Tadmor, B. and M.H. Tulloch, (eds.)., *The Business Case for Pharmacogenomics*, 2nd ed. Woburn, MA: AdvanceTech Monitor, pp. 37–49 (2001).

Snustad, D. P., et al., *Principles of Genetics*, New York: John Wiley & Sons, Inc. (1997).

Utility Examination Guidelines, United States Patent and Trademark Office, 66 *Fed. Reg.* 1092–1104 (2001).

CLINICAL APPLICATIONS

Integration of Pharmacogenomics into Medical Practice

Gilbert S. Omenn, M.D., Ph.D. and Arno G. Motulsky, M.D.

I. Introduction: The Importance of Identifying and Preventing Adverse Drug Reactions

Pharmaceuticals represent a major class of treatments for a broad array of human diseases. These chemicals are isolated from nature or, more often, synthesized chemically. They are administered to patients to kill or arrest the growth of cancer cells or infectious microbes, to stimulate faltering hearts, and to achieve a host of other specific clinical aims. It is widely recognized that not all people respond alike to these therapies, even when the diagnosis is quite specific.

Meanwhile, these chemicals—like chemical agents encountered at work or in hobbies or as pollutants in air, water, soil, or food—can also cause harm. Sometimes the known mechanisms of action permit us to predict the nature of toxicity to be expected. A meta-analysis of prospective studies from U.S. hospitals indicates that 6.7% of in-patients have serious adverse drug reactions; 0.3% have fatal reactions (Lazarou et al., 1998). In fact, estimates of 40,000 to 100,000 deaths per year attributed to errors in medical care, primarily due to adverse reactions to pharmaceuticals, make this phenomenon a major cause of death in the United States (Meyer, 2000). A tremendous

Pharmacogenomics: Social, Ethical, and Clinical Dimensions, Edited by Mark A. Rothstein.
ISBN 0-471-22769-2   Copyright © 2003 Wiley-Liss, Inc.

effort is underway across the nation to enhance patient safety in hospitals and in office practice (IOM, 2000); recognition of the reasons for adverse drug reactions is a leading component of efforts on patient safety in the broad arena of improving quality of care (IOM, 2001).

Physicians' aim in prescribing medications is to achieve a high therapeutic margin—high benefit with tolerable or negligible adverse effects. Sometimes the margin is small but the aim is compelling, as in chemotherapy for advanced stages of cancers or for life-threatening infections with resistant microorganisms. At the other end of the spectrum, the tolerance for side effects is very low when preventive interventions (vaccines, cancer chemoprevention, dietary supplements, cholesterol-lowering drugs) are introduced for large, healthy populations.

II. THE ORIGINS OF PHARMACOGENETICS AND THE NEW FIELD OF PHARMACOGENOMICS

Over the past 50 years, evidence has accumulated about the clinical significance of marked individual differences in metabolism of drugs and in responses to drugs at the sites of therapeutic and adverse actions. The pioneers—Kalow and Lehman—identified specific inherited traits that produced these marked differences in people who were entirely normal unless "challenged" with the particular drugs (Motulsky, 1957; Kalow, 1962). This new area of investigation was named "pharmacogenetics" by Vogel (1959). That term has powerful meaning today, generally referring to the actions of particular, individual genes or a few specific interacting genes on drug response or adverse drug reaction. Pharmacogenetics is a subclass of the field "eco-genetics," with drugs being one of the many environmental agents interacting with genetic (and nongenetic) variation in the host individual of various exposed populations (Omenn and Motulsky, 1978; Eaton et al., 1999). Eco-genetics, in turn, is a significant part of "public health genetics" (Khoury et al., 2000; Omenn, 2000a).

The newer term "pharmacogenomics" should be reserved, we believe, for genomic analyses employing the powerful tools of modern molecular biology: comparative genome hybridization for amplification of gene copy number, gene expression microarray analysis of levels of hundreds or thousands of messenger RNAs, and proteomic analysis of the level of expression of very large numbers of proteins. These analyses aim to detect evidence of variation in response to pharmaceutical action and in host factors influencing the absorption, distribution, metabolism, and excretion (ADME) of those chemical agents. Originally, these methods identified mostly "spots" of unknown function; now more mRNAs and proteins are identified with particular classes of genes or proteins and their known or likely functions. It is possible that automation of the new techniques will so facilitate testing and

TABLE 8.1. Classic Pharmacogenetic Traits

Variation in Biotransformation of Pharmaceuticals

Suxamethonium sensitivity, due to atypical serum pseudocholinesterase
    Frequency: 1/2000 Caucasians
       Consequence: prolonged respiratory paralysis on exposure to the drug
       Suxamethonium (succinylcholine) for muscle relaxation for anesthesia

Slow acetylator phenotype, due to mutations in liver *N*-acetylase transferase,
    NAT 2
    Frequency: 45–65% of Caucasians and African Americans; 10–15% of Asians
    Slow inactivation of drugs such as isoniazid (for tuberculosis), dapsone (for
       leprosy), and hydralazine (for high blood pressure), leading to toxicity from
       the drug at doses well tolerated in people with rapid acetylator phenotype
    Clinical consequences depend on the specific side effects of the drugs

Variation in Tissue Sensitivity to Pharmaceuticals

Glucose-6-phosphate dehydrogenase (G6PD) deficiency in red blood cells
    X-linked recessive inheritance; many specific mutations
    Common variants: mild deficiency variant in African Americans, moderately
       severe deficiency variant in Caucasians of Mediterranean descent
    Frequency: 1/10 males; 1/100 males, African American and Mediterranean,
       respectively
    Clinical consequence: hemolysis (breakdown of circulating red blood cells) from
       antimalarials, sulfonamides, nitrofurantoin, and other drugs.

α-1 Antitrypsin deficiency, due to variants in this circulating plasma protein
    Frequency: 1/3000 Northern European Caucasians
    Clinical consequence: predisposition to emphysema in early middle age,
       especially in cigarette smokers, due to failure to protect against trypsinlike
       enzymes in lung

---

clinical interpretation in reference laboratories, including those of academic medical centers, that practicing physicians will more readily use these new pharmacogenomic tests than the few previously-developed specific tests for pharmacogenetic variation.

Several classic pharmacogenetic traits are listed in Table 8.1 and discussed below. For a variety of reasons, mostly reflecting the slow penetration of genetic thinking into clinical medicine and the limited predictive value of even the best-validated tests, these tests have been significantly underutilized.

"Suxamethonium sensitivity," prolonged paralysis of respiratory muscle function caused by failure to cleave the short-acting muscle relaxant succinylcholine widely used in anesthesia, was made testable with a simple bedside assay in 1968 (Motulsky and Morrow). However, that test was not incorporated widely into practice. Most anesthesiologists felt they could simply monitor all patients and "bag" those not resuming respiratory action, without testing for a trait that would be found in only one of 2000 patients.

The anesthesiologists' response is at the heart of the difference between individual clinical decisions for individual patients and the epidemiological or public health approach required for progress in patient safety. The patient safety approach begins with millions of patients undergoing anesthesia and asks how to find the large number of individuals who constitute 1/2000 of such a large population exposed to the drug.

Acetylator phenotype is tested by administering a standard dose of isoniazid or caffeine or other drug requiring acetylation for inactivation and then measuring the ratio of acetylated metabolite to unchanged drug at a standard time (2 or 6 h) later in plasma (and sometimes urine) samples from the patient. This procedure is time-consuming for the patient compared with a straightforward in vitro blood test that is available for detection of polymorphic variants mediating drug acetylation by N-acetyltransferase (NAT2) (Meyer and Zanger, 1997). Relevant drugs (Table 8.1) are likely to cause toxicity only in slow acetylators at standard doses; peripheral neuropathy from isoniazid is a classic example. Conversely, rapid acetylators may not maintain circulating levels of the drug at sufficiently high levels to gain full therapeutic benefit; this problem arises when isoniazid is administered, especially in developing countries, on a twice-per-week schedule rather than daily to treat or prevent the emergence of tuberculosis in exposed individuals. It should be noted that slow acetylators are also at higher risk of developing cancers of the urinary bladder from industrial chemicals known to cause such cancers, namely benzidines and β-naphthylamine. On the other hand, rapid acetylators may be at higher risk of colon cancer (Hein, 1988).

Unlike suxamethonium sensitivity and slow acetylator phenotype, which are caused by a deficiency of the enzyme to biotransform or metabolize the drug itself, glucose-6-phosphate dehydrogenase (G6PD) deficiency involves an enzyme of the pentose-phosphate shunt pathway whose inherited deficiency makes red blood cells highly susceptible to hemolysis from oxidative stress. Such stress can occur with a variety of drugs and upon ingestion of fava beans. Susceptibility to fava bean-triggered hemolysis in G6PD-deficient individuals requires additional genetic variation, as well as the exposure to fava beans. G6PD deficiency has been studied from numerous points of view—clinical consequences, evolutionary advantage (in the face of malarial infections), molecular heterogeneity (hundreds of different mutations), and marker for embryological development and cancer cell lineages. Because its inheritance is X-linked recessive, the deficiency and the clinical consequences are more severe in males. Much less attention has been paid to discerning the consequences of G6PD deficiency in other tissues that are known to have active pentose-phosphate shunt pathways and that might, therefore, be susceptible to the same oxidizing drugs. These sites include mammary gland and testis.

G6PD deficiency occurs in two common types in different populations. A severe deficiency occurs in people of Mediterranean origin; a much less severe deficiency occurs in people of African origin. As this book documents, gene frequencies often vary in different populations, whether defined geographically, ethnically, or "racially." These differences arise because of original mutations with a founder effect in small populations and occasionally persist because of favorable selection that increases the prevalence of the variant gene. Important examples are the higher survival of children who have G6PD deficiency and children who have sickle cell trait in geographic areas that have a history of endemic malaria infection. These relationships are highly specific; for example, falciparum malaria is responsible for the selective survival advantage of G6PD deficiency and of hemoglobin S (sickle), whereas a different kind of malaria (vivax) is responsible for selection of the red blood cell variant Duffy negative (loss of the antigen that is a receptor for penetration of this organism into red blood cells).

III. MODERN PHARMACOGENETICS

Much more systematic investigation with molecular tools is revealing variants of key biotransformation enzymes. When these variants occur at population frequencies greater than 1%, they are classified as "polymorphisms." Some of these variants have clinically significant differences in enzyme activity (Evans and Relling, 1999). By knowing the nature of the biotransformation of specific drugs, one is able to predict which drugs should be examined for variation in metabolism via these polymorphic pathways.

This approach has been particularly fruitful for the cytochrome P450 monooxygenase known as CYP2D6. Early work with debrisoquine, a drug discontinued during development to treat high blood pressure because of a striking drop in blood pressure in a "super-sensitive" 5% of early volunteer patients, and with sparteine, a drug developed for entirely different clinical indications (antioxytocin effect) converged on the same CYP2D6 polymorphism. Later, people were differentiated into extensive versus poor metabolizers of these drugs. More than two dozen additional lipophilic bases, including antidepressants and beta-blockers, are now known to be differentially metabolized by this same enzyme. In fact, there are multiple mutations in CYP2D6 that reduce activity. In addition, there are ultrarapid metabolizers, whose variant enzyme has such high activity that ordinary drug dosages have no effect (Johnson et al., 1993). Asians only rarely have the null CYP2D6 allele that is present in 6–10% of Caucasians, yet up to one-half have a partially deficient allele (Pro34Ser) which slows their metabolism and sets up a lower dosage requirement for antidepressants and neuroleptics in this population (Bertilsson, 1995). A related enzyme, CYP2C19, metabolizes certain acidic (the anticonvulsant mephenytoin), basic (the antidepressant imipramine and the antiulcer proton pump inhibitor omeprazole), and neutral (the antianxiety drug diazepam) drugs. Many other CYP enzymes and their variants govern the metabolic pathways of a great many other therapeutic agents (Meyer, 2000; Evans and Relling, 1999).

Learning which functions and relevant proteins are "upregulated" in clinical disorders can guide drug development to new targets for pharmaceuticals. A dramatic example is overexpression of the HER2/neu receptor in the tumors of about 25–30% of breast cancer patients; a drug directed at inhibiting this receptor, Herceptin, has proved highly useful in patients with advanced breast cancer. Use of the drug should be and is limited to those who are tested before treatment and found to be overexpressing this specific gene and its receptor (Baselge et al., 1998; Goldenberg, 1999). The choice of test for overexpression is not a simple matter; recent studies indicate that the most reliable test in diagnostic laboratories is based on fluorescence in situ hybridization (Bartlett et al., 2001).

Another dramatic example of drug development targeted to a specific gene variant is Gleevec (Novartis STI-571). This drug is remarkably effective against one form of leukemia, namely chronic myelogenous leukemia (CML), because it inhibits very specifically the activation of the tyrosine kinase coded for by the bcr-abl fusion gene at the site of the pathognomonic Philadelphia chromosome 9/22 translocation in CML patients (Druker et al., 2001). It also works in blast crisis with Ph^+ acute lymphoblastic leukemia and in very uncommon gastrointestinal stromal tumors (GIST). Even this highly targeted drug, however, has encountered resistance to its action and relapse of the leukemia in CML patients. Gorre et al. (2001) found progressive bcr-abl gene amplification in 3 patients and a point mutation in the bcr-abl kinase domain in 6 other patients, among 11 patients relapsing. Binding and inactivation of the drug by α-1 acid glycoprotein may be another resistance mechanism.

One of the most intriguing findings to date about differential response to drug action has emerged from use of aminoglycoside antibiotics to treat *Pseudomonas* and other gram-negative infections in patients with cystic fibrosis (CF). In cultured respiratory epithelial cells, Bedwell and colleagues showed that antibiotics such as gentamicin could overcome the effect of stop mutations in the CFTR gene by facilitating read-through of the CFTR mRNA (Bedwell et al., 1996). Such stop mutations account for about 10% of all CF patients and about 60% of Ashkenazic Jewish CF patients. A similar read-through effect can be achieved in a mouse model for Duchenne muscular dystrophy, increasing expression of full-length mRNA from dystrophin genes with stop codon mutations (Barton-Davis et al., 1999). Clinical trials should be of interest.

IV. PHARMACOGENOMICS: EXAMINING THE GENOME, NOT JUST INDIVIDUAL GENES

New methods employing microarrays for gene expression (as mRNAs) and utilizing 2-D gels or 2-D liquid phase separations and powerful mass spectrometry for detecting variation in protein expression (proteomics) permit examination of hundreds and thousands of genes and proteins, instead of one gene or a few related genes at a time. For common diseases, multifactorial causation is the usual pattern—involving multiple genes and multiple environmental, behavioral, and nutritional modifying factors. The same must be true for the complex metabolism and actions of common drugs—with genes and environmental factors influencing absorption, distribution, biotransformation, protein binding, excretion, and interactions at various sites of action in cells and tissues.

Although these methods can detect variation in inherited genes that would influence susceptibility or resistance to genes, as described above for individual genes, the greatest application thus far has been in studies of somatic cells—especially cancer cells. During the past two years, many impressive clinical investigations for various common cancers have been published, identifying different patterns of expression of genes and/or proteins in normal tissue, benign tumors, malignant tumors, and metastatic tumors of prostate, breast, esophagus, lung, colon, lymphomas, and melanomas. Some of this new work is described in the following sections.

A. B CELL LYMPHOMAS

A landmark publication by Alizadeh et al. (2000) from Stanford and six other institutions demonstrated the usefulness of studying gene expression by gene chip methodology in differentiating patients with different prognoses for survival after chemotherapy against large B cell lymphoma. First, they characterized lymph node gene expression into certain patterns representing normal lymph nodes, the germinal centers of normal lymph nodes, proliferation zones of lymph nodes, and normal T cells. With some 10,000 oligonucleotides on the chips to hybridize and recognize expression of genes, a great many genes showed differences from background or from the normal patterns. A convention was introduced to filter the results to show those genes whose expression is increased or decreased at least 2.0- or 2.5-fold and to suppress the mass of data on many genes whose expression varies relatively little. In fact, this convention may introduce a serious problem. There are genes whose gene products are critical as rate-limiting steps for whole pathways, so that a change of much less than a factor of two or more might be quite important.

When diffuse large B cell lymphoma gene expression patterns were then examined, two distinctly different patterns were identified, one resembling closely the germinal centers of B cells in lymph nodes and the other resembling the activated proliferation zone of B cells in lymph nodes. Of great clinical interest, these two patterns were associated with strikingly different survival curves, 9 deaths among 24 patients at "low" risk and 11 deaths among 14 patients at "high" risk ($p < 0.002$) over 8–10 years of follow-up.

With additional pretreatment information, the survival rates could be stretched to 85% versus 15%. Of course, for management of individual patients, it would be desirable to move as close to 100% vs. 0% as feasible; nevertheless, these results can help identify a subgroup of patients not otherwise identifiable that can be expected to respond dramatically well to aggressive chemotherapy, whereas the other subgroup might be spared the toxicity of the treatment in light of the very limited chance of benefit. These findings are being subjected to independent confirmation in larger patient groups. Hopefully, a different treatment will now be devised and tested in clinical trials for the poorly responding subgroup. Without the molecular signatures, physicians and patients would remain committed to a therapy with efficacy in 40% of patients and toxicity in close to 100%.

B. PROSTATE CANCERS

Rubin, Chinnaiyan, and colleagues at the University of Michigan have published extensive studies of molecular profiles of prostate cancer (Dhanasekaran et al., 2001). Using methods similar to those of Alizadeh et al., they classified carefully dissected prostate tissue specimens into normal adjacent tissue, benign prostatic hypertrophy, localized prostate cancer, and metastatic prostate cancer. Two control or comparison groups were utilized: first ,a commercial pool of normal prostates from men with no cancers, and second, normal adjacent prostate from men with prostate cancers. Each type of specimen was differentiated by the microarray clustering method, combined with arrays of prostate tissue from many patients. Many of the genes with substantial up- or downregulation fell into well-recognized categories of functions, including transcription factors, cell adhesion, protease/antiprotease, phosphatase/kinase, free radical scavenger, and inflammation/immunity genes (and protein products). Furthermore, prostate cancers that are responsive to anti-testosterone (estrogenic) therapy were differentiated from cancers that are nonresponsive or refractory to such therapy. They also identified a specific biomarker (Hepsin) that seems to have prognostic value for prostate cancers.

C. CUTANEOUS MALIGNANT MELANOMAS

Bittner et al. (2001), in a study involving 11 different institutions, brought modern molecular techniques to this important type of cancer, whose incidence is rising steeply. Melanomas have no histopathologic, molecular, or immunochemical markers to differentiate subsets of patients. There are few known recurring genetic or cytogenetic changes in these tumors, and nonsurgical treatment is notoriously ineffective in cases of advanced disease. With a clever visual clustering of gene expression data for 31 patients, 19 were concentrated in the major cluster, whereas the rest were distributed rather widely. Just 22 genes accounted for much of the separation between the major cluster and the other specimens; 19 have known functions.

A substudy examined cell lines from very invasive malignant melanomas of the uvula, together with the cutaneous melanomas. The molecular assays were combined with bioassays. First, a simulated scratch wound was made in plated cells, to see whether the cells would grow out, how far they would migrate, and how rapidly they moved away from the initial site. The cells from the invasive lesions migrated dramatically from the initial position. Second, these cells actively stimulated coagulation of collagen in the gel. And, finally, the most invasive melanoma cells pushed a trabecular framework aside. In these bioassays, the 31 cutaneous melanoma specimens showed remarkable differentiation: The major cluster of 19 melanomas had downregulation of several genes related to spreading, migration, and formation of focal adhesions (integrins, syndecan 4, vinculin), whereas the mixed group of 12 other melanomas had an opposite pattern, with higher activities associated with invasiveness, notably fibronectin, a promigratory molecule. Preliminary survival data (7/10 vs. 1/5 for major vs. nonmajor clusters) showed differences. Clearly, longer follow-up and confirmatory studies are needed.

D. LUNG CANCERS

Microarray analysis for mRNA expression profiling distinguished squamous cell carcinomas from adenocarcinomas and revealed distinct subclasses of adenocarcinomas of the lung: tumors with high relative expression of neuroendocrine genes and tumors with high expression of type II pneumocyte genes (Bhattacharjee et al., 2001). Neuroendocrine tumors were further subdivided, as carcinoids had a specific set of gene markers. These findings are likely to be clinically useful, because histopathology of adenocarcinomas, including bronchioloalveolar carcinomas, generates substantial disagreement among pathologists. Morphologic and immunohistochemical evidence of neuroendocrine features is noted among high-grade small-cell lung cancers, large-cell tumors, and intermediate/low-grade carcinoid tumors. When these results were compared with an independent study with a different set of tumors and expression profiling platform (Garber et al., 2001), there was substantial similarity, plus some differences yet to be sorted out. These methods also were useful in distinguishing primary lung adenocarcinomas from metastases of nonpulmonary origin.

Proteomics studies of patients with lung cancers are leading to potentially useful circulating biomarkers, including auto-antibodies against annexins and PGP9.5 antigenic proteins (Brichory et al., 2001; Oh et al., 2001). Cell lines and microdissected tumors are useful in developing biomarkers, which must be evaluated clinically and epidemiologically. The proteome is a highly dynamic compartment, regulated both through transcription of mRNAs and the subsequent translation into proteins and then also through posttranslational modification of the proteins. Compartmentalization within the cell is critical—location matters. Subsets of proteins can be examined—based on secretion, membrane location, phosphorylation, glycosylation, antigenic properties, and other categories (Hanash and Beretta, 2002).

For decades proteins have been displayed as "spots" on two-dimensional gels, separating proteins by charge and by size. Actually, this approach was rather discouraging, because so few of the many spots visualized could be identified. Genome sequences and mass spectrometry of proteins have dramatically altered this situation. New methods are appearing rapidly; three journals dedicated to proteomics have been launched in the past year (*Proteomics, Briefings in Functional Genomics and Proteomics, Journal of Proteomics Research*; joining *Molecular & Cellular Proteomics*). Organizationally, there is now an international Human Proteome Organization (HUPO), patterned after the quite successful Human Genome Organization (HUGO) (Abbott, 2001).

Measuring both protein and gene expression is important, because evidence is accumulating that the two levels are often not closely correlated. Many other factors besides transcription of the gene are important. These factors include splicing, translation, posttranslational modifications, binding, catabolism, and clearance. Some protein biomarkers will be related to drug action.

E. BREAST CANCERS

There is tremendous potential for more specific examination of gene expression and protein expression in already-recognized forms of breast cancers. Clinically cogent comparisons include estrogen receptor positive vs. estrogen receptor negative, HER2/neu positive vs. -negative, and familial BRCA 1 vs. familial BRCA 2 vs. sporadic breast cancers. It is practical to evaluate the influence of chemotherapy agents and chemopreventive agents on such tumors by sampling cells and proteins in ductal fluid from nipple aspirates. The most instructive clinical result to date is the development of Herceptin for advanced breast cancer (above).

V. TOXICOGENOMICS

Similar methods are being used under leadership from the National Institute for Environmental Health Sciences' (NIEHS) National Center for Toxicogenomics and the Chemical Industry Institute of Toxicology (CIIT) to search for "molecular signatures" of the carcinogenic, mutagenic, teratogenic, and other toxic effects of specific chemical agents (including drugs) on various tissues. Also, pretreatment variation in the tissues can be identified; such tissue susceptibility signatures may help to explain differences in risk of adverse effects or prospects for therapeutic efficacy. Clusters of genes involved in particular mechanistic pathways underlying biotransformation, cell division, proliferation, apoptosis, and other phenomena can be examined for coordinated, coherent responses (Omenn, 2000b).

The National Center for Toxicogenomics created a ToxChip with 2090 unique human genes from 65,000 nonredundant clusters in UniGene, in turn selected from 750,000 human sequences in GenBank. These genes included 72 related to apoptosis, 90 oxidative stress, 22 peroxisome proliferator responsive, 12 Ah receptor battery, 63 estrogen responsive, 84 housekeeping, 76 oncogenes and tumor suppressor genes, 51 cell cycle control, 131 transcription factors, 276 kinases, 88 phosphatases, 23 heat-shock proteins, 30 cytochrome P450s, and 349 receptors. During 1999–2002, the NIEHS Environmental Genome Project resequenced 123 of its list of 554 environmentally responsive genes and identified more than 1700 SNPs in these genes (work at University of Washington and University of Utah Genome Centers, *www.niehs.nih.gov/envgenom/*). It is likely that some of these methods will become common in the next few years in reference laboratories for pharmacology and toxicology.

In Britain, a large-scale study of gene-environment interactions, called the UK Population Biomedical Collection is being launched (Berger, 2001). Jointly funded by the UK Medical Research Council, the Wellcome Trust, and the Department of Health, "BioBank UK" would recruit a population cohort of 500,000 volunteers 45–69 years of age as a resource for investigating genetic and enviromental infulences on a wide range of conditions, from responses to drugs to risks of cancers, heart, and other diseases. The goals and design are still under discussion. For example, in order to enhance the prospects of detecting significant interactions, Wright et al. (2002) proposed a subcohort of 20,000 sibpairs and parents.

VI. INFECTIOUS DISEASES: GENOMIC VARIATION IN HOST-PARASITE INTERACTIONS

The first free-living organism to have its genome fully sequenced was *Haemophilus influenzae* (Fleischmann et al., 1995). Genomes of many pathogenic microbes and their innocuous relatives have now been sequenced, with important insights. The genome size and gene number vary widely, reflecting diversity in ecological niches. In general, organisms that need to survive diverse environments have larger genomes with comprehensive biosynthetic pathways. In contrast, obligate parasites have smaller genomes with adaptations that facilitate an existence entirely dependent on the host. For example, *Mycobacterium tuberculosis* has genomic expansions of enzymes involved in lipid metabolism and cell wall biogenesis, which facilitate resistance to anti-TB drugs through changes in permeability and transport (Cole et al., 1998). This organism uses enzymes of the glyoxylate pathway, which seems to enable survival in lung tissue of humans. *Mycobacterium leprae* is an intracellular obligate parasite, with massive gene decay and greatly restricted metabolism, yet it has several species-specific genes and enzymes not found in the larger *M. tuberculosis* genome (Cole et al., 2001). Another example is the enterohemorrhagic *Escherichia coli* strain O157:H7, which has expansion of several pathogenic determinants and 1387 additional genes compared with the innocuous K-12 strain (Perna et al., 2001). Pathogenicity islands are prominent in pathogens like *Vibrio cholera*, *Helicobacter pylori*, and *Yersinia pestis*; these genomic regions of 10–200 kilobase pairs have distinctive structural features and encode adhesins, invasins, toxins, and protein secretion systems that are determinants of bacterial virulence. In the human host, polymorphisms in genes that modulate the immune response in macrophages, cytokines, chemokines, and Toll receptors alter susceptibility to various infections, including HIV (Littman, 1998) and *E. coli* O157:H7, as well as efficacy and safety of antimicrobial drugs.

VII. Issues of "Race" in Pharmacogenetics and Pharmacogenomics

The geographic origins of different frequencies of specific genes in different populations are sometimes converted to discussions of "racial differences." Serious, avoidable misunderstandings arise. The usual superficial physical features associated with race—especially skin color—are not pathophysiologically correlated with internal variation in biotransformation of drugs or likelihood of therapeutic or adverse effects, or with other genes predisposing to various internal diseases, for that matter. The correlations are strictly statistical, derived from common geographic origins and population movements of people of African, Hispanic, Asian, and Native American origins. Furthermore, the correlations are diluted by interracial admixture of genes, as revealed in studies of population structures.

Using a method for multilocus genotyping (Pritchard et al., 2000), Wilson et al. (2001) genotyped 30–48 individuals from each of eight human populations. They inferred a structure for the populations from genotypes of 16 chromosome 1 microsatellites and 23 X-linked microsatellites, variation not known to be related to any phenotypic differences. The structure best fit four clusters by geographic area: Western Eurasians (Norwegians, Ashkenazic Jews, Armenians), sub-Saharan Africans (South African Bantu), Chinese, and Papua New Guineans. The Ethiopians and Afro-Caribbeans had extensive admixture. Then they determined the frequencies of 11 functionally significant variant alleles of six genes coding for drug-metabolizing enzymes [CYP1A2, 2C19, and 2D6; NAT2; NAD(P):quinone oxidoreductase (DIA4); and glutathione-*S*-transferase M1 (GSTM1)], which showed only moderate correlation with the clusters determined genetically, and even less with "ethnicity" limited to Eurasian, black, and Asian.

Studies of racial factors in public health services, plus higher incidence and poorer survival rates for various clinical conditions in minority populations, have generated considerable attention to underdiagnosis, underservice, different benefit/risk equations, and other "disparities" in health status and health care (IOM, 2002). According to the Institute of Medicine, minority populations tend to receive lower-quality health care than whites do, even when insurance status, income, age, and severity of conditions are comparable (IOM, 2002). Ironically, this report refers to the large undifferentiated nonminority population as "white"; see Bhopal and Donaldson (1998) for commentary on this extremely heterogeneous descriptor introduced for administrative purposes, including census, and not for scientific or medical/public health purposes. The *British Medical Journal* has provided guidance that study populations be described for what they are (leaving aside what would be appropriate for the particular study question) and then be called "comparison" populations (*BMJ* 1994; 312:1094). Fullilove (1998) emphasized the cultural traditions and practices associated with "ethnicity" and the many covariates of geographic "place," while decrying the outdated use of the term "race" and its reliance on skin color. Finally, Rothstein and Epps (2001) provide a balanced assessment, favoring the elimination of the term and concept of "race," while utilizing the much broader and more flexible term "ethnicity."

It is ironic that the U.S. Government research funding agencies, notably the National Institutes of Health and the Centers for Disease Control and Prevention (CDC) of the Department of Health and Human Services, require explicit recruitment of self-identified underrepresented minorities into clinical trials and other studies at the same time that there is fierce resistance to "racial profiling" or "racial labeling" of patients in the clinical records and on clinical rounds. For example, a pair of articles relevant to treatment of patients with congestive heart failure appeared in the *New England Journal of Medicine* 3 May, 2001. The articles by Exner et al. (2001) and Yancy et al. (2001) had "black versus white patients" and "race," respectively, in the titles. The ACE inhibitor enalapril appears to reduce rates of hospitalization in whites more effectively than in blacks, whereas the beta-blocker carvedilol was reported to be similarly effective in reducing mortality or hospitalization in black and nonblack patients. These articles triggered strongly opposed editorials. One praised the potential usefulness of choosing drugs in light of racial subpopulation mean differences in response (Wood, 2001), whereas the other decried such studies as racial profiling (Schwartz, 2001). There are definitely group differences in frequencies of polymorphic drug metabolizing alleles (Table 8.2). On the other hand, the conclusion (Rothstein and Epps, 2001) that "observations of differences in responses to a drug in different groups that correspond with historical racial categories have been important to advances in pharmacogenomics" seems to be overstated.

TABLE 8.2. FREQUENCIES OF CERTAIN ALLELES OF PHARMACOGENETIC
SIGNIFICANCE

	Caucasians	African Americans	Asians
NAT2, slow acetylator	40–70%	50–60%	10–15%
CYP2D6/null (PM)	6–10%	5%	1%
CYP2C19/null (PM)	2%	4%, 18%	12–23%
CYP2C9/het/low (PM)	14–37%	0–8%	0–1%
β-2 adrenergic receptor (codon 16)	37%	49%	59%

PM = poor metabolizer phenotype.
From Bertilsson, 1995; Evans et al., 2001, Mancinelli, 2000; Meyer, 2000; Wood, 2001; Wilson et al., 2001.

CDC recently established three centers at schools of public health to further knowledge and utilization of genetics in public health practice. These Centers on Genomics and Public Health are at the University of Washington, the University of Michigan, and the University of North Carolina. The engagement of communities, and especially communities of color, is emphasized in this initiative. The National Institute of General Medical Sciences of NIH in 2001 offered administrative supplements to principal investigators with cooperative agreements funded as part of its Pharmacogenetics Research Network and Knowledge Base (PharmGKB) to foster "research aimed at understanding pharmacogenetic differences among racial and ethnic groups . . . either to collect scientific information or to develop a sample resource" (NIGMS, 2001).

In this context, it is interesting to recall that the Vice President for Minority Affairs at the University of Washington pointed out more than a decade ago that he declines to check the racial origin box, because the government's demand "makes me choose among my grandparents." In the year 2000 U.S. Census, individuals were given the option of indicating "multiple racial groups," without checking any specifically; 7% of respondents indicated "multiple." Another group did not respond at all. These two groups make analysis more complex. Meanwhile, the Human Genome Project has demonstrated that there is striking similarity among people, with 99.9% of genes described as "identical." Even so, with 6 billion nucleotides in the double-stranded DNA of humans, 0.1% represents 6 million differences!

VIII. PHARMACOGENETICS IN CLINICAL PRACTICE

In a response published in *The Lancet* to the question "If I had a gene test, what would I have and who would I tell," Motulsky (1999) emphasized the potential value of pharmacogenetic tests before treatment for certain specific conditions. He cited pharmacogenetic predispositions caused by CYP2D6 alleles that impair metabolism of commonly used drugs, like antidepressants or beta-blockers. Conversely, the much less frequent ultrarapid metabolizers (because of gene amplification) cause failure of drug response unless very much higher dosages are used. Other gene tests worth considering include carrier states for the common thrombophilias, such as factor V Leiden (5–8% of the population) and prothrombin variant 20210 (2–3% of the population). However, current data suggest that the risk of hemorrhagic complications from anticoagulant drugs outweighs the risk of thrombosis in these carriers, unless there is a personal or family history of thrombosis. On balance, wide-scale testing for these genes is not recommended. Finally, individualization of drug therapy for high blood pressure would be desirable; despite many mechanistically interesting, but very rare mutations affecting blood pressure through actions in the kidney or the angiotensin/renin pathway (Lifton and Gharavi, 2001), treatment currently remains entirely empirical. The key variable in control of blood pressure is net renal salt balance.

Without regard to therapy, potentially valuable diagnostic tests are available for presymptomatic evaluation of risk of breast cancer due to predisposition from BRCA 1 or BRCA 2 and of colon cancer related to familial adenomatous polyposis (APC gene) or hereditary nonpolyposis mismatch repair genes (MSH 2). Genetic predisposition to Alzheimer disease associated with ApoE4 is neither sufficient nor necessary to lead to the clinical condition, and no definitive therapy is available.

In Table 8.3, we provide illustrative combinations of genetic polymorphisms and relevant drugs by organ system or clinical specialty for prevention of adverse effects and tailoring of drug dosage. Evans and Relling (1999) have provided (as a Web-based supplement to their *Science* review) a lengthy tabulation of polymorphic variants identified in phase I (metabolism) and phase II (conjugation) reactions (at *www.sciencemag.org/feature/data/1044449.shl*). For many variants, no clinical consequences are yet known. For others with serious clinical consequences, the frequencies are highly variable, with some possibly too infrequent to warrant screening. Extraordinary success analyzing rare Mendelian defects in renal functions has generated new knowledge about molecular mechanisms of blood pressure variation and predictable properties of new pharmaceutical agents specific for these molecular targets (Lifton and Gharavi, 2001; Mukherjee and Topol, 2002).

IX. PROSPECTS FOR USE OF PHARMACOGENOMICS IN CLINICAL PRACTICE

The new methods to detect and characterize variations in gene and protein expression across the human genome in individuals will become common

TABLE 8.3. ILLUSTRATIVE GENETIC VARIATION BY CLINICAL FIELD, DRUG CLASS, AND POLYMORPHIC GENETIC TRAIT (METABOLISM OR TISSUE SENSITIVITY)

Cardiovascular Diseases:
- Dosing with beta-blockers and anti-arrhythmics (CYP 2D6)
- Dosing with ACE inhibitors (ACE variants)
- Thromboembolism increased with factor V Leiden, factor II (prothrombin), or high factor VII variants
- Increased sensitivity to Coumadin (CYP 2C9)
- Prolonged Q-T interval syndromes and dysrhythmias (potassium channel variants HERG, KvLQT1,KCNE2; late-opening sodium channels, with specific blockers)
- Mechanism-based high blood pressure treatments (numerous rare Mendelian disorders)
- Failure to respond to pravastatin (B2B2 variant of cholesterol ester transfer protein)
- Enhanced response to simvastatin (apoE4 and high Lp(a))

Cancers:
- Susceptibility to excessive myelosuppression from 6-mercaptopurine, 6-thioguanine, azathioprine in treatment of acute lymphocytic leukemia (thiopurine methyltransferase)
- Treatment of Her2/neu (EGFR2)-upregulated breast cancers with Herceptin
- Neurotoxicity from fluorouracil (dihydropyrimidine dehydrogenase)
- Mucositis (toxicity) from methotrexate (methylene tetrahydrofolate reductase TT variant)

Depressive Disorders:
- Antidepressants (CYP 2D6)
- Neuroleptics (serotonin receptor)
- Tardive dyskinesia from antipsychotic agents (dopamine D_2 and D_3 receptor variants)

Alzheimer Disease Treatment
- Anticholesterase treatment with Tacrine (apoE4)

Asthma
- Response to β-agonists (β-adrenergic-2 receptors)
- Response to nonsteroidal anti-inflammatory agents (CYP 2C9)

Pain Control
- Nonresponse to codeine (PMs of CYP2D6 fail to O-demethylate to morphine)
- Cramps and other reactions from excess codeine effect in 2D6 "ultrarapid metabolizers"

Anesthesia
- Drug-induced malignant hyperthermia (ryanodine receptor)

Adapted from Evans and Relling (1999), Meyer (2000), and Mukerjee and Topol (2002).

in the next few years in reference laboratories. Physicians will not be expected to understand the methods and the patterns. Instead, physicians will be expected to utilize the interpretations provided from these laboratories, as with classical pathology and with "exotic" nongenetic tests, to characterize the stage of disease in patients with cancers and to assist in directing therapy for a host of conditions. Patients will rely on their physicians for such interpretations. Chips already exist with numerous known mutations of the various relevant CYP enzymes; other potentially useful polymorphisms could be added to such chips for automated analyses.

The choice of drug therapy in large B cell lymphoma may be guided now by the patterns of gene expression reported by Alizadeh et al. (2000). The possibility that women with BRCA1 respond differently than women with BRCA2 to treatment with tamoxifen or other selective estrogen receptor modulators, based on very preliminary clinical epidemiological observations (King et al., 2001), seems ripe for molecular pharmacogenomic investigation. If this is true, prevention with these agents might be recommended differentially.

Physicians addressing patients with suspected environmental, occupational, or recreational exposures to particular agents may be able to submit blood samples or tissue biopsies to special laboratories of toxicogenomics to assay for evidence of molecular signatures of particular exposures and possibly for inherited susceptibility to adverse effects from such exposures. Of course, nongenetic factors—nutrition, smoking, alcohol, other pharmaceutical, nutraceutical, and chemical exposures, infections, physical activity, age, gender, menstrual cycle, and body mass index—will have to be evaluated knowledgeably to understand, let alone predict, efficacy and toxicity. The power of prediction from screening tests (see Lave and Omenn Value-of-Information Model, 1986, 1988) is always less than 100% for either positive predictive value or negative predictive value, or, generally, for both, often much less (Holtzman and Marteau, 2000; Holtzman, 2003). That deficiency is true for all kinds of tests in medicine and public health, whether biochemical or physiological, including the use of the electrocardiogram as a test for risk of heart attack.

For the near term, it is certain that pharmacogenomics will be utilized in characterizing tumors and choosing therapies for multiple cancers, as described above. Oncology is clearly leading the way among clinical fields, although many other fields and diseases will be generating evidence from microarray and proteomics research studies. Much of cancer care is managed by specialists, especially because of the complexity of treatments and the toxicity of potent drugs. There is widespread awareness of the limited efficacy of most chemotherapies in many cancer patients and, therefore, much interest among clinicians in the newer methods of improving the benefit-to-risk ratio. Inherited predispositions to specific disorders have attracted more attention in oncology as well, especially for breast cancer and colorectal cancer.

X. PROSPECTS FOR USE OF PHARMACOGENOMICS IN THE PHARMACEUTICAL INDUSTRY

There is ambivalence in the pharmaceutical industry about the likely impacts of pharmacogenomics. The possibility of developing new drugs much better targeted to well-characterized subpopulations of patients should increase the therapeutic margin (ratio of dose for toxicity to dose for efficacy) and, therefore, the benefit-to-risk ratio (which is weighted for the proportion of those treated experiencing benefit or toxicity) (Roses, 2000). If the therapeutic margin is sufficiently improved, smaller numbers of patients might suffice to demonstrate clinical efficacy, even to the satisfaction of the FDA. A most remarkable example is Novartis' STI-571 (see above) in CML. Had the drug been tested in a heterogeneous group of leukemia patients, its efficacy might have been so diluted as to be unrecognized; in fact, this drug was nearly discarded during development as ineffective against common cancers and heart disease.

Most preclinical drug development is conducted by companies themselves, with toxicology and toxicity assessed by contract laboratories. Clinical phase 1 trials tend to be done with academic centers. Increasingly, companies are returning to academic medical centers for phase 2 and, to some extent, phase 3 trials, after a tumultuous period of relying on community-based physicians. A big part of this transition is the effort by several major institutions to provide the necessary infrastructure for efficient recruitment and reliable compliance with human subject requirements and FDA regulatory requirements. Meticulous attention to full description of patient characteristics, including environmental exposures, diet, lifestyle behaviors, age, menstrual status, and evidence of clinical or metabolic heterogeneity is critical to full interpretation of drug efficacy and risk of adverse events. Such detailed data collection and such anticipation of heterogeneity are important for effective delineation of appropriate target populations.

Meanwhile, the companies are deeply concerned that the cost of drug development, regulatory approvals, and marketing will be nearly as great for new drugs that are much more narrowly targeted and, therefore, have much smaller markets. Companies have been reluctant to market existing drugs to restricted populations that can be defined with biomarker tests, as in the case of bisphosphonates for osteoporosis patients, which were marketed to all postmenopausal women, rather than to the 50% of such women who have evidence of predisposition to, or early manifestations of, osteoporosis by bone density measurement and/or N-terminal collagen cyclic peptide biomarkers (Osteomark) (Miller et al., 1999). The possibility that a drug will be effective in an individual patient, even if the likelihood is very low, may not preclude companies from marketing the drug and physicians from using it, especially if there are few other effective alternatives. Probably the drug will be aggressively marketed even if there are effective alternatives, because of the competitive nature of the business.

One constructive resolution of this situation may emerge from pharmacogenomic and proteomic analyses. One can imagine that a particular receptor or set of receptors is involved in, say, certain adenocarcinomas of the lung and of other tumor sites, like colon, pancreas, and (estrogen receptor negative) breast. The appropriate patient markets could be the sum of those patients, defined mechanistically, rather than by the "geographic" location of the tumors in the body. The present strategy of treating all lung cancer patients, or all lung adenocarcinoma (non-small cell cancer) patients, identically has been quite unproductive to date. Many other examples may emerge from broad analyses of clinical conditions currently thought to be unrelated.

XI. EDUCATIONAL NEEDS

Accurate content about genetics, genomics, and proteomics is essential for curricula for undergraduate medical students, training programs for residents, and continuing medical education programs for practicing physicians. The teaching must be understandable and empowering, for these physicians to keep up with the accelerating pace of discoveries and applications in the pharmaceutical and biotechnology worlds. Physicians must be prepared for much more knowledgeable and self-confident patients, reflecting the tremendous media coverage of medical "news," much of it involving genetics and genomics. At a Josiah Macy, Jr., Foundation conference on this subject, there was a strong endorsement for teaching and learning about the new genetics and its practical and ethical ramifications; the University of Vermont reported that its entire curriculum was being revised to give a central place to genetics and to ethics throughout the four years of medical school (Hager, 1999). A broad array of health care professionals, not just physicians, needs education about genetics, and geneticists need to become more familiar with broader aspects of health care delivery and public health.

Continuing medical education programs in every specialty already are incorporating some of the new genetic and genomic findings. However, it is well known that physicians choosing and attending these programs are insistent that content be "practical" for their current patient care responsibilities. It is a complex task to balance introduction of new methods and tests, let alone clinical decision making, before the relevant cases are frequent enough to hold the practitioner's attention and reinforce the learning. It is important to make genetics an integral part of all aspects of medical professional education, rather than treating genetic practices and ethical or policy issues as "exceptional" or "outside the medical mainstream."

Several other chapters in this book address ethical issues, which are essential for the educational programs. Pharmacogenomic analyses will multiply the challenges of explaining available tests and gaining informed consent to perform each test. In some cases, patients or relatives will be asked to consider information that was not necessarily sought by the patient but emerged from the analysis. To address patients' and families' questions about common diseases, and to satisfy ourselves as clinicians that we have credible answers about risks and treatments, we need and must teach from a solid foundation of information linking data on genetic variation with data on other key variables. No longer should we debate "nature versus nurture," genetics versus environment; nor should we presume, either actively or by omission, that genes by themselves "determine" common clinical disorders and outcomes of care. Understanding disease risks, predicting therapeutic and adverse effects of treatments, and creating comprehensive clinical and public health interventions require attention to the interactions. Nongenetic variables, as noted above, include diet, metabolism, behaviors, medical diagnoses, and various environmental exposures.

Some of these variables require data from individuals and may be available, if systematically recorded, in patient charts or administrative datasets mostly related to billing for clinical care. Even so, the information may be separate from the genetic test results. Other variables may have to be estimated from population samples and environmental monitoring. Computer links to health department epidemiological surveillance (possibly using current EPI-INFO from CDC), EPA pollution monitoring, census tract marketing, health care and pharmacy utilization, and lifestyle databases will have to be devised, all with great care to protect confidentiality and privacy and to respect cultural sensitivities.

There are many scientific and technical hurdles to establishing these linkages, conducting appropriate studies, and applying the knowledge gained in educating health care professionals, patients, and communities. However, the biggest hurdle is not scientific; it is the policy position that genetic information is uniquely different from other kinds of personal medical information, mostly because of well-justified fears about discrimination in employment and insurance. Murray (1997) has called this strategy "genetic exceptionalism."

Institutional Review Boards (IRBs) have put special requirements for re-consent on ancillary analyses of genetic markers in large clinical prevention trials (Thornquist et al., 2002). Legislative proposals to regulate or prohibit access to personal genetic information (Annas et al., 1995), let alone links among essential datasets with personal identifiers, would leave many important questions unanswerable. After all, despite the special features of family history, other kinds of medical information are highly sensitive and often involve family members—including information about sexually transmitted infections, tuberculosis, environmental chemical exposures, smoking, mental illness, and interpersonal violence. Hopefully, the hotly debated federal HIPAA privacy regulations will bridge this situation by limiting access to all medical information to those personnel with a true need to know in support of the patient's care and with the patient's knowledge. These regulations specify and facilitate exceptions for patient care, properly authorized criminal investigations, and IRB-supervised research. The Social Issues Committee of the American Society for Human Genetics has also provided good guidance on handling genetic and other sensitive medical information (ASHG, 1998). The framework now exists for building the linkages to other datasets that could energize and inform both educators and practitioners about the interactions of genes, environment, and behaviors.

REFERENCES

Abbott, A., "Workshop Prepares Ground for Human Proteome Project," *Nature*, **413**, 763 (2001).

Alizadeh, A.A., et al., "Distinct Types of Diffuse Large B-Cell Lymphoma Identified by Gene Expression Profiling," *Nature*, **403**, 503–511 (2000).

American Society for Human Genetics, Social Issues Subcommittee on Familial Disclosure, "Professional Disclosure of Familial Genetic Information," *Am. J. Hum. Genet.*, **62**, 474–483 (1998).

Annas, G.J., et al., "Drafting the Genetic Privacy Act: Science, Policy, and Practical Considerations," *J. L. Med. Ethics*, **23**, 360–366 (1995).

Bartlett, J.M., et al., "Evaluating HER2 Amplification and Overexpression in Breast Cancer," *J. Pathol.*, **195**, 422–428 (2001).

Barton-Davis, E.R., et al., "Aminoglycoside Antibiotics Restore Dystrophin Function to Skeletal Muscles of Mdx Mice," *J. Clin. Invest.*, **104**, 375–381 (1999).

Baselge, J., et al., "Recombinant Humanized Anti-HER2 Antibody (HERCEPTIN) Enhances the Antitumor Activity of Paclitaxel and Doxorubicin Against HER2/neu Overexpressing Human Breast Cancer Xenografts," *Cancer Res.*, **58**, 2825–2831 (1998).

Bedwell, D.M., et al., "Suppression of a CFTR Premature Stop Mutation in a Bronchial Epithelial Cell Line," *Nat. Med.*, **3**, 1280–1281 (1997).

Berger, A., "UK genetics database plans revealed," BMJ, **322**, 1018 (2001).

Bertilsson, L., "Geographical/Interracial Differences in Polymorphic Drug Oxidation: Current State of Knowledge of Cytochromes P450 (CYP) 2D6 and 2C19," *Clin. Pharmacokinet.*, **29**, 192–209 (1995).

Bhattacharjee, A., et al., "Classification of Human Lung Carcinomas by mRNA Expression Profiling Reveals Distinct Adenocarcinoma Subclasses," *Proc. Natl. Acad. Sci. USA*, **98**, 13790–13795 (2001).

Bhopal, R., and L.W. Donaldson, "White European, Western, Caucasian, or What? Inappropriate Labeling in Research on Race, Ethnicity, and Health," *Am. J. Publ. Hlth.*, **88**, 1303–1307 (1998).

Bittner, M., et al., "Molecular Classification of Cutaneous Malignant Melanoma by Gene Expression Profiling," *Nature*, **406**, 536–540 (2000).

BMJ, Editorial: "Ethnicity, Race, and Culture: Guidelines for Research, Audit, and Publication," *BMJ*, **312**, 1094 (1994).

Brichory, F., et al., "Proteomics-Based Identification of Protein Gene Product 9.5 as a Tumor Antigen that Induces a Humoral Immune Response in Lung Cancer," *Cancer Res.*, **61**, 7908–7912 (2001).

Broder, S., et al., "The human genome," In: *Pharmacogenomics: The Search for Individualized Therapeutics*. (in press, 2002).

Cole, S.T., et al., "Deciphering the Biology of *Mycobacterium tuberculosis* from the Complete Genome Sequence," *Nature*, **393**, 537–544 (1998).

Cole, S.T., et al., "Massive gene decay in the leprosy bacillus," *Nature,* **409**, 1007–1011 (2001).

Department of Health & Human Services. Health Insurance Portability and Accessibility Act Privacy Regulations. *Fed. Reg.*, **65**, 82463–82829 (2000).

Dhanasekaran, S.M., et al., "Delineation of Prognostic Biomarkers in Prostate Cancer," *Nature*, **412**, 822–826 (2001).

Druker, B.J., et al., "Efficacy and Safety of a Specific Inhibitor of the BCR-ABL Tyrosine Kinase in Chronic Myeloid Leukemia," *N. Engl. J. Med.*, **344**, 1031–1037 (2001).

Eaton, D.L., et al., "Genetic Susceptibility," In Rom W.N. (ed), *Environmental and Occupational Medicine*, 3rd ed. Philadelphia: Lippincott-Raven, pp. 209–221 (1998).

Evans, W.E., and M.V. Relling, "Pharmacogenomics: Translating Functional Genomics into Rational Therapeutics," *Science*, **186**, 487–491 (1999). Data also from Supplemental Table 1 at (*www.sciencemag.org/feature/data/1044449.shl.*)

Exner, C.V., et al., "Lesser Response to Angiotensin-Converting-Enzyme Inhibitor Therapy in Black as Compared with White Patients with Left Ventricular Dysfunction," *N. Engl. J. Med.*, **344**, 1351–1357 (2001).

Fleischmann, R.D., et al., "Whole-Genome Random Sequencing and Assembly of *Haemophilus influenzae Rd.*," *Science*, **269**, 496–512 (1995).

Fullilove, M., Comment: "Abandoning 'Race' as a Variable in Public Health Research—an Idea Whose Time Has Come," *Am. J. Publ. Hlth.*, **88**, 1297–1298 (1998).

Garber, M.E., et al., "Diversity of Gene Expression in Adenocarcinoma of the Lung," *Proc. Natl. Acad. Sci. USA*, **98**, 13784–13789 (2001).

Giordano, T.J., et al., "Organ-Specific Molecular Classification of Primary Lung, Colon, and Ovarian Adenocarcinomas Using Gene Expression Profiles," *Am. J. Pathol.*, **159**, 1231–1238 (2001).

Goldenberg, M.M., "Trastuzumab, a Recombinant DNA Derived Monoclonal Antibody, a Novel Agent for the Treatment of Metastatic Breast Cancer," *Clin. Ther.*, **21**, 309–318 (1999).

Gorre, M.E., et al., "Clinical Resistance to STI-571 Cancer Therapy Caused by BCR-ABL Gene Mutation or Amplification," *Science*, **293**, 876–880 (2001).

Hager, M. (ed). *Education for More Synergistic Practice of Medicine and Public Health.* New York, Josiah Macy, Jr., Foundation (1999).

Hanash, S.M., and L.M. Baretta, "Operomics: Integrated Genomic and Proteomic Profiling of Cells and Tissues," *Briefings Funct. Genomics Proteomics*, **1**, 10–22 (2002).

Hein, D.W., "Acetylator genotype and arylamine-induced carcinogenesis," *Biochim. Biophys. Acta*, **948**, 37–66 (1988).

Holtzman, N.A., and T.M. Marteau, "Will Genetics Revolutionize Medicine?," *N. Engl. J. Med.*, **343**, 141–144 (2000).

Institute of Medicine, *Crossing the Quality Chasm*, Washington, DC: National Academy Press (2001).

Institute of Medicine, *To Err is Human*, Washington, DC: National Academy Press (2000).

Institute of Medicine, *Unequal Treatment: Confronting Racial and Ethnic Disparities in Health Care*, Washington, DC: National Academy Press (2002).

Johansson, I., et al., "Inherited amplification of an active gene in the cytochrome P450 CYP2D-locus as a cause of ultrarapid metabolism of debrisoquine," *Proc. Natl. Acad. Sci. USA*, **90**, 11825–11829 (1993).

Kalow, W., *Pharmacogenetics—Heredity and the Responses to Drugs*, Philadelphia: W.B. Saunders, 1962.

Kalow, W., and A.G. Motulsky, "General Conclusions and Future directions," In: Kalow W., Meyer U.A., Tyndale R.F., *Pharmacogenomics*, New York: Marcel Dekker, pp. 389–395 (2001).

Khoury, M.J., et al., (eds), *Genetics and Public Health in the 21st Century: Using Genetic Information to Improve Health and Prevent Disease*, New York: Oxford University Press (2000).

King, M.C., et al., "Tamoxifen and Breast Cancer Incidence Among Women with Inherited Mutations in BRCA1 and BRCA2," *JAMA*, **286**, 2251–2256 (2001).

Kleyn, P.W., and E.S. Vesell, "Genetic Variation as a Guide to Drug Development," *Science*, **281**, 1820–1821 (1998).

Lave, L.B., and G.S. Omenn, "Cost-Effectiveness of Short-Term Tests for Carcinogenicity," *Nature*, **324**, 29–34 (1986).

Lave, L.B., et al., "Information Value of the Rodent Bioassay," *Nature*, **336**, 631–633 (1988).

Lazarou, J., et al., "Incidence of Adverse Drug Reactions in Hospitalized Patients: a Meta-Analysis of Prospective Studies," *JAMA*, **279**, 1200–1205 (1998).

Lifton, R.P., and A.G. Gharavi, "Molecular Mechanisms of Human Hypertension," *Cell*, **104**, 545–556 (2001).

Liggett, S.B., "The Pharmacogenetics of Beta2-Adrenergic Receptors: Relevance to Asthma," *J. Allergy Clin. Immunol.*, **105**, 487–492 (2000).

Littman, D.R., "Chemokine Receptors: Keys to AIDS Pathogenesis?," *Cell*, **93**, 677–680 (1998).

Mancinelli, L., et al., "Pharmacogenomics: The Promise of Personalized Medicine," *AAPS PharmSci*, **2** (1) Article 4 (*http://www.aapspharmsci.org/*) (2000).

Martin, E.R., et al., "SNPing Away at Complex Diseases: Analysis of Single-Nucleotide Polymorphism around APOE in Alzheimer Disease," *Am. J. Hum. Genet.*, **67**, 383–394 (2000).

Meyer, U.A., "Pharmacogenetics: Clinical Viewpoints," In Kalow W., Meyer U.A., Tyndale R.F., *Pharmacogenomics*, New York: Marcel Dekker, pp. 135–150 (2001).

Miller, P.D., et al., "Practical Clinical Application of Biochemical Markers of Bone Turnover," *J. Clin. Densitometry* **2**, 323–342 (1999).

Motulsky, A.G., "Drug Reactions, Enzymes and Biochemical Genetics," *JAMA*, **165**, 835–837 (1957).

Motulsky, A.G., "If I Had a Genetic Test, What Would I Have and Who Would I Tell?" *Lancet*, **354** (suppl 1), 35–37 (1999).

Mukherjee, D., and E.J. Topol, "Pharmacogenomics in cardiovascular diseases," *Prog. Cardiovasc. Dis.*, 44, 479–498 (2002).

Murray, T.H., "Genetic Exceptionalism and 'Future Diaries': Is genetic information different from other medical information?" In: Rothstein MA (ed), *Genetic Secrets: Protecting Privacy and Confidentiality in the Genetic Era*. New Haven, CT: Yale University Press (1997).

National Institute of General Medical Sciences (NIGMS). "Availability of administrative supplements to the pharmacogenetics research network to address health disparities." Notice NOT-GM-01-004, 3 April, 2001.

Oh, M., et al., "A Database of Protein Expression in Lung Cancer," *Proteomics*, **1**, 303–1319 (2001).

Omenn, G.S., "Public Health Genetics: An Emerging Interdisciplinary Field for the Post-Genomic Era," *Annu. Rev. Publ. Hlth.*, **21**, 1–13 (2000a).

Omenn, G.S., "The Genomic Era: A Crucial Role for the Public Health Sciences," *Environ. Hlth. Perspect.*, **108**, 204–205 (2000b).

Omenn, G.S. and A.G. Motulsky, "Eco-genetics: genetic variation in susceptibility to environmental agents," In: Cohen B.H., Lilienfeld A.M., and Huang P.C. (eds), *Genetic Issues in Public Health and Medicine*. Springfield, IL: CC Thomas, pp. 83–111 (1978).

Perna, N.T., et al., "Genome Sequence of Enterohaemorrhagic *Escherichia coli* O157:H7," *Nature*, **409**, 529–533 (2001).

Priori, S.G., et al., "Genetic and Molecular Basis of Cardiac Arrhythmias: Impact on Clinical Management," *Circulation*, **99**, 518–528 (1999).

Pritchard, J.K., et al., "Inference of Population Structure Using Multilocus Genotype Data," *Genetics*, **155**, 945–959 (2000).

Ramaswamy, S., et al., "Multiclass Cancer Diagnosis Using Tumor Gene Expression Signatures," *Proc. Natl. Acad. Sci. USA*, **98**, 15149–15154 (2001).

Roses, A.D., "Pharmacogenetics and the Practice of Medicine," *Nature*, **405**, 857–865 (2000).

Rothstein, M.A., and P.G. Epps, "Pharmacogenomics and the (Ir)Relevance of Race," *Pharmacogenomics J*, **1**, 104–108 (2001).

Schwartz, R.S., "Racial Profiling in Medical Research," *N. Engl. J. Med.*, **344**, 1392–1393 (2001).

Su, A.I., et al., "Molecular Classification of Human Carcinomas by Use of Gene Expression Signatures," *Cancer Res.*, **61**, 7388–7393 (2001).

Thornquist, M.D., "Streamlining IRB review in multisite trials through single-study IRB Cooperative Agreements: experience of the β-Carotene and Retinol Efficacy Trial (CARET)," *Controlled Clinical Trials*, **23**, 80–86 (2002).

Ulrich, C.M., et al., "Pharmacogenetics of Methotrexate: Toxicity Among Marrow Transplantation Patients Varies with the Methylenetetrahydrofolate Reductase C677T Polymorphism," *Blood*, **98**, 231–234 (2001).

Vogel, C.L., et al., "First-line Herceptin® Monotherapy in Metastatic Breast Cancer," *Oncology*, **61** Suppl S2, 37–42 (2001).

Vogel, F., "Moderne Probleme der Humangenetik," *Ergeb. Inn. Med. Kinderheilk.*, **12**, 52–125 (1959).

Weber, W.W., and M.T. Cronin, "Pharmacogenetic Testing," In: R.A. Meyers (ed), *Encyclopedia of Analytical Chemistry*, Chichester, UK: Wiley & Sons, pp. 1–25 (2001).

Weinshilboum, R.M., et al., "Methylation Pharmacogenetic Catechol O-Methyltransferase, Thiopurine Methyltransferase, and Histamine N-Methyltransferase," *Annu. Rev. Pharmacol. Toxicol.*, **39**, 19–52 (1999).

Wilson, J.F., et al., "Population Genetic Structure of Variable Drug Response," *Nat. Genet.*, **29**, 265–269 (2001).

Wood, A.J., "Racial Differences in the Response to Drugs—Pointers to Genetic Differences," *N. Engl. J. Med.*, **344**, 1394–1396 (2001).

Wright, A.F., et al. "Gene-environment interactions—the BioBank UK study," *Pharmacogenomics J.*, **2**, 75–82 (2002).

Yancy, C.W., et al., "Race and the Response to Adrenergic Blockade with Carvedilol in Patients with Chronic Heart Failure," *N. Engl. J. Med.*, **344**, 1358–1365 (2001).

CLINICAL UTILITY OF PHARMACOGENETICS AND PHARMACOGENOMICS

NEIL A. HOLTZMAN, M.D., M.P.H.

I. INTRODUCTION

Scientists claim that genetic research will explain "interindividual variation of the therapeutic effectiveness and toxicity of drugs" (Meyer, 2000).
Genetic tests will, they say, result in "safer and efficient pharmacotherapy"
(Ingelman-Sundberg, 2001) by enabling physicians "to more precisely select
medications and dosages that are optimal for individual patients" (Evans
and Relling, 1999). Scientists at a leading pharmaceutical company state that
genetic "profiling of patients' risks may help reduce the number of admissions to hospital that arise because of adverse drug reactions and interactions" (Fears et al., 2000). A scientist at another large pharmaceutical
company predicted, "We will soon be able to profile variations between individuals' DNA to predict responses to a particular medicine.... [This] will
change the practice and economics of medicine" (Roses, 2000b). As the
"genetic revolution" comes to fruition, writes one representative of a biotechnology company, there will be no need to wait until a patient gets sick:
"Shortly after a person is born, her genotype is recorded at her physician's
office.... Assisted by a decision support system, her physician may prescribe a personal immunization and screening schedule or recommend specific preventive measures" (Sander, 2000). The editor of this volume wrote:

Pharmacogenomics: Social, Ethical, and Clinical Dimensions, Edited by Mark A. Rothstein.
ISBN 0-471-22769-2   Copyright © 2003 Wiley-Liss, Inc.

"the prescription of medications based on genotypic information will become the standard of care for physicians" (Rothstein and Epps, 2001).

Despite these claims, few genotypes (or haplotypes) predict patients' drug-handling[1] with high fidelity. This raises the importance of requiring evidence of clinical validity and utility before marketing of pharmacogenetic tests. In this chapter, I first consider how the role of genes in drug handling is established. I then examine the importance of pharmacogenetics in explaining adverse drug responses and therapeutic failures in terms of both the frequency of genetic variants and the strength of the association between genetic variants and altered drug handling. We see that the predictive values of putative pharmacogenetic tests are seldom very high, and we examine several explanations, as well as implications, of this observation. In the last section, I consider the need for regulation of pharmacogenetic tests.

II. FINDING GENETIC VARIANTS IN DRUG HANDLING

For drugs already on the market, the discovery of inherited variants[2] utilizes two unwanted effects of drugs: adverse drug reactions (ADRs) and therapeutic failures.

A. ADVERSE DRUG REACTIONS

Physicians encountering an adverse event in a patient receiving a newly marketed drug cannot be certain that it is drug related, particularly when the event does not occur immediately after therapy is started. To aggregate data to establish whether an event is associated with the administration of a drug, physicians need to report possible adverse drug reactions (ADRs) to the manufacturer, the Food and Drug Administration, (FDA) or both. Underreporting occurs frequently (Heeley et al., 2001) and can delay recognition of the hazards of a drug.

Once an adverse event is associated with a drug, comparisons between patients with ADRs and those who have taken the drug but have not suffered an adverse reaction can be made. Possible differences include dosing, duration of treatment, other diseases present (comorbidity), other drugs

---

[1] Throughout this paper I use "drug handling" to signify all of the possible steps from absorption of the drug to the excretion of it or its metabolites. In addition to metabolism of the drug to active or inactive forms, these steps include transport to tissues or organs and the cells therein and binding by receptors to which they are targeted and that influence their pharmacodynamic effect.

[2] Unless otherwise stated, I use the terms "genetic" and "inherited" interchangeably to refer to the alleles that a person receives from his or her parents, and which could be transmitted to his or her children. I do not mean acquired somatic cell mutations or mutations in infectious agents.

being used, diet, environmental exposures, and demographic factors (e.g., older people taking certain drugs are more likely to have ADRs). When researchers find such differences, changes in dosing or in the indications and contraindications of the drug can reduce the incidence of ADRs.

If the differences cannot be explained, further study is needed. One approach is to gain clues from the nature of the ADR. A small fraction of patients receiving the muscle relaxant succinylcholine had prolonged apnea, suggesting the possibility that they were not inactivating the drug as rapidly as the vast majority in whom the effect was short-lived. Those with the ADR were found to have much lower concentrations of plasma cholinesterase, the enzyme that inactivates the drug (Kalow and Grant, 1995). Some individuals taking the antimalarial primaquine developed hemolytic anemia very shortly after therapy was started. This was traced to their having a form of the enzyme glucose-6-phosphate dehydrogenase (G6PD) that was less able than the common form to protect red blood cells from the oxidative side effects of the drug (Luzzatto and Mehta, 1995).

Another approach is to administer a standard dose of the drug to those who have had ADRs and to those who tolerate the drug and compare their blood concentrations. An association between high blood concentrations of isoniazid and peripheral neuritis was found among patients being treated for tuberculosis with standard doses of isoniazid. Most of those with high blood concentrations were found to be slow acetylators of the drug (Evans and Clarke, 1961). To achieve a therapeutic effect of the anticoagulant warfarin without excessive bleeding the dosage may have to be varied over a 16-fold range (Linder, 2001). Those who require very low doses may have a variant form of an enzyme, cytochrome P450-2C9, that does not inactivate the drug properly.

B. THERAPEUTIC FAILURES

Some failures will be due to the presence of variants in drug handling. Patients who are rapid acetylators of isoniazid have a slower antituberculous response than slow acetylators (Evans and Clarke, 1961). Asthmatics who do not respond well to β_2-agonist bronchodilators may have fewer functioning β_2-adrenergic receptors (Drysdale et al., 2000). Variations in the synthesis or structure of the serotonin transporter protein, which is involved in selective reuptake of serotonin by presynaptic neurons, may explain why some patients with depressive disorders respond to selective serotonin reuptake inhibitors and others do not (Steimer et al., 2001).

Other failures will result from differences in disease etiology, although responders and nonresponders may have the same or very similar symptoms and signs before treatment is started. Some asthmatics do not respond to drugs that block the synthesis of leukotrienes because leukotrienes do not account for their asthma (Drazen et al., 1999).

A combination of altered drug handling and etiologic differences may account for some therapeutic failures. Even when hypertensive patients attain the same blood concentrations of an antihypertensive drug, they do not all respond by a reduction in blood pressure. Some hypertensive patients will respond better to drug A with one mechanism of action than to drug B with a different mechanism, but others will respond better to drug B than to drug A (Turner et al., 2001).

C. ARE THE VARIANTS INHERITED?

Finding that the concentration or activity of proteins differs between those who handle drugs as expected and those who do not proves neither that the associations are genetic nor that they are causally related to the altered response. Differences in the function of enzymes or proteins that affect drug handling may result from posttranslational (nongenetic) modification (Banks et al., 2000), which might be influenced by other drugs or other exogenous factors. The persistence of variation in cultured cells makes the role of exogenous factors less likely.

An inherited basis is suggested by studies in healthy first-degree relatives of patients with altered drug handling. When some of these relatives have the same abnormality in protein concentration or activity as the patient, and the proportion with the abnormality is consistent with Mendelian inheritance, a single gene of high penetrance is probably responsible. This is also suggested when studies of populations of those using a drug indicate that the distribution of its concentration in the blood, or of a metabolite or of one of the enzymes that metabolizes the drug, is bimodal (Evans, 1963; Turner et al., 2001). One peak is usually due to heterozygotes and the less frequent homozygote and the other to the more common homozygote. If the population surveyed is sufficiently large (several hundred), the less frequent homozygotes may constitute a third peak, provided that the enzyme activity differs between them and heterozygotes. Bi- or trimodal distributions have been observed for G6PD activity (Harris, 1980), isoniazid acetylation (Evans, 1963), cholinesterase activity (Kalow and Grant, 1995), debrisoquine (an antihypertensive agent) hydroxylation (CYP2D6) (Kalow and Grant, 1995), and thiopurine methyltransferase (TPMT) activity (low TPMT activity results in life-threatening myelosuppression in patients treated with mercaptopurines for leukemia and other disorders) (Weinshilboum, 2001).

A unimodal distribution, which is observed more often, suggests that within the population there are many different alleles at the same gene locus whose diploid combinations each affect drug handling slightly differently or that there are many different gene loci that affect drug handling.

Either bimodal or unimodal distribution curves may result from environmental, dietary, and behavioral factors, the bimodal curve indicating different exposures between the two classes and the unimodal curve indicating that several different combinations of these factors affect the distribution.

Confirmation of a genetic basis for variation due to the predominant effect of an allele, or alleles, at a single gene locus comes from sequencing its DNA. One would not undertake these costly procedures without preliminary evidence that inherited factors are operating. If the gene encoding the protein under investigation has already been identified, then patients with ADRs or nonresponsiveness should share a variant sequence significantly more often than people who tolerate the drug or who are responsive to it. Causality is suggested when the variant sequence is likely to have a functional effect on the protein product or influences the transcription or translation of the gene by changes in the promoter region or splice junctions as well as in exons (Drysdale et al., 2000; Furlong et al., 1998; Kuehl et al., 2001). Some sequence changes will not have any effect on the synthesis, structure, or function of the encoded protein. As we consider below, this may often be the case when the variant sequence occurs in 1% or more of the population. Such frequently occurring changes are known as polymorphisms.[3]

Even when a direct causal relationship is lacking, polymorphisms are extremely useful in finding genes related to drug handling. Scientists look for polymorphic markers along the genome, one of whose forms is shared by those with altered drug handling more often than by those who tolerate the drug or are responsive to it. When "linkage" is found, it is likely that the causal sequence is on the same chromosome as the marker and close to it. This approach is most likely to succeed when a single mutation, occurring many generations ago, accounts for a large proportion of all patients with the altered drug effect. In that case, the mutation occurred on a chromosome carrying the polymorphic variant.[4] This approach works less well when dif-

---

[3] Polymorphisms may involve as little as a single nucleotide change (single-nucleotide polymorphism or SNP), although they also may involve longer sequences. They occur in genes or in nongene parts of the genome. The term is sometimes used for rare variants (<1%), but not here.

In this paper, "polymorphism" is sometimes used synonymously with one allelic form of the polymorphism.

[4] Occasionally, the mutation might be the polymorphic variant itself. Such a variant could have attained polymorphic frequency ($\geq$1%) if it conferred a selective advantage in our ancestors, for example, by providing enzymatic protection against some environmental hazard. Drugs, a very recent innovation, today serve as substrates for these enzymes, converting beneficial forms into possibly harmful ones (Nebert, 1997). Although there are a few examples of genes that are advantageous in one environment being disadvantageous in another, I am not aware of any proven examples involving genes whose expression affects drug handling. The allele for sickle cell (Hgb S) protects against malaria but has no beneficial effect when the threat of malaria is removed. The so-called "thrifty genotype" was advantageous in cultures that went through alternate periods of feast and famine but is disadvantageous when the surfeit of food is continuous.

ferent mutations account for the alteration; not all of these mutations would have occurred on the chromosome with the polymorphism or necessarily reside at only one gene locus. When multiple mutations play a role, one or more of them are likely to be found only by studying a very large number of patients with the altered response.

III. How Important Are Genetic Variants in Predicting Drug Handling?

Serious ADRs were estimated to occur in over 2.2 million hospitalized patients in the United States in 1994 (6.7% of all hospitalized patients) and were fatal in 106,000 (0.32% of hospitalized patients), making them between the 4th and 6th leading cause of death (Lazarou et al., 1998). They may account for 7–13% of hospitalizations in two European countries (Ingelman-Sundberg, 2001). No comprehensive data are available on how often patients fail to respond to drugs on the market. If genetic variants contributed significantly to ADRs and nonresponsiveness, then tests for them before drug use could reduce the frequency of ADRs and nonresponsiveness. The extent of this reduction depends, first, on the population frequency of genetic variants that might play a role in altered drug handling, and second, on the strength of the effect.

A. Frequency of Genetic Variants That Might Explain ADRs

Rare variants (frequency of less than 1%) cannot account for a high proportion of frequently occurring ADRs, but they may play an important role in unusual ADRs, such as prolonged apnea, which occurred in fewer than 1% of patients receiving succinylcholine (Evans, 1963), or myelosuppression in about 0.3% of Caucasians receiving mercaptopurines (Weinshilboum, 2001). Polymorphic variants may account for ADRs to seldom-used drugs but will have little effect on the magnitude of ADRs. On the other hand, polymorphic variants associated with ADRs that follow the use of frequently prescribed drugs could make a substantial contribution to all ADRs. Phillips et al. recently gathered information on ADRs that were reported in two or more studies. Not surprisingly, 24 of the 27 were observed among the 200 most frequently prescribed drugs. Sixteen of the 27 (59%) drugs are metabolized by enzymes encoded by genes with known polymorphisms (Phillips et al., 2001). They could account for a high proportion of the adverse reactions to these drugs in Caucasians (Table 9.1), particularly when, as is often the case, the polymorphisms occur with a higher frequency in some population groups than the rate of ADRs.

TABLE 9.1. DRUGS FREQUENTLY INVOLVED IN ADRs FOR WHICH POLYMORPHISMS IN DRUG METABOLISM ARE KNOWN

Drug	Therapeutic category	Polymorphic enzyme in the major pathway for metabolism of the drug	Frequency of poor metabolizer genotypes, %	
			Caucasians	Other
Fluoxetine	SSRI	CYP1A2	2–6	Unknown
	antidepressant	CYP2D6	3–10	<2 (J,C,A)
Imipramine	Tricyclic	CYP1A2	12	Unknown
	antidepressant	CYP2C19	2–6	16 (C); 20 (J)
		CYP2D6	3–10	<2 (J,C,A)
Isoniazid	Antituberculosis	NAT2	50–59	41(A),20(C), 9(J)
Metoprolol	Beta blocker	CYP2D6	3–10	<2 (J,C,A)
Naproxen	NSAID	CYP2C9	2–6	Unknown
Phenytoin	Anticonvulsant	CYP1A2	12	Unknown
Piroxicam	NSAID	CYP1A2	12	Unknown
S-Ibuprofen	NSAID	CYP1A2	12	Unknown
S-Warfarin	Anticoagulant	CYP2C9	2–6	Unknown
Theophylline	Bronchodilator	CYP1A2	12	Unknown

SSRI = Selective serotonin reuptake inhibitor; NSAID = Nonsteroidal anti-inflammatory drug; CYP = cytochrome P-450; NAT2 = N-acetyl transferase type 2; J = Japanese; C = Chinese; A = African American. Poor metabolizers are not always at increased risk of ADR.
Table is drawn from Tables 1, 2, and 3 of Phillips et al. (2001).

B. STRENGTH OF EFFECT: POSITIVE AND NEGATIVE PREDICTIVE VALUES OF PHARMACOGENETIC TESTS[5]

Unless a genetic variant is highly predictive of altered drug handling, a test for the variant may not be very useful in making clinical decisions. Surprisingly, I could find no studies that report the predictive values of pharmacogenetic tests. A few studies provided sufficient data to calculate them (Aithal et al., 1999; Arranz et al., 1998; Chen et al., 1996; Layton et al., 1999; Poirier et al., 1995; Evans and Clarke, 1961; Taube et al., 2000).[6] One might expect

[5] In the discussion that follows positive predictive value (PPV) is defined as the proportion of people with a positive test result who have or will develop an unwanted response when the drug is administered. Negative predictive value (NPV) is the proportion of people with a negative test who will not have or develop an unwanted response when the drug is administered.
[6] Sufficient data were not included in other reports to calculate predictive values. For instance, Drysdale et al. (2000) reported over a twofold difference between responses to bronchodilators in asthmatics who differed in haplotypes of the β_2-adrenergic receptor gene, but predictive values could not be calculated because the number of asthmatics in each response and haplotype category were not given. Marked differences in mean absorption of digoxin between indi-

that high relative risks (>2) would indicate high predictive values. This is not necessarily the case (Holtzman, manuscript in preparation). As shown in Table 9.2, only for a polymorphism that diminishes transcription of the ALOX5 gene was the PPV greater than 90%. The study involved a small number of subjects. If the findings are confirmed, it is likely that asthmatic patients who have the polymorphism will not respond to leukotriene inhibitor drugs. However, with an NPV of 48%, about half of the patients who test negative for these polymorphisms will not respond to the drug.

Although the NPVs of the other six tests shown in Table 9.2 are higher, ranging from 73% to 97%, their PPVs are all lower, ranging from 16% to 83%. Thus the presence of a polymorphism associated with altered drug handling is usually not a powerful predictor of altered drug handling even when the enzyme encoded by the gene is in the major pathway for metabolism of the drug (Table 9.1).

Although not included in Table 9.2, the PPV and NPV of tests for G6PD deficiency in males to predict hemolytic anemia following the administration of primaquine are presumably both high. Among polymorphic variants, G6PD remains exceptional in this respect. Even for the well-established risk of polyneuritis in isoniazid users who are slow acetylators, the predictive value of testing is rather weak. Strong associations between rare variants and ADRs do occur, as discussed above.

For only three of the eight drugs in Table 9.2, isoniazid, oral contraceptives, and warfarin, have the susceptibilities conferred by polymorphisms been replicated. Arranz et al. used meta-analysis of studies subsequent to their original study (shown in Table 9.2) to combine results but found that without including their original study, the association between nonresponse to clozapine and polymorphic variants in the serotonin 2A receptor gene was no longer significant (Arranz et al., 1998). Subsequent to the study on the negative effect of apolipoprotein e4 on response of patients with Alzheimer disease to tacrine, an acetylcholinesterase inhibitor (Table 9.2), a larger study found that male patients with one or two apo e4 alleles responded as well to the drug as non-apo e4 patients, but female patients with an apo e4 allele did not respond compared to women without apo e4 (Farlow et al., 1998). In two other studies of newer acetylcholinesterase inhibitors no effect of apo e4 polymorphisms on response was observed (Farlow et al., 1999; Wilcock et al., 2000).

Ioannidis et al. recently showed that after initial reports of associations between polymorphic variants and disease, subsequent studies had lower odds ratios, which were often insignificant, even when meta-analysis

viduals homozygous for two different polymorphisms in the human multidrug resistance gene were observed by Hoffmeyer et al. (2000), but no cutpoint was given to separate normal from abnormal absorption, precluding further calculations. In a study on the effect of cholesteryl ester transferase polymorphisms on response to the lipid-lowering drug pravastatin, only differences in the means of coronary artery before and after use of the drug were given (Kuivenhoven et al., 1998).

showed significant results (Ioannidis et al., 2001). Clozapine response in schizophrenia was the only drug included in their study.

IV. WHY ARE MOST PHARMACOGENETIC TESTS OF LIMITED CLINICAL UTILITY?

The low predictive values of pharmacogenetic tests for most polymorphic variants means there will be false positive (when PPV is low) and false negative (when NPV is low) test results. Both reduce the clinical utility of the tests. There are a number of reasons for the reduced predictive values.

A. THE POLYMORPHISMS ARE NOT THE MUTATIONS THAT RESULT IN ALTERED DRUG HANDLING

Rather than being the variant that accounts for altered drug handling, the polymorphism identified by a genetic test may be in linkage disequilibrium with it. This is particularly likely when the polymorphism does not alter the rate of transcription or the structure or function of the gene.[7] For example, a polymorphism in exon 26 of the multidrug resistance (MDR)-1 gene is associated with low activity of the membrane P-glycoprotein even though it does not change the amino acid sequence (Hoffmeyer et al., 2000). A polymorphism in the pseudogene linked to CYP3A5 cannot have any effect on transcription of the actual gene. It is linked to mutations in the CYP3A5 gene itself that do alter translation of the protein (Kuehl et al., 2001).[8] It is also likely when several linked polymorphisms (haplotypes), but not any one of them alone, are associated with altered drug handling. This is the case for the β_2-adrenergic receptor described above (Drysdale et al., 2000). The strength of the association between a linked marker or haplotype and altered drug handling will vary inversely with the frequency of recombination between the marker(s) and the inherited mutation responsible for the alteration.[9] Even

---

[7] Genetic variants that do not affect protein structure, function, or synthesis could have attained polymorphic frequency by chance propagation (genetic drift) aided by a genetic bottleneck in which the population size was reduced.

[8] Two mutations were identified: a splice variant that eliminated an exon and a frameshift mutation that truncated the protein. The linkage between the pseudogene polymorphism and these mutations was much stronger in Caucasians than African Americans. Thus a test for the pseudogene polymorphism in African Americans would have lower predictive values than in Caucasians.

[9] Recombination will occur more frequently the greater the number of nucleotides that separate the marker from the mutation. Some segments of the genome, "hot spots," are more prone to recombination than others (Goldstein, 2001).

The greater the number of generations that have passed since the mutation occurred the weaker the linkage with a marker will be as there will have been more opportunity for recombination. Persistence of linkage could result from either genetic drift—the chance propagation of the 2 together in the absence of any selective advantage—or a selective advantage of the marker, the mutant protein or both.

TABLE 9.2. RELATIVE RISKS (RR) AND PREDICTIVE VALUES OF TESTS FOR POLYMORPHIC FORMS FOR ADRS OR DRUG RESPONSIVENESS

Drug	Polymorphic gene/enzyme	No. of subjects	Problem (% of subjects in which problem observed)	Variant enzyme, allele, or genotype fx, %	RR	PPV %	NPV %
Anti-depressants* (Chen et al., 1996)	CYP2D6	74	Nervousness, agitation, lethargy (24)	11	3.2	63	80
Clozapine, study 1 (Arranz et al., 1998)	Serotonin (5HT) neurotransmitter	164	Schizophrenia a nonresponse (36)	38	1.9	51	73
Isoniazid (Evans and Clarke, 1961)	N-acetyl transferase A	158	Polyneuritis (16)	56	4.2	24	94
Leukotriene inhibitor (Drazen et al., 1999)	ALOX5	114	Asthma nonresponse (56)	11	1.8	91	48
Oral contraceptives (Spannagl et al., 2000)	Factor V Leiden	165	Thromboembolism (10)	6	3.2	27	92

Penicillamine (Layton et al., 1999)	Glutathione S-transferase M1	Rheumatoid arthritis nonresponse (22)	81	53	2.3	30	87
Tacrine (Poirier et al., 1995)	non-apolipoprotein e4	Alzheimer improvement (74)	40	80**	2.1**	83	60
Warfarin (Aithal, 1999; Taube, 2000)	CYP2C9	Low dose needed to prevent bleeding (7)	697	34	5.9	16	97

PPV, NPV, and RR were calculated from the following data, which were provided or could be estimated: (1) the proportion of people receiving the drug who had altered drug handling (A); (2) the frequency of the variant enzyme or polymorphism among all those receiving the drug (G); and (3) the proportion of people who have *both* the variant and altered drug handling (a). Then:

$$PPV = a/G; \qquad NPV = [(1 - G) - (A - a)]/(1 - G); \qquad RR = PPV/(1 - G)$$

In each case, the population being studied consisted of consecutive patients receiving the drug regardless of whether or not they had altered drug-handling. In the leukotriene inhibitor report, no subjects with "mutant genotypes" responded to the leukotriene inhibitor. To estimate PPV, I assumed that ~10% of patients with "mutant genotypes" would respond if a larger number of patients were studied.

*Tricyclics and other antidepressants known to be metabolized by CYP2D6.

**Combined frequency of e2 and e3 phenotypes. The RR is the probability of response of non-apo e4 subjects compared to apo e4 subjects.

Data in this table will be included in Holtzman (manuscript in preparation).

when no recombination has occurred, the polymorphism would still not be a perfect predictor because it is likely to have occurred generations before the causal mutation. Consequently, some people will possess the polymorphism but not the linked mutation.[10] They will have false positive test results.

B. OTHER MUTATIONS IN THE SAME GENE

In addition to the mutation in linkage disequilibrium with the polymorphic marker (or haplotype) being tested, other mutations that affect drug handling could have occurred in the same gene but on chromosomes not containing the polymorphism. Individuals who test negative for the polymorphism would then be at risk for altered drug handling (false negative results).

The problems that arise by using tests for polymorphisms as predictors of altered drug-handling could be circumvented by the use of DNA microarrays ("chips") that can detect many different mutations. This technology may still fail to detect all causal mutations. Another approach is to assay for the protein responsible for altered drug handling. Unfortunately, the enzymes or proteins involved may not be present in readily accessible tissues. One way around this is to amplify the DNA of the gene in readily accessible cells, for instance, lymphocytes from blood, for proteins that are not expressed in blood. The amplified DNA could then be transcribed and translated into protein whose activity could be examined. This might still not detect problems in promoter and other regulatory segments. An approach that has been used for a number of years is to administer a "probe drug" that is metabolized in the same way as the drug about to be prescribed but that does not have adverse effects if the patient has altered drug handling (Streetman et al., 2000; Zhu et al., 2001). Thus dextromethorphan can be used to test for the presence of CYP2D6-related altered drug handling and an erythromycin breath test for CYP3A-related variants.

C. MUTATIONS IN GENES DIFFERENT FROM THE ONE CONTAINING (OR LINKED TO) THE POLYMORPHISM

There are many steps in drug transport and metabolism and, possibly, more than one target receptor for a drug. Thus mutations in any one of several different genes could affect drug handling. Until scientists devise tests to detect alterations in the gene or protein for each of the steps, a test for mutations in only one gene will yield false negative results.

---

[10] If the polymorphism occurred generations *after* the causal mutation, some people would have the responsible mutation but not the polymorphism. A test for the polymorphism in them would fail to predict altered drug handling. Mutations leading to loss of function of some CYP2D6 alleles may have occurred before the origin of some CYP2D6 haplotypes observed today (Kalow and Grant, 1995).

Roses has suggested the use of a "SNP linkage disequilibrium profile" to predict altered drug handling (Roses, 2000b). The profile for each drug would be established by comparing haplotypes throughout the genome of patients receiving the drug who display altered drug handling to patients who respond appropriately to the drug. His unstated assumption is that such a profile will predict altered drug handling with few false positives or false negatives. From the relatively weak associations so far reported this seems questionable. It is true that by not limiting the profile to one particular gene, interacting genes will be detected, resulting in greater predictive value. It is also possible that more than one profile will have the same predictive effect. In that case, the number of subjects needed to establish the relationships will be enormous.

D. Gene-Gene and Gene-Environment Interactions

Some mutations will not alter drug handling unless mutations in genes at other loci are simultaneously present. This may be the case when a mutation results in lowered (but not absent) enzyme activity in a step that is not rate-limiting in a drug-metabolizing pathway; an ADR may not be observed until a mutation elsewhere in the pathway further lowers throughput by creating a new rate-limiting step. Similarly, a mutation in a receptor gene may not affect drug responsiveness until a mutation in a gene for an alternative receptor precludes it also binding the drug. Considering the number of steps in drug handling, mutations in more than two independently segregating genes may well be involved, and there may be different combinations of mutations that will alter drug handling. This is akin to the problem that has already been encountered in finding susceptibility-conferring genotypes for the common forms of common complex diseases (Holtzman, 2001).

A mutation in a drug-related gene may only be harmful in certain environments, including the presence of particular dietary constituents or other drugs. Many patients receive more than one drug at a time. Positive test results for alterations in any one gene will correctly predict altered drug handling only when other genetic and environmental alterations are simultaneously present.

V. Are Pharmacogenetic Tests Needed?

As the PPV and NPV of pharmacogenetic tests approach 100% they become more useful, but this ideal is seldom approached with tests for common polymorphisms and haplotypes. In those cases, will predictive pharmacogenetic testing help physicians decide on treatment? If there are no other drugs with lower risks or a greater margin of safety, they may have no choice but to administer the drug and monitor the patient closely. The nature of the

ADR or therapeutic failure must also be considered. Despite the availability of a test for factor V Leiden and the confirmed increased risk of thromboembolism, several groups have argued against screening for the polymorphism before the administration of oral contraceptives (Middeldorp et al., 2001; Sarasin and Bounameaux, 1998; Spannagl et al., 2000). As shown in Table 9.2, the test has a low positive predictive value for thromboembolism. If women who had the factor V Leiden polymorphism were not to be administered contraceptives their risk of pregnancy would increase, and the risk of thromboembolism with pregnancy is much greater than with oral contraceptives. It might, however, be appropriate to perform the test if a family history of thromboembolism were elicited.

A. COMPENSATORY MECHANISMS

Some ADRs are self-limiting, and others can be compensated by other drugs. Although the hemolytic anemia following primaquine treatment of patents with G6PD deficiency can be severe, it is self-limited. As the older blood cells are destroyed and the patient's red blood cell count reaches a new equilibrium with a younger population, hemolysis subsides because the younger population of red blood cells can withstand the oxidative shock. The polyneuritis sometimes observed in slow acetylators of isoniazid can be prevented by administering vitamin B_6 (pyridoxal phosphate). Moreover, slow acetylators' tuberculosis often responds more rapidly to isoniazid than that of rapid acetylators; the active form of the drug has a longer half-life. Thus, rather than screening for G6PD or isoniazid polymorphisms, the necessary drugs could be administered and the patients monitored carefully.

Therapeutic failure can sometimes be corrected by safely administering higher doses. Patients with gastroesophageal reflux, who have the extensive metabolizer polymorphism of CYP2C19, do not raise their gastric pH with standard doses of lansoprazole, a proton pump inhibitor, but can do so with more frequent doses of the drug (Furuta et al., 2001). On the other hand, heterozygotes for a poor metabolizer polymorphism have increased serum gastrin levels, which may increase the risk of atrophic gastritis (Sagar et al., 2000). Thus maintaining high levels of proton pump inhibitors may have harmful as well as beneficial effects.

B. WHEN TESTS WITH LOW PREDICTIVE VALUES WILL BE USEFUL

Tests with relatively low predictive values may be useful in a few situations. First, testing may be valuable when the ADR in those with a testable genetic susceptibility at standard therapeutic dosages occurs suddenly and becomes irreversible or life-threatening before the drug can be stopped. Testing before administration of the standard dose would prevent these ADRs but would also exclude use of the drug in some people with the susceptibility who would not develop an ADR and would benefit from the drug. Rather than use a genetic test before treatment, all candidates for the drug could be started at a lower dose and monitored carefully for ADRs. This would not be possible in situations in which rapid drug response is needed; many people would not respond at the lower dose. Second, testing may be valuable when the efficacious response takes a long time to appear. This is the case for some psychiatric drugs. When the patient's status may deteriorate before the drug exerts a therapeutic effect, knowing in advance that one drug is unlikely to work but another will avoids a trial that is likely to end with a poor outcome.

In light of the low predictive values of many pharmacogenetic tests, two other situations in which pharmacogenetic testing has been heralded need closer inspection: their use in drug development and as a substitute for prescribing on an ethnic-racial basis.

C. POLYMORPHISMS AND DRUG DEVELOPMENT

If the only subjects enrolled in phase I and II clinical trials had genotypes (discovered by pharmacogenetic testing) that indicated that a new drug was likely to be safe and/or efficacious in them, the size of the clinical trial could be reduced, lowering costs and making it more likely that fewer subjects would respond adversely or not at all (Roses, 2000a; Shi et al., 2001). The question is, how would researchers know which genotypes are more likely to make a *new* drug safe and efficacious? Predictions based on the importance of a drug-metabolizing enzyme may not be realized. A minor enzyme in the metabolism of a drug (e.g., CYP1A2) may play a major role in ADRs and another that is estimated to be in the major metabolic pathway of several drugs (e.g., CYP2D6) may not be involved in as many ADRs because of the lower frequency of the harmful polymorphisms (Phillips et al., 2001) or the presence of compensatory pathways (Kalow and Grant, 1995). If tests used to screen subjects for clinical trials do not have high positive predictive values, some people in whom the drug would prove to be efficacious and safe will be eliminated from the trial.

Rather than screen before phase I and II trials, the "pharmacogenetic profiles" can be obtained afterwards. Those who tolerated and responded to a drug can be compared with those with ADRs or therapeutic failure. The ensuing phase III trials, which enroll many more patients than phase I and II trials, could recruit only those patients with a "favorable" pharmacogenetic profile. "[T]rials could be smaller and faster, and less expensive" (Roses, 2000b). This approach poses two problems. First, a favorable profile is unlikely to be found only in drug responders and never in nonresponders; subject-patients who lack the profile could still benefit from the drug. Second, by reducing the number of subjects in phase III trials there will be less of a chance of detecting the full range of ADRs associated with the drug. The enzymes responsible for drug metabolism will be different from those responsible for drug efficacy except when a slow metabolizer variant increases toxicity as well as efficacy. Roses suggests that, once efficacy is demonstrated in phase III trials for those with a certain pharmacogenetic profile, "provisional marketing approval" be given for those in whom pharmacogenetic tests reveal the favorable profile. The "first few hundred thousand . . . who receive prescriptions could have blood spots stored on filter paper in an approved location. As ADRs are documented, DNA could be extracted for patients with a particular drug-related adverse event and compared with well-matched patients who took the drug but did not experience the ADR. The abbreviated SNP profile for these ADRs could then be added to the abbreviated SNP profile for efficacy" (Roses, 2000b).

It is easy to see why pharmaceutical companies would like this approach. It reduces the size, and hence the cost, of clinical trials. Moreover, by allowing marketing, albeit provisional, on pharmacogenetic testing and collection of the blood spot, not only does it greatly expand sales of the drug but it also permits manufacturers to include a profit in their price for the drug; FDA regulations do not allow this in clinical trials. Unfortunately, there is no assurance that physicians or other drug dispensers will obtain the blood spot or report all ADRs. With such an approach many more people will suffer ADRs than would be the case if there were a larger, less restrictive clinical trial.

Removal of a drug from the market after adverse reactions occur because of polymorphic differences—as was the case for debrisoquine—is a much more expensive proposition than never marketing such a drug in the first place. Based on the suspicion (sometimes ill-founded) that drug handling will be affected by a common polymorphism in the major population to whom the drug will be marketed, drug developers could either abandon development of the drug or design drugs for only that segment of the population that they assume will handle the putative drug safely and efficaciously. Either strategy will leave people bereft of therapy.

D. ETHNIC-RACIAL EFFECTS

If physicians decided not to prescribe drugs in an ethnic-racial group in which altered drug handling occurred more frequently than in other groups, they might be denying some patients a safe and beneficial drug. Although frequencies of drug-related polymorphisms vary among ethnic-racial groups (Table 9.1), no group is homogeneous. In a recent analysis in which polymorphisms for drug-metabolizing enzymes were examined in populations ethnically labeled Africans (Bantu, Ethiopian, Afro-Caribbean), Caucasians (Norwegians, Ashkenazi Jews, Armenians), or Asians (Chinese, New Guineans), no polymorphism associated with altered drug handling occurred in more than 74% in any population. Of the six different genes studied in each of the three groups, the polymorphism altering drug handling occurred in less than 50% in 11 of the 18 possibilities (Wilson et al., 2001).[11]

Rather than simply avoiding the use of certain drugs in all members of an ethnic-racial group, physicians could use pharmacogenetic tests before prescribing these drugs. However, even then, the predictive value of the tests may be low. As an editorial accompanying the analysis of ethnic-racial differences noted, "[T]he authors have looked at the association between population genetic structure [ethnic-racial group] and polymorphisms in drug metabolizing enzymes; they have not examined the response to drugs which is likely to involve, in most cases, the interaction of several genes as well as other factors" (Editorial, 2001).

VI. NEED FOR REGULATION OF PHARMACOGENETIC TESTS

The Task Force on Genetic Testing was created by the NIH-DOE Working Group on Ethical, Legal, and Social Implications (ELSI) of Human Genome Research to make recommendations to ensure the development of safe and effective genetic tests. Many of its recommendations were subsequently adopted by the Secretary's Advisory Committee on Genetic Testing (SACGT) created by the Secretary of Health and Human Services in 1998. Here I present the Task Force's recommendations that are especially relevant to pharmacogenetic testing.

A. REPLICATION OF FINDINGS

"The genotypes to be detected by a genetic test must be shown by scientifically valid methods to be associated with the occurrence of a disease. The observations must be independently replicated and subject to peer review" (Holtzman and Watson, 1998, p. 25).

The Task Force considered replication a necessary first step *"before either linked markers or putative disease-related mutations are used as a basis of genetic tests"* (Holtzman and Watson, 1998, p. 25). Reported associations between

[11] By genotyping several neutral polymorphisms on chromosome 1 and the X chromosome in each of the population groups, the authors found that some genetic similarities among people assumed to be of the same ethnic-racial group broke down. Thus 62% of Ethiopians but only 4% of Bantus shared the same genetic structure as white Europeans, and only 2% of Papua New Guineans, considered Asians, had the same genetic structure as 85% of Chinese. Except for rare, isolated populations, fewer and fewer of which exist in our shrinking world, the amount of genetic commingling has been so great since humans moved out of Africa over 50,000 years ago that there are few if any genes that are the exclusive property of any ethnic-racial group.

genetic variants and altered drug-handling have not always been confirmed. The number of subjects in published studies is often small. Moreover, publication bias may inhibit the reporting of negative findings, at least until a positive result is reported first (Ioannidis et al., 2001). Thus significant associations should be confirmed before a pharmacogenetic test is recommended before administration of the relevant drug or to assess ADRs or nonresponse.

B. ANALYTICAL VALIDITY

"Analytical sensitivity and specificity of a genetic test must be determined before it is made available in clinical practice" (Holtzman and Watson, 1998, p. 25).

An analytically sensitive test always detects analytes when they are present in specimens. An analytically specific test does not detect analytes when they are absent. The Task Force also recommended that laboratories providing a test for routine clinical use (after it had been developed) demonstrate their ability to provide analytically valid tests. These laboratories are regulated under the Clinical Laboratories Improvement Amendments of 1988 (Holtzman, 2000).

C. CLINICAL VALIDITY

"Data to establish the clinical validity of genetic tests (clinical sensitivity, specificity, and predictive value) must be collected under investigative protocols" (Holtzman and Watson, 1998, p. 28).

Since Congress gave it authority to regulate in vitro diagnostic tests in 1976, the FDA has included assessment of clinical validity based on data made available by the test's developer (Holtzman, 2000). In view of the paucity of published data on predictive values of pharmacogenetic tests, and the likelihood that many of them will be low (Table 9.2), data on clinical validity should be required before pharmacogenetic tests are used clinically.[12]

D. CLINICAL UTILITY

"Before a genetic test can be generally accepted in clinical practice, data must be collected to demonstrate the benefits and risks that accrue from both positive and negative results" (Holtzman and Watson, 1998, p. 29).

---

[12] On the recommendation of SACGT(Secretary's Advisory Committee on Genetic Testing, 2000), the Food and Drug Administration is developing plans to increase its scrutiny of all genetic tests (see Chapter 6 by David Feigal). Previously, it actively regulated only tests marketed as kits. Now it is preparing to regulate tests marketed as clinical laboratory services as well.

The benefits and risks of a pharmacogenetic test may differ depending on whether it is used to predict a person's response or to explain an ADR or a therapeutic failure. The Task Force recognized the problem of trying to set one cutpoint for all uses of a test. *"Review panels could become enmeshed in endless debate if they attempt to set cutpoints for sensitivity and PPV; these should vary depending on the particular test, its use, options for treatment, and other factors. Even for a particular test, reasonable people will differ on how much test uncertainty they can tolerate. It is more important for external reviewers to ensure that the data have been appropriately collected and analyzed than to attempt to set cutpoints"* (Holtzman and Watson, 1998, p. 37). The Task Force recommended, *". . . test developers should provide information to professional organizations and others in order to permit informed decisions about routine use"* (Holtzman and Watson, 1998, p. 36). Both the Task Force and the SACGT recognized that some tests will need more stringent scrutiny than others.

VII. DO GENETIC TESTS FOR PREDICTING RISK OF DISEASE REQUIRE MORE REGULATION THAN PHARMACOGENETIC TESTS?

Roses argues that pharmacogenetic tests that predict altered drug handling should not be regulated in the same way as genetic tests that predict risks of disease. Pharmacogenetic tests "will not specifically 'test' the patient for the presence or absence of a disease gene-specific mutation, nor will they provide any other significant disease-specific predictive information about the patient or family members. . . . The genetics of response to the medicine will be the only data generated using an abbreviated SNP linkage disequilibrium profile and, practically, could be easily designed, edited and safeguarded to be totally meaningless with respect to any known disease-specific mutation. . . . It is therefore incumbent that medical guidelines for Mendelian- or susceptibility-gene testing do not extend automatically to discussion of other types of genetically based profiles in pharmacogenetics" (Roses, 2000b). Roses refers specifically to the Task Force and SACGT recommendations.

Given the sparse and not particularly impressive data on the clinical validity and utility of many pharmacogenetic tests, I see no reason why the same standards of validity should not be applied to pharmacogenetic tests as to genetic tests for disease. Moreover, the same possibilities of discrimination based on the results of genetic tests for disease arise for pharmacogenetic tests. First, if a health insurer is asked to cover a pharmacogenetic test for drugs known to be used for certain diseases, the insurer can infer that the patient either has or is at risk of disease. Second, if the result of the pharmacogenetic test reveals that a patient will not benefit from a drug, the insurer or employer may discriminate against the patient, particularly when alternative drugs are not available. Finally, regardless of whether the pharmacogenetic test is measuring an inherited mutation that directly affects drug handling or a polymorphism or haplotype in linkage disequilibrium with the mutation, first-degree relatives will also be at increased risk of carrying the mutation. Thus pharmacogenetic tests require at least the same level of confidentiality as other genetic tests.

VIII. CONCLUSIONS

Many nongenetic factors affect drug handling. Even when significant associations are found between inherited variants on the one hand and adverse drug reactions or therapeutic failure on the other, they may not translate to tests that will be clinically useful. Seldom is a single genetic variant the sole cause of an adverse drug reaction or a therapeutic failure. Consequently, testing for one variant—whether it be a direct test for the mutation or one for linked polymorphisms or haplotypes—will yield false negative results. False positive results are particularly likely in tests for polymorphisms.

Many in vitro diagnostic tests have been introduced into clinical practice without adequate validation. The result has been unnecessary testing and, sometimes, erroneous decision making. Pharmacogenetic tests on which prescribing decisions will be based need the same level of regulation as genetic tests to predict risks of disease. In addition to subjecting the patients to potentially harmful medical interventions, both types of tests may subject the patients and their relatives to unjust discrimination.

REFERENCES

Aithal, G.P., et al., "Association of Polymorphisms in the Cytochrome P450 CYP2C9 With Warfarin Dose Requirement and Risk of Bleeding Complications," *Lancet*, **353**, 717–719 (1999).

Arranz, M.J., et al., Meta-Analysis of Studies on Genetic Variation in 5-HT2A Receptors and Clozapine Response. *Schizophr. Res.*, **32**, 93–99 (1998).

Banks, R.E., et al., "Proteomics: New Perspectives, New Biomedical Opportunities," *Lancet*, **356**, 1749–1756 (2000).

Chen, S., et al., "The Cytochrome P450 2D6 (CYP2D6) Enzyme Polymorphism: Screening Costs and Influence on Clinical Outcomes in Psychiatry," *Clin. Pharmacol. Ther.*, **60**, 522–534 (1996).

Drazen, J.M., et al., "Pharmacogenetic Association Between ALOX5 Promoter Genotype and the Response to Anti-Asthma Treatment," *Nat. Genet.*, **22**, 168–170 (1999).

Drysdale, C.M., et al., "Complex Promoter and Coding Region Beta 2-Adrenergic Receptor Haplotypes Alter Receptor Expression and Predict In Vivo Responsiveness," *Proc. Natl. Acad. Sci. U.S.A.*, **97**, 10483–10488 (2000).

Editorial, "Genes, Drugs and Race," *Nat. Genet.*, **29**, 239–240 (2001).

Evans, D.A.P., "Pharmacogenetics," *Am. J. Med.*, **34**, 639–653 (1963).

Evans, D.A.P. and C.A. Clarke, "Pharmacogenetics," *Br. Med. Bull.*, **17**, 234–240 (1961).

Evans, W.E. and M.V. Relling, "Pharmacogenomics: Translating Functional Genomics Into Rational Therapeutics," *Science*, **286**, 487–491 (1999).

Farlow, M. R., et al., "Metrifonate Treatment of AD: Influence of APOE Genotype," *Neurology*, **53**, 2010–2016 (1999).

Farlow, M.R., "Treatment Outcome of Tacrine Therapy Depends on Apolipoprotein Genotype and Gender of the Subjects with Alzheimer's Disease," *Neurology*, **50**, 669–677 (1998).

Fears, R., et al., "Rational or Rationed Medicine? The Promise of Genetics for Improved Clinical Practice," *BMJ*, **320**, 933–935 (2000).

Furlong, R.A., et al., "Analysis and Meta-Analysis of Two Serotonin Transporter Gene Polymorphisms in Bipolar and Unipolar Affective Disorders," *Am. J. Med. Genet.*, **81**, 58–63 (1998).

Furuta, T., et al., "Effect of High-Dose Lansoprazole on Intragastric pH in Subjects Who Are Homozygous Extensive Metabolizers of Cytochrome P4502C19," *Clin. Pharmacol. Ther.*, **70**, 484–492 (2001).

Goldstein, D.B., "Islands of Linkage Disequilibrium," *Nat. Genet.*, **29**, 109–111 (2001).

Harris, H., *Human Biochemical Genetics*, 3rd ed. Amsterdam: Elsevier (1980).

Heeley, E., et al., "Prescription-Event Monitoring and Reporting of Adverse Drug Reactions," *Lancet*, **358**, 1872–1873 (2001).

Hoffmeyer, S., et al., "Functional Polymorphisms of the Human Multidrug-Resistance Gene: Multiple Sequence Variations and Correlation of One Allele with P-Glycoprotein Expression and Activity In Vivo," *Proc. Natl. Acad. Sci. U.S.A.*, **97**, 3473–3478 (2000).

Holtzman, N.A., "FDA and the Regulation of Genetic Tests," *Jurimetrics*, **41**, 53–62 (2000).

Holtzman, N.A., "Putting the Search for Genes in Perspective," *Int. J. Hlth. Serv.*, **31**, 445–461 (2001).

Holtzman, N.A., and M.S. Watson (eds), *"Promoting Safe and Effective Genetic Testing in the United States. Final Report."* Baltimore: Johns Hopkins University Press (1998).

Ingelman-Sundberg, M., "Pharmacogenetics: An Opportunity for a Safer and More Efficient Pharmacotherapy," *J. Intern. Med.*, **250**, 186–200 (2001).

Ioannidis, J.P. et al., "Replication Validity of Genetic Association Studies," *Nat. Genet.*, **29**, 306–309 (2001).

Kalow, W., Grant, C. M., "Pharmacogenetics,". In: Scriver CR et al. (eds), *The Metabolic and Molecular Bases of Inherited Disease*, 7th ed. New York: McGraw-Hill (1995).

Kuehl, P., et al., "Sequence Diversity in CYP3A Promoters and Characterization of the Genetic Basis of Polymorphic CYP3A5 Expression," *Nat. Genet.*, **27**, 383–391 (2001).

Kuivenhoven, J.A., et al., "The Role of a Common Variant of the Cholesteryl Ester Transfer Protein Gene in the Progression of Coronary Atherosclerosis," *N. Engl. J. Med.*, **338**, 86–93 (1998).

Layton, M.A., et al., "The Therapeutic Response to D-Penicillamine in Rheumatoid Arthritis: Influence of Glutathione S-Transferase Polymorphisms," *Rheumatology* (Oxford), **38**, 43–47 (1999).

Lazarou, J., et al., "Incidence of Adverse Drug Reactions in Hospitalized Patients: A Meta-Analysis of Prospective Studies," *JAMA*, **279**, 1200–1205 (1998).

Linder, M.W., "Genetic Mechanisms for Hypersensitivity and Resistance to the Anti-coagulant Warfarin," *Clin. Chim. Acta*, **308**, 9–15 (2001).

Luzzatto, L., and A. Mehta, "Glucose 6-Phosphate Dehydrogenase Deficiency," In: Scriver CR et al. (eds), *The Metabolic and Molecular Bases of Inherited Disease*, 7th ed. New York: McGraw-Hill (1995).

Meyer, U.A., "Pharmacogenetics and Adverse Drug Reactions," *Lancet*, **356**, 1667–1671 (2000).

Middeldorp, S., et al., "A Prospective Study of Asymptomatic Carriers of the Factor V Leiden Mutation to Determine the Incidence of Venous Thromboembolism," *Ann. Intern. Med.*, **135**, 322–327 (2001).

Nebert, D.W., "Polymorphisms in Drug-Metabolizing Enzymes: What is Their Clinical Relevance and Why Do They Exist?," *Am. J. Hum. Genet.*, **60**, 265–271 (1997).

Phillips, K.A., et al., "Potential Role of Pharmacogenomics in Reducing Adverse Drug Reactions: A Systematic Review," *JAMA*, **286**, 2270–2279 (2001).

Poirier, J., et al., "Apolipoprotein E4 Allele as a Predictor of Cholinergic Deficits and Treatment Outcome in Alzheimer Disease," *Proc. Natl. Acad. Sci. U.S.A.*, **92**, 12260–12264 (1995).

Roses, A.D., "Pharmacogenetics and Future Drug Development and Delivery," *Lancet*, **355**, 1358–1361 (2000a).

Roses, A.D., "Pharmacogenetics and the Practice of Medicine," *Nature*, **405**, 857–865 (2000b).

Rothstein, M.A., and P.G. Epps, "Ethical and Legal Implications of Pharmacoge-nomics," *Nat. Rev. Genet.*, **2**, 228–231 (2001).

Sagar, M., et al., "Omeprazole and CYP2C19 Polymorphism: Effects of Long-Term Treatment on Gastrin, Pepsinogen I, and Chromogranin A in Patients with Acid Related Disorders," *Aliment. Pharmacol. Ther.*, **14**, 1495–1502 (2000).

Sander, C. "Genomic Medicine and the Future of Health Care," *Science*, **287**, 1977–1978 (2000).

Sarasin, F.P. and H. Bounameaux, "Decision Analysis Model of Prolonged Oral Anti-coagulant Treatment in Factor V Leiden Carriers with First Episode of Deep Vein Thrombosis," *BMJ*, **316**, 95–99 (1998).

Secretary's Advisory Committee on Genetic Testing. *Enhancing the Oversight of Genetic Tests: Recommendations of the SACGT* (2000). *http://www4.od.nih.gov/oba/sacgt/gtdocuments.html*

Shi, M.M., et al., "Pharmacogenetic Application in Drug Development and Clinical Trials," *Drug Metab. Dispos.*, **29**, 591–595 (2001).

Spannagl, M., et al., "Are Factor V Leiden Carriers Who Use Oral Contraceptives at Extreme Risk for Venous Thromboembolism?" *Eur. J. Contracept. Reprod. Hlth. Care*, **5**, 105–112 (2000).

Steimer, W., et al., "Pharmacogenetics: A New Diagnostic Tool in the Management of Antidepressive Drug Therapy," *Clin. Chim. Acta*, **308**, 33–41 (2001).

Streetman, D.S., et al., "Phenotyping of Drug-Metabolizing Enzymes in Adults: A Review of In-Vivo Cytochrome P450 Phenotyping Probes," *Pharmacogenetics*, **10**, 187–216 (2000).

Taube, J., et al., "Influence of Cytochrome P-450 CYP2C9 Polymorphisms on Warfarin Sensitivity and Risk of Over-Anticoagulation in Patients on Long-Term Treatment," *Blood*, **96**, 1816–1819 (2000).

Turner, S.T., et al., "Antihypertensive Pharmacogenetics: Getting the Right Drug into the Right Patient," *J. Hypertens.*, **19**, 1–11 (2001).

Weinshilboum, R., "Thiopurine Pharmacogenetics: Clinical and Molecular Studies of Thiopurine Methyltransferase," *Drug Metab. Dispos.*, **29**, 601–605 (2001).

Wilcock, G.K., et al., "Efficacy and Safety of Galantamine in Patients with Mild to Moderate Alzheimer's Disease: Multicentre Randomised Controlled Trial," *BMJ*, **321**, 1445–1449 (2000).

Wilson, J.F., et al., "Population Genetic Structure of Variable Drug Response," *Nat. Genet.*, **29**, 265–269 (2001).

Zhu, B., et al., "Assessment of Cytochrome P450 Activity by a Five-Drug Cocktail Approach," *Clin. Pharmacol. Ther.*, **70**, 455–461 (2001).

MEDICAL LIABILITY FOR PHARMACOGENOMICS

LARRY I. PALMER, LL.B.

I. INTRODUCTION

As pharmacogenomics develops new ways of specifying the risks of drugs to particular patients, medical liability law will be used to determine whether there is legal culpability for adverse effects of drugs and, if so, how to allocate responsibility between health care professionals and pharmaceutical companies. Medical liability law includes three broad categories of doctrines developed by judges, and sometimes modified by legislatures, to balance the benefits and risks of medical intervention: (1) the special standard of care judges and juries use to determine whether licensed health care professionals should compensate for medical mishaps ("medical malpractice"); (2) the doctrine used to determine whether patients understood the risks of medical intervention ("informed consent"); and (3) those doctrines developed to assess whether the manufacturers of medical products have reduced the risks of their products to acceptable levels ("product liability").

Critics have long complained about the ineffectiveness of medical liability law both as a means of reducing the risks of injuries and as a system of compensation for injuries. So far, none of these critiques has led policy makers to jettison our fault-based medical liability system and to replace it with some type of no-fault system as proposed by some scholars. Thus some form of medical liability is going to be a feature of the social and regulatory

Pharmacogenomics: Social, Ethical, and Clinical Dimensions, Edited by Mark A. Rothstein.
ISBN 0-471-22769-2   Copyright © 2003 Wiley-Liss, Inc.

landscape as health care undergoes fundamental shifts caused by the genomic revolution (Palmer, 2002).

Policy makers, practitioners, and scholars from a variety of disciplines have recently embraced a new approach to risk reduction in health care—a "systems approach"—without proposing any specific reforms of medical liability law. The Institute of Medicine (IOM) placed its imprimatur on this approach in its recent reports (Kohn et al., 2000; IOM, 2001). In its simplest form, a systems approach to risk reduction in health care posits that an injury to a patient is often the manifestation of a latent error in the system of providing care. In other words, a medical mishap is the proverbial "accident waiting to happen" because the injury-preventing tools currently deployed, including medical liability law, are aimed at finding the individuals at fault rather than the systemic causes of error. Coexistence of a systems approach to error reduction and medical liability law as a conceptual framework for policy makers implies that the latter is likely to evolve in an incremental fashion as the former makes more visible different aspects of the medical error problem.

The prototypical form of error in the health care system that could be reduced by a systems approach is medication error. The kind of error identified in the literature—overdose of chemotherapy, injection of the wrong drug, etc.—sometimes leads to either injury or death, the kinds of harm that are the central concern of after-the-fact medical liability adjudication. Pharmacogenomics introduces not only another conception of harm—genetic risks—but also new ways of developing and prescribing drugs.

My basic thesis is that the introduction of pharmacogenomics requires a reexamination of the justification of medical liability doctrines. The traditional justification of balancing the risks and benefits of medical intervention to individual patients should be replaced by a new theory of how after-the-fact adjudication balances the risks and benefits to groups defined by genotype or their social-ethnic or racial group. To allow for appropriate balance of the risks and benefits of pharmacogenomics, liability doctrines must incorporate a systems approach in imposing or withholding liability. In other words, endorsement of a systems approach to error reduction does not mean the elimination of the compensation function of medical liability law. Rather, the underlying rationale of medical liability must be rethought with the primary goal of providing some incentives for those involved in the development, distribution, and dispensing of the products of pharmacogenomics to incorporate systemic approaches to risk reduction.

I am not asserting that judges or legislatures have in fact developed these systemic approaches to issues associated with genetic risk or injury. Rather, I argue that older cases dealing with a form of inherited injury—liability for the risk of diethylstilbestrol ("DES")—should be analyzed in a systemic fashion to prepare us for our genomic futures. The implicit systems view developed in those lines of cases suggests that physicians rather than manufacturers of DES should bear primary responsibility for the adverse consequences of DES. Were this rudimentary systems view of medical liability applied to pharmacogenomics, one might expect the producers of new drugs to be immune from liability suits brought by patients/consumers.

I further propose that litigation about new genetic products over the past decade provides a framework for determining how pharmacogenomics should alter medical liability doctrines. A California Supreme Court ruling that physician-scientists have a duty to disclose their research and financial interests in a patient's unique DNA is the starting point for a systemic analysis of medical liability doctrine (*Moore v. Regents of the University of California*, 1990). Thus duty analysis begins with the physician-scientist rather than the organizational entity that might develop and market the drug. Judges, however, will increasingly use duty analysis to link physicians and organizational actors—that is, those engaged in pharmacogenomics. A pending case in the Federal District Court in Illinois involving a dispute over the distribution of genetic knowledge and technology associated with a rare genetic disorder illustrates how a liability-focused duty analysis could be used by groups—defined by either genotype or racial or ethnic background—to reach organizational actors who own patents on genetic products (*Greenberg v. Miami Children's Hospital Research Institute, Inc.*, 2000).

Once a duty analysis links physician-scientists, product developers, and patients in a new systems view of informed consent, medical liability should shift to class action suits against manufacturers. However, plaintiffs' success with such a strategy will be highly dependent on courts' views of the liability of physicians for pharmaceutical mishaps. Another pending case in Pennsylvania involving a claim that a drug manufacturer failed to warn people with a certain genotype of the risks of a vaccine illustrates the challenges courts will increasingly face (*Cassidy v. SmithKline Beecham*, 1999). Although this is a novel case, the plaintiffs must deal with the medical liability doctrines developed by that state's courts, which at first glance appear far removed from the systemic approach that can incorporate the newer developments that pharmacogenomics will bring. This case demonstrates that although genomics may be national or even international in terms of its scope, liability doctrines remain primarily local or state-based.

Finally, I argue that changing methods of distributing drugs—direct advertising to consumers of prescription drugs—has already led some courts to suggest that manufacturers may have legal duties to warn patients of the risks of their products, in addition to the manufacturers' warnings to physicians. This crack in the implicit systems view embedded in what has been called the "learned intermediary" defense challenges the underlying notion that the physician-patient dyad is the appropriate point of analysis in developing a system for the balancing of risks and benefits. A general theory of how medical liability doctrine will evolve in light of pharmacogenomics must therefore propose a theory of how the professional-patient dyad fits within the innovation process driven by pharmacogenomics.

II. DRUGS AND GENETIC HARMS

Traditional medical liability doctrines seek to establish that the plaintiffs were in fact "injured," usually meaning that they sustained some physical harm as a result of the medical intervention. But in addition to "adverse reactions," drugs also have the ability to create genetic harm, a modification of the germ cells that damage one's progeny or leave one's progeny more susceptible to disease. Medical liability law's traditional response to increased risk of disease is best illustrated by the manner in which courts reacted to the increased intergenerational risks of various forms of cancer associated with DES.

The California Supreme Court ruled that the daughters of women who had taken DES to prevent miscarriages could not hold the manufacturers liable under product liability doctrine because DES injured them *in utero* (*Brown v. Superior Court*, 1988). The plaintiff-daughters' increased risk of cancer and other reproductive abnormalities meant that more medical monitoring and procedures to prevent malignancies as well as some actual cancers were not the legal responsibility of the manufacturers of DES. In formulating a rule for prescription drugs that set them apart from general product liability doctrines, the court noted that the manufacturer had a duty to properly prepare the drug and warn of its "dangerous propensities that were either known or reasonably scientifically knowable at the time of manufacture" (*Brown v. Superior Court*, 1988). This latter duty has been characterized as one arising out of negligence, whereas the plaintiff's proposed product liability theory has been called a theory of "strict liability" or "liability without fault."

In their action based in strict liability, the plaintiff sought a ruling that the drug was defectively designed because it created the risk of intergenerational harm. If DES were labeled a "defective product," the plaintiffs could have established a legal link between the product and the increased risk of cancer. In rejecting the applicability of strict liability, the California court reduced the pharmaceutical company's legal duty to warn of the dangerous propensity of DES *at the time of its manufacture* (a negligence standard). Because the harm complained of was not apparent until the 1970s, it was impossible for the plaintiffs to prove that the manufacturer knew of the drug's risks in the 1950s, when the plaintiffs' mothers ingested DES. The California litigation, however, was only one of several cases involving DES filed after a 1971 article described how a rare form of vaginal cancer in a 15-year-old was linked to her mother's ingestion of DES during pregnancy.

A group of women who were a part of a double-blind test of the effectiveness of DES filed a lawsuit in federal district court against a hospital and one of the manufacturers (*Mink v. University of Chicago*, 1978). The federal court, applying the liability doctrine of Illinois, ruled that the women could potentially recover against the doctors, employees of the hospital, who prescribed DES on the theory that the physician committed a battery—a form of strict liability—when the patients took DES without any knowledge of what it was. But the court also ruled that the DES mothers could not hold the manufacturer liable for an allegedly defective product. The plaintiffs' lack of success on the product liability claim was a function of the lack of "injury" to themselves as opposed to the risks of cancer to their offspring. Plaintiffs' claims that the defendants had a duty to warn them of the inherent risks—at least after 1971—also failed in the court's view, because of the lack of injury to the plaintiffs themselves.

The Illinois and California cases together provide a snapshot of how the traditional medical liability analysis handles the risks of intergenerational harm from drugs. The DES daughters cannot recover from the drug manufacturer on a "strict liability" theory, but the DES mothers might recover under a "battery" action for unconsented touching against the doctors administering the drug without the patients' knowledge or consent. Otherwise, the plaintiffs are relegated to various theories of negligence, most of which are not provable under the facts of the DES cases. The startling—at least to our modern ears—facts about DES are that a study in the 1950s mentioned in the Illinois federal case actually demonstrated that DES *caused* rather than *prevented* miscarriages. Even more surprising is that the FDA regulatory structure did not require that DES be proven safe or effective for preventing miscarriages when it was first prescribed for this purpose in 1947 (Weitzner and Hirsh, 1981). Given the present climate of a "gene rush," the clearest indication that DES is from a different era is the fact that the drug was never patented!

This analysis assumes implicitly that prescription drugs are different from most products because a physician will weigh the risks and benefits of prescribing the drug to a particular patient. This traditional analysis further supports the exemption of drugs from the strict liability doctrine by asserting that, today, the FDA is required to assess both the safety and efficacy of drugs before they are allowed in the market place (Henderson and Twerski, 2001). The drug manufacturer's responsibility for risks and injury is primarily determined by using standards of negligence. Negligence is also the standard by which the physician's relationship with a patient is usually analyzed. So the use of a negligence standard for most of the DES-related problems, except for the experimentation, is premised on the theory that the physician-patient relationship is the appropriate framework for weighing the various risks and benefits of drug intervention. The hindsight perspective of the intergenerational harm of DES could not undermine the primacy of physician risk-benefit analysis in prescribing drugs—which is what a negligence standards boils down to.

Recall that liability for malpractice requires proof that the defendant-physician violated not the jury's standards, but the standards of fellow practitioners when prescribing DES. The allocation of the liability risks in the DES cases between manufacturer and physician is only illustrative rather than prescriptive of the form of allocation that will occur for the products of pharmacogenomics. The main thrust of the DES cases is to illustrate that the standard of care for physicians undergoes a major shift when physicians are operating on the frontiers of medical practice or engaged in clinical research. This allows for the manufacturer's responsibility to be derivative of the physician's duty to deal "fairly" with patients with regard to the risks.

III. LIABILITY FRAMEWORK FOR GENETIC PRODUCTS

One significant shift in thinking about medical liability occurred in the 1970s, about the time of the first DES cases. Some courts began to rule that physicians might be liable for physical harm to patients if they failed to "inform" the patient of the risks of the procedure before undertaking the medical intervention (*Canterbury v. Spence*, 1972; *Cobbs v. Grant*, 1972). When informed consent was first introduced, there was a debate as to whether it was in fact based in theories of negligence or strict liability for unlawful touching. In other words, was it to implement the ideal of the autonomous patient as epitomized by the oft-quoted, "Every human being . . . has the right to determine what should be done with his or her own body" and be based on battery? Or was the lack of informed consent simply a judicial development of some rules for disclosing risks of medical intervention consistent with the notion that physicians' duties were defined by a special professional standard of care?

The latter view prevailed but generated yet another set of issues as to whether disclosure rules should be evaluated from the perspective of professionals or patients. Some courts took the narrower view of duty by ruling that professional standards should be used to determine what should be disclosed to patients. Although this theory of "lack of informed consent" was distinct from whether the health care provider had violated the standard of care, most courts, and many legislatures, confined the doctrine's operation to a very narrow set of circumstances consistent with the negligence standard underlying the standard of care in medical cases. Other courts took what is called the broader view and leaned toward the patient's perspective while requiring juries to impose the reasonable patient's view of risk rather than the particular patient's view of risks.

The more expansive view of informed consent was used by the California Supreme Court in *Moore v. Regents of the University of California* in 1990 to establish a fiduciary duty on the part of physician-scientists to disclose their research and financial interests in a patient's DNA during the clinical research stage of product development.

A. FIDUCIARY DUTIES AND GENETIC PRODUCT DEVELOPMENT

We need to revisit *Moore* with the view of determining how one leading appellate court used liability doctrines to allocate the risks and burdens of genetic product development among patients/subjects, physicians, and manufacturers. John Moore, a person with hairy cell leukemia, was a patient of Dr. David Golde at the University of California at Los Angeles Medical Center in 1976. Dr. Golde recommended the removal of Moore's spleen as part of the successful treatment of his leukemia. Over the course of the next seven years, Moore returned to the UCLA Medical Center from his home in Seattle for tests, providing Dr. Golde with blood, skin tissue, bone marrow, and sperm. There is a suggestion in the voluminous legal literature surrounding this case that Dr. Golde encouraged Moore to continue his treatment at UCLA rather than in Seattle, his home, by obtaining grant funds to pay Moore's transportation costs to Los Angeles (Bergman, 1992).

This interest in having direct access to Moore's body is explained by the fact that Dr. Golde apparently discovered on examination of Moore's excised spleen that his DNA was unique in that he overproduced proteins that regulate the immune system. Dr. Golde and his laboratory assistant developed a cell line from tissue, blood, and other body parts Moore had provided. Dr. Golde filed a patent on the cell line, granted a license to a biotechnology company, and received stock options and a consultancy with the biotechnology company. The university was paid an amount to cover a portion of Dr. Golde's salary over a period of years. Once Moore discovered that "Mo cell line" (U.S. Patent Number 4,438,032, 1984) had been used to develop powerful drugs for the treatment of several forms of cancer, he sued Dr. Golde, the University of California, the assignee of the patent, and the biotechnology and drug firms involved in the development and distribution of the drugs on various grounds.

Moore argued numerous theories of liability, including the claim that all the defendants—including the biotechnology and drug firms—had misappropriated his "property" when they used his DNA to develop the cell line and the drugs. The court rejected Moore's property claim based on the theory that the various defendants had interfered with Moore's possessory interest in his spleen, tissue, blood, DNA, etc. By rejecting the liability claim based on Moore's alleged property interest in his unique DNA, the California Supreme Court dismissed the claims against the assignee of the patents—namely, the biotechnology companies.

The court did, however, hold that a physician-scientist has a duty to disclose his research and financial interests in the patient's cells, tissue, and DNA, citing its earlier ruling on informed consent in *Cobbs v. Grant* (*Moore*, 129; 1992). Presumably, this means that Dr. Golde should have told John Moore of his interest in developing the cell line and pursuing the patent, as well as of his financial arrangements with the companies developing the drugs. Although the case was settled after this ruling, the court's result uses the judicially developed doctrine of lack of informed consent to balance the interests of patients and physician-scientists. The result in *Moore* protects scientific innovation because the duty to disclose established by the court exempts the companies—those that might bring successful products to market—while protecting the patient's interest in autonomy by granting a theoretical right not to participate in the research.

So viewed, *Moore* fits neatly with older cases of liability based in negligence dealing with genetic health—wrongful birth and wrongful life claims. (Wrongful birth claims are brought by parents of a child born with an inherited condition, whereas wrongful life claims are brought on behalf of a child born with an inherited condition.) Those cases deal almost exclusively with the standard of care for physicians and providers (Palmer, 2002). If we look at more recent developments in genetic health and genetic harm, it is apparent that the likely line of legal development is through a *Moore*-based expansion of the standard of information disclosure line of cases at the pharmacogenomics frontier. It was not, however, until a recently filed case involving the duties of a physician-scientist to those who furnished DNA that led to the discovery of the gene for an inherited disease that genomic medicine's focus on groups became apparent. How those groups are described within medical liability doctrines has great implications for how racial or ethnic groups are considered for purposes of medical liability.

B. GENETIC RESEARCH AND SOCIAL GROUPS

In October 2000, less than a year after the announcement of the completion of the first complete draft of the human genome, a group of plaintiffs filed suit against the developer and assignee of a patent on a test for a genetic disease (*Greenberg v. Miami Children's Hospital Research Institute, Inc.*, 2000). In this case, the plaintiffs, who are donors of human body parts and tissue, allege that the defendants, a physician-scientist and a hospital, should have disclosed their intention to patent the gene and the diagnostic test for Canavan disease, a progressive fatal neurological disease of the central nervous system affecting persons of Eastern European Jewish ancestry more frequently than members of other ethnic groups. The issue is whether any medical liability doctrines can be used by the plaintiffs to control the actions of the health care providers who extracted genetic information from their human body parts and tissue to develop a product that could be patented.

The primary plaintiffs in the case are a group of parents whose children either have or have died from Canavan disease. They allege that over a seven-year period they provided the defendant and his research team with tissue samples from their dead children's autopsied bodies, blood from themselves, and other family data. Furthermore, along with the nonprofit organizational plaintiffs—the Canavan Foundation, National Tay–Sachs and Allied Disease Association, and Dor Yeshorim, a group providing screening and counseling services to members of the Jewish community—these parents claim that they helped the defendant and his research team collect data and body tissue, blood, and other samples from families afflicted with Canavan disease from around the world. Their efforts led to the defendants' development of a test to determine whether prospective parents carry the recessive gene for Canavan disease and whether a fetus will be affected with the disease. In 1994, the defendant-physician filed for a patent on the diagnostic test and the gene itself (U.S. Patent No. 5,679,635, 1997). Shortly after the patent was granted in 1997, the defendants started to notify testing centers, including those of the organizational plaintiffs, of their intentions to vigorously defend their "intellectual property rights."

The plaintiffs are not claiming that the patent is invalid. Rather, relying on theories implicit in *Moore*, they seek to use informed consent liability doctrines to ensure low-cost genetic screening and testing for Canavan disease. The lawsuit is based on what I will call a breach of covenant, not a breach of promise. The term breach of covenant is appropriate, despite the contractual view in some of the literature (Schuck, 1994), because the relationship between the parties was not contractual. Breach of promise implies that their agreement was contractual in nature, which creates doctrinal problems (Palmer, 1989).

The individual and organizational plaintiffs allege that they supplied tissue, blood, urine, autopsy information, and money for research to the physician-scientist with the *implicit* understanding that the physician-scientist shared their goal of developing an affordable and accessible diagnostic test "modeled after the Tay–Sachs testing" program. This model involves active cooperation by donors of tissue and other DNA samples with researchers and, eventually, free screening and testing. It is worth noting that the patent for the Tay–Sachs gene and its associated diagnostic test is assigned to the United States government, because a researcher at the National Institutes of Health made the discovery (U.S. Patent No. 5,679,635, 1997). This ownership by the federal government is the crucial factor to ensure that the Tay–Sachs disease test is both inexpensive and widely available. The defendant-hospital in the Canavan disease case claimed that it hoped to attract one large company to do all the testing by granting it an exclusive license. Therefore, it began informing institutions that were performing the Canavan disease test that it would cost $25 per test in addition to a steep licensing fee. It later lowered the cost per test to $12.50 (Gorner,

2000). In contrast, the Tay–Sachs disease test patented by the Department of Health and Human Services costs about $100 per test, and the screening test for two breast cancer genes patented by a biotechnology firm costs $2,400 per test (Krimsky, 1999).

The defendants' motion to dismiss the complaint essentially alleges that the duty to disclose is limited to the doctor-patient relationship and was never meant to govern the researcher-subject relationship. As a consequence, the defendants have asked the court to rule on whether physicians/scientists engaging in genetic research have a duty to disclose their intentions to patent their discovery to those who volunteer to be human subjects. The plaintiffs must get past this motion to dismiss to explain to the court what they mean by the "Tay–Sachs model."

Because there are very few precedents that are directly relevant, the resolution of the claim will involve two fundamental issues.

1. Are the rules of disclosure different in therapeutic and research settings?

For the plaintiffs in *Greenberg* to survive the motion to dismiss, they must argue that informed consent has a different function in research settings than in therapeutic settings. This difference is due to the fact that information flow in the two settings must accomplish different goals. In the research setting, the purpose of information flow is to allow subjects to determine whether they should participate in the research. Although informed consent is necessary, it is not sufficient to determine whether proceeding with the particular clinical research project is ethical (Emanuel et al., 2000). Conflicts of interest, financial or otherwise, may be a significant issue in clinical research (Miller, 2001). In contrast, in the therapeutic setting, the purpose of information flow is to allow patients to determine whether the risks of a medical procedure, drug, test, or other treatment, as well as its possible health benefits, are worth undertaking.

This determination requires an examination of both an old and a new line of cases. In the new line of cases, plaintiffs have argued—in the therapeutic context—that health care providers have a duty to provide comparative data about the effectiveness not only of various *treatment* alternatives, but also about the qualifications, competence, and experience of the *provider*. In the one case in which plaintiffs have so far succeeded in surviving a motion to dismiss, the Wisconsin Supreme Court held that a neurosurgeon had a duty to disclose his level of experience with a procedure and the morbidity and mortality differences between himself and more experienced neurosurgeons (*Johnson v. Kokemoor*, 1996). The Wisconsin court's ruling has been labeled the "second revolution" of the informed consent doctrine (Twerski and Cohen, 1999; Ketler, 2001).

Even in the strictly therapeutic context, scholars have argued that courts should expand the informed consent doctrine to include a duty to disclose "provider-risk information" because of the growth of such information in today's health care system. With the growth of managed care, much more data are available about the relative effectiveness of providers in performing certain procedures. The bottom line of this argument is that forcing disclosure of readily available data increases consumer/patient choice. In effect, the argument is that in our form of health care with increased competition, advertisement by all kinds of providers, and multiple health care plans from which to choose, information flow (advertisement) to consumers and patients (informed consent liability) is crucial to the law's respect for autonomy. The argument for applying this theory to pharmacogenomics is strong, even though rejected by the Pennsylvania courts (*Duttry v. Patterson*, 2001), provided judges gain some rudimentary understanding of the science and technology behind the term "genomics."

There are, in fact, a few cases before 1990 that have a bearing on the informed consent theory in the research setting put forth in *Greenberg*. These cases involve clinical experiments where there is a risk of physical harm to the subject who was not a patient (*Haluska v. University of Saskatchewan*, 1965). Through this line of cases, the courts suggested that the lack of informed consent is similar to a form of strict liability based on notions of fraud and battery. This battery-based theory of informed consent in the research context suggests that its purpose is to allow potential human subjects the opportunity to refuse to participate in or to withdraw from the project at any time. The *Greenberg* plaintiffs, of course, would prefer for the court to view their case in line with this older line of cases because the corresponding emphasis on strict liability would increase their likelihood of winning.

The plaintiffs must convince the court that the answer to my first question of whether the rules of disclosure are the same in research and therapeutic settings is "No." The plaintiffs should, in my view, argue that the duty to disclose in the research context means that the researchers must not only disclose any physical risks of harm but, in the genomics world, must also disclose, without the plaintiffs' asking, their intentions regarding the patenting of genetic knowledge and data they obtain. This more modern approach is based on theories of negligence rather than older notions of battery and strict liability (*Grimes v. Kennedy Kreiger Institute*, 2001).

2. Does the Nuremberg Code provide a basis for liability?

For many commentators, the duty to disclose in the research context has its origins in the judgments against Nazi physicians who conducted experiments on concentration camp inmates without their consent (Katz, 1993). Because several German physicians were executed and others received long prison terms for "war crimes" and "crimes against humanity," some form of civil liability for violation of the requirements of consent during the course of research appears to be a reasonable extension of the legal principles established in the international criminal law context.

Some of the most ethically troublesome human experiments, such as the Tuskegee Study of Untreated Syphilis in the Negro Male and the Human

Radiation Experiments (Final Report Tuskegee Syphilis, 1973; Final Report Human Radiation, 1995), seem to scream out for an application of the Nuremberg Code. The Human Radiation Experiments came to light in the mid-1990s, when it was revealed that the United States government had sponsored various experiments employing patients and institutionalized children to study the effects of radiation on the human body during the Cold War, in which no consent was either sought or obtained. The Tuskegee Syphilis Study was a 40-year effort to understand the effects of untreated syphilis in African Americans; the study continued long after penicillin was discovered as a cure for the disease. The subjects were in fact patients of the United States Public Health Service who believed they were receiving treatment for their "bad blood" over several years, when in fact no treatment was ever provided (Jones, 1993). Yet, the few United States judges who have ever cited the Nuremberg Code in a human experimentation case, such as the lawsuit filed over one of the human radiation experiments (*In re Cincinnati Radiation*, 1995), never dealt explicitly with the main question: Is the Nuremberg Code's requirement of informed consent in fact a part of the *domestic* law of the United States (Clark, 1990)?

The Tuskegee Syphilis and Human Radiation studies' discussions of consent in the research context are examples of ethically horrific cases that invoke rhetorical charges of violating the Nuremberg Code, but they are *not* legal precedents that clearly establish the need for a different standard of informed consent in the research and therapeutic contexts. Law suits following these studies ended in settlements without full adjudications, and they were filed against governmental agencies, not private individuals.

Despite these doubts about whether, on a purely technical basis, the Nuremberg Code applies, some institutional lessons from those judgments remain and might be helpful in resolving *Greenberg* as well as other cases of genetic disease management (Palmer, 1997). Most of the scholarly attention on the Nuremberg judgment fails to note the legal distinction, even in the context of international law, between experiments that were war crimes and those that were crimes against humanity. The Nazi malaria experiments, which, like their American counterparts, involved prisoners, were aimed at resolving problems of infection arising on the battlefield. The criminal convictions for these experiments are more properly thought of as war crimes convictions. In effect, the Nazi researchers violated the rules of war by contributing to the excessive deaths and needless infliction of suffering resulting from their attempts to develop more effective treatments for malaria. The war crimes convictions of the Nazi physicians and scientists, in my view, do not turn on the lack of consent. There is, in fact, some evidence to suggest that Dr. Andrew Ivy, the American Medical Association's expert consultant for the prosecutors, misrepresented the process of obtaining consent for America's own wartime experiments with prisoners regarding malaria (Harkness, 1996; Harkness and Schuster, 1998).

The sterilization experiments, such as the tests of the effectiveness of a powerful drug and powerful X rays as sterilization devices on concentration camp inmates, are more properly thought of as crimes against humanity because their goal was the elimination of civilian populations—Russians, Poles, Jews, and other groups—by the "most scientific and least conspicuous means" available. One purpose of maintaining the distinction between war crimes and crimes against humanity is that the latter doctrine might be developed in the civil context to deal with experiments that are performed on populations selected on the basis of their race or ethnic status.

I am not suggesting by any means that the goal of the nationwide screening of Eastern European Jews, their subsequent genetic counseling, fetal testing, and decisions about terminating pregnancy, is a plot to eliminate that population. The irony of this particular ethnic group being subject to so much scrutiny in our quest for genetic health does, however, suggest the need for a more cautious approach to denying claims for participation by laypersons in the decision making process—and that is in fact the essence of the plaintiffs' claim in *Greenberg*.

The crimes against humanity doctrine in the Nuremberg judgment provides analytical support for the notion that liability law should maintain a distinction between research and therapy even as genomics blurs the distinction. In the context of *Greenberg*, the Nuremberg Code does not give the plaintiffs a trump card in the pending motion to dismiss. Rather, the important lesson from the Nazi physicians' trials is that a total professionalization of the decision making process about scientific advancement can set in motion dangerous social forces in our quest for genetic health.

Greenberg has two implications for pharmacogenomics. First, as the development of pharmacogenomics moves into the clinical testing stage with humans, medical liability doctrines could become operative by redefining the nature of information that should be disclosed to human subjects who might be donors of tissue or blood. Second, to the degree that subjects are recruited from identifiable social, racial, or ethnic groups, courts will consider imposing duties of disclosure on clinical researchers that will force producers of genetic pharmaceuticals to consider the impact on population groups (Dolgin, 2000). In other words, courts are likely to use a systemic analysis in defining the contours of medical liability doctrine in approaching pharmacogenomics, and thus we might expect some innovations in legal doctrines.

C. PHARMACOGENOMICS AND GENOTYPES

Pharmacogenomics will increase the knowledge base about risks in relation to particular genotypes and potentially change the manner of disease management. Rather than start with the notion that the patient was injured and try to allocate the compensation burden between the physician and the producer of the pharmaceutical, plaintiffs will attempt to use pharmacogenomics-based knowledge to impose new types of legal duties on the producers of drugs.

A recently filed case in Pennsylvania illustrates a possible new face for liability (*Cassidy v. SmithKline Beecham*, 1999). The plaintiff claims that the manufacturer of a vaccine for Lyme disease should have warned physicians and patients in its advertisements that 30% of the population ran the risk of developing "treatment-resistant Lyme arthritis" because their particular genotype interacted adversely with the vaccine. This duty to warn is based not in product liability doctrine, but on negligence.

The relief sought by the plaintiff is aimed not at compensation for past injury but for protection against the risks of future injury. Thus the plaintiff is asking the court to allow the suit to proceed on behalf of a class of persons with a certain genotype, presumably because the manufacturer could have known of the risks to the vulnerable population group. The plaintiff is also seeking an order requiring the defendant to establish a medical monitoring program and a trust fund to pay for future medical expenses of the plaintiff class. Finally, the plaintiff wants a court order requiring the manufacturer to warn physicians and the general public to be screened through a simple blood test before taking the vaccine.

At first blush, the increased risk of harm suit against a drug manufacturer on a negligence theory that pharmacogenomics could have been used to determine the increased risks to a portion of the population appears to be a logical extension of the doctrinal shifts in liability law signaled by *Moore*. The likelihood of success of this particular lawsuit is in fact small because of the nature of existing medical liability doctrine in Pennsylvania.

Unlike many states, Pennsylvania's courts have developed a very limited notion of informed consent. For instance, the Pennsylvania Supreme Court views the lack of informed consent doctrine as based in battery rather in negligence. It has also explicitly rejected the idea proposed by a few courts that an increased amount of information means that the provider should have an obligation to disclose more information to consumers/patients (*Duttry v. Patterson*, 2001). Thus, in viewing this new style of lawsuit, the Pennsylvania Supreme Court might take the view that the only warnings a drug manufacturer must give are those required by the Food and Drug Administration. Without explicitly holding that the regulation insulates the drug manufacturer, Pennsylvania courts' systems view might be that the regulatory agency is in the best position to make the analysis of risks and benefits of pharmaceutical innovation. Such a view is implicit in the position that prescription drugs are different from other products because courts should rely on the physician to manage the risks of any drug, including vaccines, that is beneficial for some patients (Henderson and Twerski, 2001).

IV. MANUFACTURER LIABILITY FOR HARMS TO PATIENTS

If the physician remains the primary mode of distributing prescription drugs, a protective stance toward drug manufacturers might make sense. But we know that the traditional notion that drug manufacturers market their products to physicians is rapidly giving way to "direct to consumer" advertisement of prescription drugs. Although this practice has not led to an abandonment of the restrictive application of product liability doctrines to prescription drugs or narrow views of manufacturers' duty under negligence doctrine, it has led at least one court to remove one of the drug manufacturers' defenses—the learned intermediary doctrine—in some limited circumstances.

The New Jersey Supreme Court held in 1999 that the drug manufacturer's duty to warn the consumer/patient was not satisfied by warnings to physicians when the drug manufacturer advertised directly to consumers (*Perez v. Wyeth Laboratories*, 1999). In those cases, the manufacturer must also warn the consumer. This crack in the so-called learned intermediary doctrine is not simply a matter of a more "liberal" or pro-liability doctrine protecting plaintiffs instead of drug manufacturers. Rather, it is illustrative of a particular systems view. Instead of starting with the assumption that prescription drugs and devices are different kinds of products because physicians prescribe and the FDA regulates their use, the *Perez* court looked outside of liability doctrines to practices in the marketplace.

Put in the form of a question, one might ask, What is the systems view implicit in *Perez*, and why does the drug manufacturer's liability turn on the duty to warn? The simple answer is found in the notion that once we are talking about a prescription drug, the ultimate consumer is a patient. A patient of whom? A physician who must first write the prescription authorizing a pharmacist to dispense the drug to the patient. For a prescription drug to be on the market, the FDA must approve it and allow the physician to write the prescription and the pharmacy to deliver the drug to the patient.

But the competitive marketplace and increased consumer knowledge are changing the method by which physicians prescribe drugs. Lest we get too excited by direct to consumer advertisement by drug manufacturers, we should recall that all types of health care providers—physicians, hospitals, and managed care entities—now engage in direct to consumer advertising, something they did not do twenty years ago. This change in practice is explained by at least two forces, the changing and more competitive aspects of health care and the removal of restrictions on advertising by professionals. The legal controversy has been fought through a series of law suits challenging the legality of professional codes of ethics' restrictions on advertising.

The patient in the Pennsylvania case discussed above, for instance, saw the advertisement for the vaccine and apparently asked for the vaccine, assuming that the risks of contracting Lyme disease were greater than the risks of the vaccine. As knowledge about genetic variation grows, one can imagine that patients will begin to ask physicians about the genetic risks of drugs. In this world, one can well imagine that plaintiffs' lawyers might push to require that the tools of pharmacogenomics be used to actually reduce the risks of these new drug products.

One potential fertile ground of litigation in the future might be the common practice of "off-label" use. Before pharmacogenomics, it may have been easy to agree with the view that a physician's use of a drug to treat a disease or condition not specifically approved by the FDA is not a "risk" under lack of informed consent doctrine. Assuming that the medical literature indicates that the particular "off-label" use is in accordance with prevailing professional standards, the regulatory status of the drug is not "material" to the patient's decision (Beck, 1999). However, once a drug is developed for a specific group of individuals with a particular genotype or racial or ethnic background (Exner et al., 2001), physicians using the drug for other groups not specified by the manufacturers could subject physicians to the risk of liability. The risk of some form of medical liability might make "off-label" use uncommon for the products of pharmacogenomics or at least force professionals to treat the use of the drug in nonspecified groups as experimental.

Medical liability doctrines, of course, put the levers of control in the hands of individuals rather than regulators. Liability and regulation both have their limitations as means of optimizing the risks and benefits of innovation in disease management. A comparative institutionalist approach does not necessarily endorse adjudication associated with liability over regulations. Rather, the approach only points out the relevant advantages and disadvantages of each approach that ultimately will be tested empirically (Komesar, 1994). The one social advantage of medical liability over regulation at this point in time is that it keeps visible the social and historical challenges that pharmacogenomics must overcome.

Although science wants to speak of human variation in terms of genotype, the social reality is that those genotypes are often associated with socially and historically constructed ethnic and racial groups. Ironically, the overall implication of the Human Genome Project is that, from a genetic perspective, human beings are essentially the same; pharmacogenomics is interested in that small variation (Rothstein and Epps, 2001). For it is in the variation that the safety and risks of drugs can be determined in the era of pharmacogenomics. Given our social construction of race and ethnicity, it will be the social groups who are interested in genetic health that will most likely want to monitor the pharmacogenomics revolution. Or put another way, knowing which proteins react adversely with other molecules in our bodies does not yet have much social meaning. But knowing that the protein or adverse reaction is more common among members of a social, ethnic, or racial group has a great deal of social meaning. One might expect that judges will be aware of the risks of scientific misadventure as medical liability responds to pharmacogenomics.

V. CONCLUSION

The localized nature of liability doctrines should make one cautious in predicting the course of innovations in liability doctrines. One of the vectors of change in medical liability law will be the field of pharmacogenomics itself, as it forces judges to see risks to patients in new ways. Another factor already influencing courts to change is the manner in which drug manufacturers and health care have changed in response to market pressures of managed care and the gene rush. The first point at which these innovations in liability doctrine will be seen is probably in clinical research. While the courts' rhetoric will speak of the duties of physician-scientists, the underlying systems view of courts will increasingly have implications for manufacturers' responsibility for engaging in research to reduce the genetic risks of their products. In other words, the division of responsibility between manufacturers and providers for genetic risks will be determined by judges in response to population groups, rather than individuals, insisting that the liability system operate as a system of social control over the development and deployment of pharmacogenomics.

Caution does not mean we must shy away from addressing the ethical and legal quandaries that are just over the horizon, but we should be explicit about our assumptions as we move forward. My four basic assumptions about the relationship of liability doctrines to pharmacogenomics are as follows. First, the professional-patient relationship as viewed in liability, as opposed to contract doctrines (Schuck, 1994), is foundational to the ethical and social obligations of all participants in the institution of medicine, including those who produce, distribute, and prescribe the products of pharmacogenomics. Second, the liability of manufacturers of the products derived from pharmacogenomics will depend upon how courts, and perhaps legislatures, view the systemic relationship of professionals and pharmacogenomics producers. Third, the initial introduction of pharmacogenomics will blur the distinction between liability doctrines associated with "clinical treatment" and "research." Fourth, scientific descriptions of groups by their genotypes in the short run will not replace descriptions of individuals as members of distinct if not insular minorities or members of particular racial or ethnic groups.

It is apparent that pharmacogenomics creates a new way of assessing the risks and benefits of drugs of all kinds at a point in time when systemic approaches to risks are being introduced into the health care system. We must work to develop a theory that allows the professional-patient dyad to be reformulated in the evolving innovation process that pharmacogenomics is propelling. This challenge is one that, ultimately, is a shared social responsibility, because we are, all of us, potential patients.

REFERENCES

Beck, J.M., and E.D. Azari, "FDA, Off-Label Use, and Informed Consent," *Food Drug Law J.*, **53**, 71 (1998).

Bergman, H.R., "Case Comment: Moore v. Regents of the University of California," *Am. J .L. Med.* **18**, 127–144, 129 (1992).

Brown v. Superior Court, 751 P.2d 470 (Cal. 1988).

Canterbury v. Spence, 464 F.2d 772 (D.C.Cir. 1972).

Cassidy v. SmithKline Beecham, No. 99-10423 (C.P. Chester County, Pa., filed 1999).

Clark, R.S., "Crimes Against Humanity." In: G. Ginsburg and V.N. Kudriavtsev (eds), *The Nuremberg Trial and International Law*, pp. 177–199 (1990).

Cobbs v. Grant, 502 P.2d 1 (Cal. 1972).

Dolgin, J., "Personhood, Discrimination, and the New Genetics," *Brooklyn L. Rev.*, **66**, 755, 763–798 (2000).

Duttry v. Patterson, 771 A.2d 1255 (Pa. 2001).

Emanuel, E., et al., "What Makes Clinical Research Ethical?" *JAMA*, **283**, 2701–2711 (2000).

Exner, D.V., et al., "Lesser Response to Angiotensin-Converting-Enzyme Inhibitor Therapy in Black as Compared with White Patients with Left Ventricular Dysfunction," *New Engl. J. Med.*, **344**, 1351 (2001).

Final Report of the Advisory Committee on Human Radiation Experiments (1995).

Final Report of the Tuskegee Syphilis Ad Hoc Panel to the Department of Health, Education and Welfare (1973).

Gorner, P., "Parents Suing Over Patenting of Genetic Tests: They Say the Researchers They Assisted Are Trying to Profit from a Test for a Rare Disease," *Chicago Tribune*, p. C1 (Nov. 19, 2000).

Greenberg v. Miami Children's Hospital Research Institute, Inc., No. 00-CV-6779 (N.D. Ill., filed Oct. 30, 2000).

Grimes v. Kennedy Kreiger Institute, 782 A. 2d 807 (Md. 2001).

Haluska v. University of Saskatchewan, 53 D.L.R.2d 436, 444 (Can. 1965).

Harkness, J.M., "Nuremberg and the Issues of Wartime Experiments on US Prisoners," *JAMA*, **276**, 1672–1675 (1966).

Harkness, J.M. and E. Shuster, "The Significance of the Nuremberg Code," (Correspondence), *New Engl. J. Med.*, **338**, 995, 995–996 (1998).

Henderson, J. and A. Twerski, "Drug Designs *Are* Different," *Yale L. J.*, **111**, 151–181 (2001).

In re Cincinnati Radiation, 874 F. Supp. 796 (S.D. Ohio 1995).

Institute of Medicine, *Crossing the Quality Chasm: A New Health System for the 21st Century* (2001).

Johnson v. Kokemoor, 545 N.W.2d 495 (Wis. 1996).

Jones, J.H., *Bad Blood: The Tuskegee Syphilis Experiment*, New York: Simon & Schuster, Inc. (1993).

Katz, J., "Human Experimentation and Human Rights," *St. Louis U. L. J.*, **38**, 7–54 (1993).

Ketler, S.K., Note, "The Rebirth of Informed Consent; A Cultural Analysis of the Informed Consent Doctrine After *Schreiber v. Physicians Insurance Co. of Wisconsin*," *Nw. L. Rev.*, **95**, 1029–1056 (2001).

Kohn, L.T., et al., *To Err is Human: Building a Safer Health Care System*, Institute of Medicine, Washington, DC: National Academy Press (2000).

Komesar, N.K., *Imperfect Alternatives: Choosing Institutions in Law, Economics, and Public Policy*, Chicago: University of Chicago Press 1994.

Krimsky, S., "The Profit of Scientific Discovery and its Normative Implications," *Chi.-Kent L. Rev.*, **75**, 15–39, 36 (1999).

Miller, F.H., "Trusting Doctors: Tricky Business When it Comes to Clinical Research," *B. U. L. Rev.*, **81**, 423, 436–37 (2001).

Mink v. University of Chicago, 460 F.Supp 713 (N.D. Ill. 1978).

Moore v. Regents of the University of California, 271 Cal. Rptr. 146 (Cal. 1990).

Palmer, L., "Disease Management and Liability in the Human Genome Era," *Villanova L. Rev.*, **47**, 1–26 (2002).

Palmer, L.I., *Law Medicine and Social Justice* 34–36, Louisville, KY: Westminster/John Knox Press, 1989.

Palmer, L.I., "The Problem of Human Experimentation," *Md. L. Rev.*, **56**, 604, 604–618 (1997).

Perez v. Wyeth Laboratories, 734 A.2d 1245 (N.J. 1999).

Rothstein, M.A. and P.G. Epps. "Ethical and Legal Implications of Pharmacogenomics" *Nat. Genet.*, **2**, 228–231 (2001).

Schuck, P.H., "Rethinking Informed Consent," *Yale L. J.*, **103**, 899— (1994).

Twerski, A.D. and N.B. Cohen, "The Second Revolution in Informed Consent: Comparing Physicians to Each Other," *Nw. L. Rev.*, **94**, 1–49 (1999).

U.S. Patent Number 4,438,032 (issued March 20, 1984), available at
http://164.195.100.11/netacgi/nph-Parser?Sect1=PTO2&Sect2=HITOFF&u=/ netahtml/search- adv.htm&r=11&p=1&f=G&l=50&d=ft85&S1=Quan.INZZ.&OS= in/Quan&RS=IN/Quan

U.S. Patent No. 5,679,635 (issued Oct. 21, 1997) available at
http://164.195.100.11/netacgi/nph-Parser?Sect1=PTO1&Sect2=HITOFF&d= PALL&p=1&u=/netahtml/srchnum.htm&r=1&f=G&l=50&s1='5,679,635'.WKU.&O S=PN/5,679,635&RS=PN/5,679,63. See, http://www.nih.gov/sigs/bioethics/ genepatenting.html

U.S. Patent No. 5,679,635 (issued Oct. 21, 1997) available at
*http://164.195.100.11/netacgi/nph-Parser?Sect1=PTO1&Sect2=HITOFF&d=
PALL&p=1&u=/netahtml/srchnum.htm&r=1& f=G&l=50&s1='5,679,635'.WKU.
&OS=PN/5,679,635&RS=PN/5,679,63.*

Weitzner, K. and H.L. Hirsh, "Diethylstilbestrol—Medicolegal Chronology," *Med.
Trial Technique Q.,* **28**, 145–170 (1981).

THE CHALLENGES OF PHARMACOGENOMICS FOR PHARMACY EDUCATION, PRACTICE, AND REGULATION

DAVID B. BRUSHWOOD, R.PH., J.D.

I. INTRODUCTION

The widespread application of pharmacogenomic principles in standard drug therapy will dramatically increase the pace of change in the pharmacy profession, where rapid change already represents the norm. A profession that has, until relatively recently, dedicated itself foremost to ensuring the accuracy of pharmaceutical product distribution now finds itself responding to opportunities for patient-related clinical care that can add significantly to the product-related duties at the core of the profession. Uniform adoption of the Pharm.D. degree has elevated the expertise of pharmacists to a level that warrants public reliance on pharmacists for their competence in therapeutic decision making. The complexity of modern pharmaceuticals has led to physician collaboration with pharmacists as trusted colleagues who share with physicians the responsibility for initiation, monitoring, and modification of drug therapy. As experts in drug therapy management, pharmacists who embrace the practice of "pharmaceutical care" focus their efforts on the outcomes of drug therapy for specific patients and not simply on the structures or processes of medication use (Hepler and Strand, 1990). The patient-specificity of pharmacogenomics promises to foster the development of new

Pharmacogenomics: Social, Ethical, and Clinical Dimensions, Edited by Mark A. Rothstein.
ISBN 0-471-22769-2   Copyright © 2003 Wiley-Liss, Inc.

patient-oriented pharmaceutical care practices and thus to enhance the value of pharmacists in drug therapy management.

II. UNCERTAINTY IN DRUG THERAPY

Under the law, prescription pharmaceuticals have traditionally been considered "unavoidably unsafe" (Conk, 1990). Standard pharmaceutical products are not just risk-reducing, they are also risk-producing. Because of inherent risks, pharmaceutical products may fail to produce a desired therapeutic effect and/or they may cause an undesired adverse effect. Harm from drug use may occur even when the product is appropriately designed and manufactured; when it is accompanied by adequate warnings; and when it is prescribed and dispensed consistent with the professional standard of care. These adverse drug effects have generally been considered regrettable and perhaps tragic, but they have also been viewed as nonpreventable and legally have been deemed the fault of no one. The harm caused by such nonpreventable adverse drug effects has been viewed as the unfortunate but necessary cost of scientific uncertainty (Noah, 1990). The only way to avoid nonpreventable adverse effects would be to remove high-risk drugs from the market, denying all potential benefit for patients who are willing to accept the risk. Whether a drug is harmful or beneficial usually depends on the unique characteristics of human physiology that make patients react differently to drugs.

A. REDUCING THE UNCERTAINTY OF DRUG USE

Pharmacists have, with frustration, long observed the idiosyncratic effects of usually safe and effective drugs that fail to help some patients and actually harm other patients. Only with hindsight has it become possible to recognize outlier patients whose reaction to drugs differs from that of the majority of patients, and by that time the harm of adverse effects is often irreparable. Some of these peculiar, individualized patient reactions to drugs will soon be addressed by anticipated advances in pharmacogenomics. Pharmacists look forward to predicting patient variations in response to drugs based on individual genetic differences. Just as they have done through their development of practices based on pharmacokinetic modeling over the past two decades, pharmacists will use expertise in pharmacogenomics to forecast how a patient will respond to a particular drug before it is taken, eliminating the inefficient "trial and error" approach to determining which therapeutic option is best for a specific individual patient.

Consider, for example, the drug 6-mercaptopurine, one of the mainstays of treatment for acute lymphoblastic leukemia, a common type of childhood cancer (Rioux, 2000). Researchers have discovered that the reason some patients suffer serious adverse effects from this drug is that they have an unusually low level of thiopurine methyltransferase (TPMT), an enzyme that helps the body metabolize and eliminate the drug. A blood test has been developed to facilitate adjustments in dose, but a quicker and more efficient DNA test that identifies the gene for producing TMPT may soon replace the blood test. Approximately 10% of people have inherited a bad copy of the TPMT gene, which makes them sluggish metabolizers of 6-mercaptopurine and necessitates a reduction in dose. A small percentage of people have inherited two bad copies of the gene, making them exquisitely sensitive to the drug and requiring as much as a 95% reduction in dose. There are other drugs for which advances in knowledge have led to similarly greater understanding of idiosyncratic reactions.

B. Opportunities for Pharmacy

This new knowledge represents huge opportunities for the pharmacy profession. Pharmacists have developed drug therapy management practices that facilitate medical care through collaborative response to formalized consults issued by medical staff (Mitrany and Elder, 1999). These practices have been primarily located in hospital or medical clinic settings, but they have begun to expand into community pharmacies as well. Anticoagulation, diabetes, hyperlipidemia, and asthma clinics have proven successful when managed by pharmacists. They have improved therapeutic outcomes and they have reduced health care costs. Pharmacogenomics will expand the ability of pharmacists to provide drug therapy management services. It will further integrate pharmacists into the mainstream of service provision to physicians and patients.

Many of the applications of pharmacogenomic principles to pharmacy practice are based on individual variation in cytochrome P450 isoenzymes (Guengerich, 2000). One such enzyme, cytochrome p4502d6 (2d6) is necessary to metabolize many commonly used drugs. Learning whether a patient is a good or poor 2d6 metabolizer can enable a pharmacist to predict how the patient will respond to beta blockers, antidepressants, antipsychotics, codeine, and tamoxifen, as well as to several other drugs. A genetic test will soon be available to test for variations in the gene for 2d6.

One potential application of knowledge regarding variation in the gene for 2d6 is directly relevant to enhanced relief of suffering due to pain. It is possible that some patients for whom codeine has been ineffective in usual doses will be identified through this test. These patients may have been inaccurately labeled as "drug abusers" or as "drug seekers" because of their insistence on receiving higher doses of the drug to treat their pain. Pharmacists may soon identify these patients as having an inactive form of the 2d6 gene and thus recognize them as being unable to metabolize codeine into its desired metabolite, morphine. Within the important role of pharmacists as guardians of the nation's drug supply, this development, and others similar to it, could enable them to better prevent drug diversion while enhancing the quality of pain management.

An inactive form of the 2d6 gene also may lead to adverse drug effects, such as the toxicity caused by the overaccumulation of fluoxitine. This drug is metabolized by 2d6, and excessive levels of the drug caused by impaired metabolism may lead to hypertension or other preventable side effects. Pharmacists have accepted the responsibility of warning patients of side effects and of monitoring patients for side effects after drug therapy has begun. The ability to prevent side effects, however, has been limited to early recognition and immediate action on recognition. The success of this warning and recognition method is dependent on the patient's ability to understand and interpret signs of toxicity. It is not a consistently effective method, because patients often do not understand what to look for, or they fail to recognize an adverse effect in its early stages. Through the use of pharmacogenomic principles, a patient who is a poor 2d6 metabolizer can be identified by a pharmacist before the start of drug therapy and the possibility of side effects can be greatly reduced without depending on an unreliable system of patient education and self-monitoring.

III. PHARMACY EDUCATION

Pharmacy educators are poised to adapt to the changes in practice that will result from the widespread adoption of drug therapies based on pharmacogenomics. Pharmacy education has been in a constant state of evolution since the early twentieth century, when the education of pharmacists consisted almost entirely of basic chemistry applied to those compounds thought to be useful as medications. At that time, many academic pharmacy programs were housed in departments of chemistry. Schools and colleges of pharmacy now are independent academic units, many of which are fully integrated into major academic health science centers.

The change in the focus of pharmacy practice from product to patient has led to an emphasis by pharmacy educators on drug therapy and not just on drugs. Pharmacy students learn how the body acts on drugs and how drugs act on the body. They are taught how to apply what they know about drugs to the care of patients. Continuing education of practicing pharmacists similarly stresses the application of drug knowledge to therapy for patients. Pharmacy students and practicing pharmacists learn to participate as clinicians in collaboration with other members of the health care team.

The initial education of pharmacists requires four years of study within the professional program. At least two years of college are required to enter the professional program. Once licensed, pharmacists are required to continue their education through conferences and other activities that reflect the evolving content of academic programs. There are five academic disciplines that comprise the standard pharmacy curriculum: medicinal chemistry, pharmacology, pharmaceutics, clinical pharmacy, and pharmacy administration. The advent of pharmacogenomics-based drug therapy will lead to significant changes in the curriculum offered by the faculties within each discipline. The requirements for continuing education will likely include specific learning objectives in pharmacogenomics, just as they have in the past for emerging focus areas such as HIV/AIDS, medication errors, risk management, and palliative care/pain management.

Learning objectives within course work offered by medicinal chemists have already begun to reflect an emphasis on the structure and function of nucleotides. Pharmacy students are taught that genetic mutations are responsible for many inherited diseases and that most genetic abnormalities are discernible only through generalized analysis of DNA sequencing. Students learn that their knowledge of single-nucleotide polymorphisms (SNPs) may lead to identification of a gene responsible for a specific trait. SNPs may also affect a drug's mechanism of action. This basic science serves as the foundation for patient care in pharmacy practices that incorporate the principles of pharmacogenomics.

Pharmacologists teach their pharmacy students how drug efficacy and toxicity can be better understood through pharmacogenomics. Toxicity issues usually develop because of a SNP modifying amino acids so that an enzyme does not metabolize a drug and the drug accumulates in toxic amounts. Dose response is at the core of pharmacology instruction regarding efficacy. Variations in enzyme activity can produce different patient responses to the same drug dose. An assessment of a patient's genetic profile is an invaluable tool for a pharmacist in screening for potential drug therapy problems due to overprescribing or underprescribing. Pharmacists can improve outcomes from drug therapy by identifying patients who are not likely to respond well to a prescribed drug.

The competencies taught by pharmaceutics faculty enable pharmacy students to predict how a drug will be absorbed into a patient's body, then be distributed through the body, and eventually eliminated from the body. Pharmacokinetics is the basis for predicting a human response to a drug. Pharmacogenomics is important to pharmaceutics because some enzymes produce transformations in the active drug molecule. Drugs metabolized in the liver may be excreted from the body at a higher rate because of enzyme activity. Enzymatic changes that increase the solubility of a drug in water may increase elimination through the kidney. Drugs may be converted to a new molecular form that is either more efficacious or is of increased toxicity. Principles of pharmaceutics also teach how to deliver a drug molecule to a site within the body where the drug will exert a positive therapeutic effect.

In courses offered by clinical pharmacy faculty, students are taught the diagnostic criteria for the application of basic science principles. Because

DNA is constant throughout a patient's lifetime, the clinician can consider genetic risk factors, along with environmental factors and individual factors (such as general health and nutrition), in developing a recommendation for a patient's drug therapy. Clinical pharmacists can incorporate genetic information into the evidence-based and consensus-driven clinical care algorithms they develop for patient care. As they observe the effects of drugs in a population of users, clinical pharmacists can record and report patient outcomes so that scientists will know how closely the predicted effect of a new drug matches the actual effect when used by large numbers of patients.

The purpose of pharmacy administration course work is to teach pharmacy students how to design and manage medication use systems that produce optimal results for patients. In pharmacy administration courses, students learn how to conduct medication use evaluations that measure patient outcomes. They are taught how to communicate effectively with patients and with other health care providers. Pharmacy administration courses also teach ethical and legal responsibilities to monitor drug therapy and to protect patients from problems with drug therapy.

IV. Controlling Drug Risks

The protection that patients are afforded from adverse drug effects does not begin with pharmacist monitoring. In fact, pharmacists are the final professional risk evaluators in a long chain of careful decisions about risk that precede dispensing of a medication to a patient. The highly paternalistic drug regulatory process constructs formidable barriers to prevent patients from making foolish choices in the use of medications. A molecule may be developed and otherwise available for use as a pharmaceutical product, but unless that product has been scientifically shown to be safe and effective for the population of users, and unless a highly trained medical professional has determined that the drug will be appropriate for a specific individual within the population, the opportunity to use the molecule will be denied. New drugs must be proven safe and effective before being placed into interstate commerce, and most newly approved drugs are limited to prescription-only status.

The cautious approach to drug approval, as overseen by the federal Food and Drug Administration (FDA), has been described in detail elsewhere (Findlay, 1999). The risks and benefits of a new drug submitted for FDA approval are rigorously studied through randomized, controlled clinical trials. Despite the enormous effort put forth to study new drugs, and the huge financial investment required to support clinical trials, many questions are often left unanswered by the relatively small-scale preapproval studies that may fail to detect, or fully describe, low-incidence adverse effects. These adverse effects may become evident within the months or years immediately following drug approval, leading to a reassessment of the initial decision to approve the drug. Pharmacists play a significant role in postmarketing surveillance activities (Noah and Brushwood, 2000). They are able to discern adverse drug effects that were not identified in clinical trials of a newly approved drug. They can also discover an increased incidence of previously identified adverse effects that were initially thought to be of acceptably low incidence. In addition, pharmacists can detect the occurrence of adverse effects that initially were considered preventable based on clinical trials with a small number of subjects but in fact were not consistently prevented after release of the drug for widespread use in clinical practice.

V. PHARMACY RESPONSES TO SOCIAL MANDATES

The morbidity and mortality that results from drug use may be due either to direct toxicity or to failure of therapy in a patient for whom the drug has been prescribed. Adverse patient outcomes can also result from the failure to use a drug that should have been prescribed for a patient. Empirical data suggest that medication misuse may account for as much as 25% of hospital admissions (Nelson and Talbert, 1996). The annual direct cost of treating drug-related morbidity and mortality has been estimated to be $177.4 billion (Ernst and Grizzle, 2001). This figure does not include the cost of lost productivity, or the value of the unrelieved suffering. A significant percentage of drug-related morbidity and mortality is preventable. In one study, for example, it was discovered that 76% of drug-related hospital readmissions could have been prevented by improved drug therapy monitoring (Bero et al., 1991). Because the concept of preventability will soon change with adoption of pharmacogenomics in drug therapy, the costs of drug-related morbidity and mortality will likely be lowered significantly.

Pharmacists are well positioned in the health care community to control preventable drug-related morbidity and mortality. The opportunities for pharmacists' involvement with pharmacogenomic applications in drug therapy correspond directly with the evolution of the practice of pharmacy in America. Initially serving society as extemporaneous compounders of medications, pharmacists have, over many decades, accepted additional responsibilities for the accurate distribution of manufactured dosage forms, the reduction of drug costs through the substitution of equivalent but less expensive products, the monitoring of medication use, and the assurance of optimal therapeutic outcomes for patients. Each new role adoption in the history of pharmacy has occurred in response to a specific social mandate. The acceptance of each new role has added to pharmacists' responsibilities without replacing existing responsibilities. Each added social mandate creates an opportunity for participation by contemporary pharmacists in drug therapies based on the principles of pharmacogenomics. Traditional and emerging responsibilities remain at the core of the profession, but pharmacogenomics applications in drug therapy will create new opportunities for public service by pharmacists.

A. THE ACCESS MANDATE

Through history pharmacists have served the important role of preparing raw chemicals into dosage forms that can be ingested, inserted, or applied to the human body for medicinal purposes. Without extemporaneous compounding by pharmacists, patients of the past would have been denied access to available but otherwise unusable drugs. On a smaller scale, pharmacists continue to provide valuable compounding services to patients, although pharmaceutical manufacturers have largely taken over responsibility for drug product formulation. Rarely do pharmacists today roll pills or press tablets or encapsulate powders for use by large numbers of patients, but they do occupy niche markets that large manufacturers spurn.

Compounding practices in the community pharmacy today focus primarily on the customization of dosage forms to meet special patient needs. For example, a pharmacist may crush tablets, or empty the contents of a capsule, and suspend the drug in liquid for a patient who cannot swallow and whose therapy requires the use of a drug that is commercially unavailable in liquid form. Or perhaps the pharmacist must create a low-strength pediatric dosage form of a drug that is commercially available only in adult strengths. Physicians rely on pharmacists to help meet the unique needs of those special patients for whom the commercially available product lines provide no suitable therapy. In a very real sense the patients who are "orphaned" by the industry are "adopted" by compounding pharmacists.

Institutional pharmacies have far more ambitious compounding practices. Intravenous and intrathecal solutions are routinely prepared in hospital pharmacies pursuant to orders by physicians for individual patients. Although the types of drugs used in compounding may remain fairly constant from one day to the next, the combinations and doses used will vary greatly. Drugs do not have doses; patients have doses. For drug therapy to be successful it is necessary to customize drug dosage forms to meet individual patient needs, and it is simply impossible for pharmaceutical manufacturers to anticipate the needs of every patient. In institutional pharmacies the theory of compounding can be technically complex and the technology used can be very advanced. This is particularly true of nuclear pharmacy and other specialty compounding pharmacy practices. Institutional pharmacy compounding facilities have much the same general appearance as pharmaceutical manufacturer production facilities, except for their smaller scale.

Pharmacy compounding is specifically allowed under the federal Food, Drug and Cosmetic Act (FDCA) (21 U.S.C. § 353A). Pharmacists who compound a drug product for an identified individual patient based on the unsolicited receipt of a valid physician's order are not subject to misbranding, adulteration, or new drug approval provisions that could otherwise apply. There are limits on the authority of pharmacists to compound under the FDCA. These limits restrict the types of drugs that may be compounded and the bulk substances that may be used in compounding. Within these limits, the FDCA enables small-scale, individualized patient care by pharmacists and physicians. The FDCA restricts large-scale manufacturing businesses that masquerade as pharmacies. Exemptions from misbranding, adulteration, and new drug approval provisions do not apply to licensed pharmacies that make large amounts of drugs for use by people who they do not know and may never be able to identify.

Because many pharmacogenomics therapies will be intended for use in small groups of patients, it is likely that they will be unavailable commercially. Through their compounding practices pharmacists will be able to offer to patients those pharmacogenomic drugs that are not commercially available but are capable of being compounded. Particularly in hospitals, where the infrastructure of equipment and computer technology is readily available, compounding practices will be vastly expanded with the advent of drug therapies based on pharmacogenomics principles. Compounding of pharmacogenomics-based drug therapies will be technically complex, but education will prepare pharmacists for this complex task, and technology will be made available through institutional resources.

The oversight of traditional compounding activities by pharmacy regulators has been a source of controversy within the pharmacy profession. Compounding pharmacists have insisted that it is their responsibility to meet the needs of patients who otherwise would do without those necessary medications that are unavailable from large-scale pharmaceutical manufacturers. Pharmacists have, at times, viewed pharmacy regulators as overly restrictive and intolerant of practices that result in the creation of new drug products without the scientific testing that pharmaceutical manufacturers use before widespread distribution of a new drug. It is simply impossible for a pharmacist, presented with a prescription requiring immediate compounding for a patient, to study the safety and efficacy of the drug. Not only have pharmacists and pharmacy regulators had disagreements over the appropriate limits on the regulation of extemporaneous compounding, state and federal pharmacy regulators have disagreed as well.

The key problem in the regulation of pharmacy compounding is the lack of uniform standards for compounding. Some extemporaneous compounding is necessary for the promotion of good therapeutic outcomes, whereas other compounding unnecessarily places the patient at risk of harm without any realistic expectation of success in therapy. The widespread application of pharmacogenomics principles in drug therapy has the potential to exacerbate this existing problem. The pharmacy profession will have to work with regulators to determine which pharmacogenomics products are appropriate for extemporaneous compounding and what ingredients and processes must be used in their compounding. Standards must also be developed for monitoring clinical outcomes from the use of compounded pharmacogenomic drugs, so that problems in the use of these drugs will become apparent when they occur.

B. THE ACCURACY MANDATE

Pharmacists pride themselves on their ability to get the right drug, in the right strength, to the right patient, with the right directions, at the right time. No pharmacist wants to make a mistake. The core responsibility in pharmacy is technical accuracy in order processing. Pharmacists suffer loss of self-esteem, as well as professional and social censure, when an error occurs. Yet, despite the many incentives to avoid error in pharmacy, the occurrence of pharmacy error appears to be on the increase (U.S. General Accounting Office, 2000). Because there is no comprehensive database of pharmacy errors, it is difficult to know whether the perceived increase in error reflects an increase in prescription volume alone, or whether the rate of error is actually increasing along with the increase in prescription volume. It is quite clear that pharmacy errors are a problem warranting significant attention from the profession and its regulators. Any tendency within the profession to move beyond technical accuracy in order processing as a core product-oriented responsibility, and to move instead into more challenging patient-oriented areas of practice is misplaced. Technical accuracy in order processing continues to be a necessary ingredient of success in pharmacy, although it may no longer be sufficient.

Studies of pharmacy error disclose that dispensing the wrong drug is the most frequent error made by pharmacists (Brushwood and Maultsby, 1999). Wrong-strength and wrong-directions errors also occur with significant frequency. These are overt errors, and they are intellectually straightforward. They are obvious and fundamental variations from a generally unambiguous physician's order, about which little judgment need be exercised by a pharmacist. The apparent simplicity of this problem belies the difficulty of the solution. Complex pharmacy systems contain many variables, all of which may coincide to "set pharmacists up" for failure. Problems related to practice sites, such as workload, distractions, and poor facility design contribute to the challenges pharmacists face with problems related to the product, such as look-alike and sound-alike drug names, as well as virtually identical packaging of very different drugs. The challenge for pharmacy is to cope with system complexity. Pharmacogenomics will increase complexity by introducing new and different drugs. Whereas a specific drug today may be available in only a single dosage form and a very few strengths, with standard directions applicable to most patients, drugs derived through pharmacogenomics principles will likely have numerous variations in dosage form, strength, and directions. Each new variation presents a new opportunity for pharmacy error.

An emerging consensus in pharmacy has changed the focus in error prevention from personal failures to system-related problems (Grasha, 2000). Even the most competent and caring pharmacist will fail within a pharmacy practice system that is not conducive to success. It is certainly necessary for pharmacists to be competent and caring, and it is well recognized that no pharmacy practice system can be designed well enough to produce positive outcomes by incompetent or uncaring pharmacists. However, effective pharmacy practice systems can be implemented to detect and absorb errors that even the best and most conscientious pharmacist will inevitably make.

The malpractice standard for pharmacy has begun to reflect a requirement that pharmacy managers implement sufficient institutional controls over the practice of pharmacy at all locations where pharmacy is practiced (*Harco Drugs, Inc. v. Holloway*, 1995). Before this development in the law, pharmacy corporations had been held vicariously liable for the failings of their employees, but this new standard reflects primary responsibility of the pharmacy owner to control the quality of pharmacy at the practice site. At a minimum, pharmacy owners are required to design systems that are anticipated to be effective and to document threats to quality for subsequent review and system redesign. The fact of error is less significant than is the lack of systematic effort to prevent error.

Pharmacy regulators have struggled to develop an appropriate response to the public health threat presented by the apparent increase in pharmacy order processing error. The advent of pharmacogenomics, and the attending increased complexity that will likely produce increased opportunity for pharmacy error, should serve as a strong incentive to work out a firm plan for regulation to prevent order processing error. The traditional approach of pharmacy regulators has been either to ignore error or to assess punishment (fines or license suspensions) for the commission of an error. But punishment does nothing to prevent the error that has already occurred, and it likewise is ineffective as a deterrent to inadvertent error. The threat of punishment cannot prevent one from making an error that one is already desperately trying to avoid and which one makes nonvolitionally. Public health protection is most effective when it prevents errors, rather than when it reacts to them.

As an alternative to punishment, some pharmacy regulatory agencies have begun to establish a requirement for pharmacies to conduct continuous quality improvement (CQI) programs (Nau and Brushwood, 1998). The intent of this requirement is to anticipate potential problems with pharmacy practice systems, and to correct those problems before a pharmacist's error results in harm to a patient. It is proactive regulation as opposed to reactive regulation. Inherent in the CQI requirement is recognition that there is no one-size-fits-all "right" way to practice pharmacy. Each pharmacy must discover what system works well for its patients and its personnel. There is no pretense that all errors can be prevented through mandatory CQI, but complaints of error can be evaluated more reliably by regulators. Errors occurring in pharmacies that have adopted and appropriately utilized CQI programs might appropriately be forgiven as unpreventable, though tragic. However, similar errors that occur in pharmacies that have failed to adopt or use CQI programs can be seen as preventable. Administrative discipline may be appropriate for those pharmacies that fail to prevent errors through the adoption of a CQI program.

C. THE EFFICIENCY MANDATE

The growing expense of pharmaceutical products, both in actual dollars and as a percentage of health care costs, has led to concerted efforts to develop more efficient uses of pharmaceuticals and to reduce their cost without adversely affecting the quality of patient care. As expensive as they may be, pharmaceutical products are generally considered to be a cost-effective means for reducing the utilization of more expensive health care procedures. Yet, despite the general cost-effectiveness of pharmaceuticals, there is widespread emphasis on developing programs to ensure that the least costly alternative is chosen from among several equivalently safe and effective pharmaceutical options. Pharmacists have played a significant role in oversight of the process through which this choice is made.

Generic substitution is now a widely recognized role for community pharmacists, although at its inception it was controversial (Keating, 1998). After a period of time permitted for exclusive marketing under the patent rights of a new drug's developer, other manufacturers are permitted to market generic versions of the innovator product by providing assurance that their product is essentially identical to the innovator product. The generic product is not subjected to the extensive and costly clinical trials required of the innovator product. The assumption is made that if the innovator product is safe and effective, and if the generic product is essentially identical to the innovator product, then the generic product is safe and effective. This assumption is subject to theoretical criticism, but in practice it has withstood the test of time. Physicians are permitted by state laws to insist that a specific product be dispensed, but in the absence of such an instruction the pharmacist may select a less expensive, equivalent product and dispense it in place of the prescribed product. Some insurance plans will pay only for a generic product, which creates a dilemma if the physician insists on the innovator product. The law may require dispensing of the innovator product, but if the patient has no money to pay for it, then the pharmacist is placed in a quandary. It is the pharmacist's responsibility to find a way to intervene on the patient's behalf to make available necessary and affordable medications, within the framework of applicable laws.

In hospital settings, the pharmacy department generally implements a therapeutic interchange program as adopted by the Pharmacy and

Therapeutics Committee (Schachtner et al., 2002). This committee produces an evidence-based, consensus-driven guide to drug therapy known as the formulary system. Through therapeutic interchange, a pharmacist dispenses a formulary drug, when a nonformulary drug has been ordered for a patient. Unlike generic substitution, which involves the dispensing of a different brand of the same drug, therapeutic interchange results in the dispensing of an entirely different drug that is within the same class of drugs as the ordered drug. Only when an order for a more expensive medication can be firmly justified will it be dispensed in place of the formulary drug.

The pressures for efficiency in the use of pharmaceuticals can be expected to increase with the advent of pharmacogenomics. These will be expensive drugs. Those who pay for them will want to be certain that there is value in them. Pharmacists will be asked to oversee the use of expensive drugs produced through the application of pharmacogenomics, and pharmacy regulators will ensure that this oversight is done responsibly. Simply because a drug is expensive does not mean that its use should be denied. But it can be expected that any order for an expensive drug will be closely scrutinized, and the order will be filled only if the characteristics of the patient justify use of the drug. Just as they have done with generic substitution and therapeutic interchange of traditional drug products, pharmacists will interface between the prescribers of pharmacogenomics-based drugs and the third-party payers. It will be the pharmacist's responsibility to determine that expensive drug therapies based on pharmacogenomics are used only when sound science warrants their use.

Decisions about the availability of expensive drug therapies are best made when data about therapeutic outcomes can be readily accessed. To make good decisions about the appropriateness of using expensive drug therapies, pharmacists need to know the outcomes of a therapy for patients in other institutions and in their own institutions. Clinical practice guidelines and decision assistance algorithms can further facilitate decisions about the appropriate use of expensive drug therapies. It would be wasteful of resources to use an expensive drug with no realistic expectation of benefit for the patient. It would be heartless and unethical to deny appropriate drug therapy on the basis of cost alone. Pharmacy regulators have adopted workable standards for generic substitution and therapeutic interchange. They will have to adopt similar standards for decision making about the availability of expensive drug therapy based on the principles of pharmacogenomics. Such regulations will have to be similarly based on valid interpretation of available scientific data, and they must recognize the necessary trade-offs in any decision that balances cost and quality.

D. THE SAFETY MANDATE

As recently as two decades ago, a pharmacist who accurately dispensed a medication pursuant to a physician's prescription could not be held legally liable for errors in the prescription. Technical accuracy in order processing was a virtual insurance policy against pharmacist liability, no matter how obviously the physician may have erred in the prescription and no matter how easily the pharmacist may have been able to rectify the error. It was the physician alone who bore responsibility for correctness in prescription orders, and the physician's responsibility superseded the pharmacist's responsibility.

Legal responsibilities of pharmacists have changed dramatically in recent years. A lengthy body of case law now recognizes a firm responsibility of pharmacists to detect and rectify obvious errors in a physician's prescription (Hornish, 2000). When a pharmacist observes that a physician has prescribed a medication that may lead to a significant drug-drug interaction, an overdose, or an allergic reaction, the pharmacist has a legal duty to contact the physician. The attention paid to medical error in the widely publicized Institute of Medicine Report entitled "To Err is Human" (Kohn et al., 2000) has emphasized the importance of this prescription-screening function. Pharmacists cannot guarantee that all prescribing has been done without error, but they are responsible for ensuring that an obvious mistake by a physician is brought to the physician's attention. The pharmacist functions as a "safety net" in drug therapy, ensuring that patients receive what their physician intends and not what the physician has ordered in error. This role is not adversarial to the physician, because the goal is to supplement, rather than supplant, the responsibility of the physician.

The case of *Happel v. Wal-Mart Stores* (2002 Ill. Lexis 296), decided by the Supreme Court of Illinois in 2002, serves as an example of how courts have begun to expand the standard of practice for pharmacists. The plaintiff alleged that she had been dispensed the drug ketorolac, despite knowledge by the dispensing pharmacist that she was allergic to this type of drug. The prescribing physician indicated that he had not known of the contraindication of this drug for patients with allergies such as the plaintiff's. The trial court granted the defendant pharmacy's motion for summary judgment, concluding that a pharmacy has no duty to warn either the prescriber or the patient under such circumstances. In reversing this ruling, the appeals court noted that both the likelihood and the reasonable foreseeability of injury to the patient were great. The court concluded that any negative consequences of recognizing a duty to warn were far outweighed by the substantial reasons favoring such a duty.

Pharmacogenomics will further expand the responsibility of pharmacists to screen prescriptions, because physicians will have more opportunities for error in prescribing, just as pharmacists will have more opportunities for error in dispensing. Knowing whether the physician has written a prescription correctly for a patient will require that pharmacists evaluate patient-specific genetic information in addition to the other patient-related information that pharmacists currently have available for analysis. This may mean that in the absence of specific genetic information a pharmacist cannot make a decision about the appropriateness of a prescribed drug therapy, and that the pharmacist will have to acquire the relevant genetic information before dispensing.

Pharmacists also perform an important role as educators of patients (Fleischer, 1999). This is a safety role, because many drugs are safe if used as intended but unsafe if used differently from the way they are intended. Pharmacists routinely inform patients about the need to ingest medications at certain times of the day to ensure steady blood levels. They caution about drowsiness or other manageable side effects. Psycho-social issues are also important in patient education by pharmacists, because every patient develops a unique medication use behavior, and sometimes these behaviors can interfere with the effectiveness of drug therapy (Bloom, 1996).

In 2002, the Supreme Judicial Court of Massachusetts joined a growing number of state courts that have recognized a responsibility of pharmacists to warn patients of possible side effects from prescribed medications. In the case of *Cottam v. CVS Pharmacy* (764 N.E.2d 814), that court affirmed a jury verdict in favor of a patient who had not been warned of the risk of priapism by the pharmacist who dispensed trazodone to him. The facts of the case disclosed that the pharmacy provided a short list of warnings to the patient but priapism was not included on the list. The court ruled that when a patient can reasonably conclude that a list of side effects is a complete and comprehensive list, the pharmacy has undertaken a legal duty to provide complete warnings and information.

The patient education role of pharmacists will be critical when dispensing and monitoring drug therapies based on pharmacogenomics, because of the unfamiliarity of patients with these therapies. The clinical aspects of these drugs will require explanation, particularly because general information about drug therapy may be inapplicable to a patient's unique needs based on genetic information. Patients receive information about drug therapy from many sources, and it will be critical to ensure that patients do not generalize from information about other patients when those patients have significant genetic variations that make their experiences inapplicable to the patient's expected experience. Some patients may become concerned about the perceived "unnatural" aspects of the therapy, and pharmacists may find themselves challenged to dispel myths about drug therapies based on pharmacogenomics.

As the safety responsibilities of pharmacists evolve in the field of pharmacogenomics-based drug therapy, pharmacy regulators will struggle to evaluate the conduct of pharmacists with reference to the evolving standard of care. Although the Internet (for community pharmacists) and integrated hospital computer systems (for hospital pharmacists) promise to bring pharmacy into the mainstream of health care, the fact remains that communication barriers often interfere with the ability of pharmacists to interface effectively with physicians and patients. Pharmacy regulators will be asked to resolve complaints alleging that pharmacists ought to have detected a prescribing error, or ought to have educated a patient on drug therapy, but that the pharmacist did not do so. "Ought" implies "can". Regulators will have to decide how much information about pharmacogenomics-based drug therapy it is reasonable to expect pharmacists to know in their prescription screening and patient education roles. General information about pharmacogenomics will disseminate rapidly to the profession, but some outlier pharmacists will be slow adopters. Specific information about patients and their genetic profile may be less well disseminated through information systems that are readily accessible by pharmacists. Pharmacy regulators will have to be sure that the expectations they place on pharmacists are realistic; that they reflect the state of knowledge actually available to pharmacists, and that the regulatory requirements protect the public health without imposing impractical burdens on the profession.

E. THE QUALITY MANDATE

The increasing importance of outcome-oriented health care has created opportunities for pharmacists to collaborate with physicians in the monitoring of drug therapy to create a cyclical, rather than linear, medication use process (Hepler and Grainger-Rousseau, 1995). Using the principles of pharmacokinetics, pharmacists can predict the absorption, metabolism, and excretion of a drug in a patient's body, based on observed characteristics of the patient and relevant laboratory values (Johnson, 2000). Pharmacists can individualize drug therapy for patients based on the predicted safety and efficacy of a drug as determined by pharmacokinetic parameters. Trial and error in drug therapy has been replaced by a system that constantly produces measured outputs leading to improved inputs. The learning is organized rather than haphazard. The goal is to produce positive outcomes, not just avoid negative outcomes.

Collaborative drug therapy management has been authorized through legislation or regulation in a majority of states (McDonough, 2001). Although the specific approach to pharmacist-physician collaboration differs from state to state, the general pattern in every adopting state is to require a written agreement through which a physician shares responsibility for a patient's care with a pharmacist. The physician performs the diagnostic procedures and develops a general treatment plan, often pursuant to a protocol that has been designed through consensus within a group practice. The pharmacist follows the plan in the initiation, monitoring, modification, and discontinuation of drug therapy. The arrangement works particularly well when a physician decides to pursue a general course of therapy but relies on the pharmacist to make detailed decisions about drugs, doses, and length of therapy.

There is tremendous potential for pharmacist-physician collaboration in drug therapy based on pharmacogenomics principles. It is the detail of the decisions that will make pharmacogenomics-based therapy challenging. Just as physicians have begun to delegate to pharmacists those patients who require antibiotic or anticoagulation therapy, leaving pharmacists to make decisions about what medications to use and how to use them, physicians will also be able to rely on pharmacists to individualize pharmacogenomics-based therapy, once the decision to treat the patient has been made.

Role definition will be the primary responsibility of pharmacy regulators as the practice of collaborative drug therapy management increases with pharmacogenomics-based drugs. Pharmacists will want to avoid encroachment into the practice of medicine, and physicians will want to guard their professional turf. Patients have a right to expect that professional roles within collaborative practice are well defined and that each person responsible for their care is a person in whom they have confidence. Firm regulatory guidance can prevent misunderstandings of role and responsibility. The activities of pharmacists in pharmacogenomics-based drug therapy will be significant, but they will be circumscribed by clear boundaries established through regulation.

VI. CONCLUSION

Pharmacogenomics will be a significant advance in therapy, but it will not replace the traditional care that pharmacists have provided to patients. Although genetics is an important component of patient variability in drug therapy, factors such as organ status and nutrition can also affect patient response to drugs, and the competent pharmaceutical care patients receive will continue to require a comprehensive approach. Every advance in drug therapy brings with it solutions to old problems, along with an array of new problems that require creative solutions. Pharmacogenomics will be no exception to this general rule.

Pharmacists will be prominently involved with pharmacogenomics-based decisions, and they will, at times, fail to meet their responsibilities. The task of the regulatory agency is to define for the regulated profession what is expected of it, so that practitioners are not left to guess what they should do in their practice. Regulators also have the ability to empower practitioners to avail themselves of productive advances in care, such as pharmacogenomics. It is regulators who must ensure the competence of licensed practitioners by acting to provide opportunities for practitioners to learn how quantum advances in therapy, such as pharmacogenomics, can be applied productively to patient care. Perhaps most importantly, pharmacy regulators must educate themselves about pharmacogenomics, so that when the time comes they will be well positioned to provide leadership to the profession regarding the role pharmacists must play in the improvement of outcomes and reduction of costs from drug therapy based on pharmacogenomics.

REFERENCES

Bero, L.A., et al., "Characterization of Geriatric Drug-Related Hospital Readmissions," *Med. Care*, **29**, 989–1003 (1991).

Bloom, D.L., "Facing the Next Challenge of Pharmaceutical Care: Patient Noncompliance," *Med.-Interface*, **9**, 67–72 (1996).

Brushwood, D.B. and J.H. Maultsby, "Judging Pharmacy Practice: An Evaluation of Pharmacy Case Law and of Incident Reports from a Major Pharmacy Chain: Can Records of Past Problems Facilitate Improvement in the Future?," *Pharm. L. Annual*, **5**, 73 (1999).

Conk, G.W., "The True Test: Alternative Safer Designs for Drugs and Medical Devices in a Patent-Constrained Market," *UCLA L. Rev.*, **49**, 737 (1990).

Cottam v. CVS Pharmacy, 764 N.E.2d 814 (Mass. 2002).

Ernst, F.R. and A.J. Grizzle, "Drug-Related Morbidity and Mortality: Updating the Cost-of-Illness Model," *J. Am. Pharmaceut. Assoc.*, **41**, 192–199 (2001).

Findlay, R.J., "Originator Drug Development," *Food Drug L. J.*, **54**, 227–232 (1999).

Fleischer, L., "From Pill-Counting to Patient Care: Pharmacists' Standard of Care in Negligence," *Fordham L. Rev.*, **68**, 165–207 (1999).

Grasha, A.F., "Into the Abyss: Seven Principles for Identifying the Causes of and Preventing Human Error in Complex Systems," *Am. J. Health-System Pharm.*, **57**, 554–564 (2000).

Guengerich, F.P., "Pharmacogenomics of Cytochrome P450 and Other Enzymes Involved in Biotransformation of Xenobiotics," *Drug Devel. Res.*, **49**, 4–16 (2000).

Happel v. Wal-Mart Stores, 766 N.E.2d 1118 (Ill. 2002).

Harco Drugs, Inc. v. Holloway, 669 So. 2d 878 (Ala. 1995).

Hepler, C.D. and L.M. Strand, "Opportunities and Responsibilities in Pharmaceutical Care," *Am. J. Hosp. Pharm.*, **47**, 533–543 (1990).

Hepler, C.D. and T.J. Grainger-Rousseau, "Pharmaceutical Care Versus Traditional Drug Treatment: There is a Difference," *Drugs*, **49**, 1–10 (1995).

Hornish, M.L., "Just What the Doctor Ordered—Or Was It?," *Missouri L. Rev.*, **65**, 1075–1100 (2000).

Johnson, J.A., "Predictability of the Effects of Race or Ethnicity on Pharmacokinetics of Drugs," *Intl. J. Clin. Pharmacol. Therapeut.*, **38**, 53–60 (2000).

Keating, E.J., "Maximizing Generic Substitution in Managed Care," *J. Managed Care Pharm.*, **4**, 557–563 (1998).

Kohn, L.T., et al., (Eds), *To Err is Human: Building a Safer Health System*, National Academy Press, Washington, DC (2000).

McDonough, R.P., "Developing Collaborative Working Relationships Between Pharmacists and Physicians," *J. Am. Pharmaceut. Assoc.*, **41**, 682–692 (2001).

Mitrany, D. and R. Elder, "Collaborative Pharmacy Practice: Idea Whose Time Has Come," *J. Managed Care Pharm.*, **5**, 487 (1999).

Nau, D.A. and D.B. Brushwood, "State Pharmacy Regulators' Opinions on Regulating Pharmaceutical Care Outcomes," *Ann. Pharmacotherapy*, **32**, 642–647 (1998).

Nelson, K.M. and R.L. Talbert, "Drug-Related Hospital Admissions," *Pharmacotherapy*, **16**, 701–707 (1996).

Noah, L., "Pigeonholing Illness: Medical Diagnosis as a Legal Construct," *Hastings L. J.*, **50**, 241–306 (1999).

Noah, B.A. and D.B. Brushwood, "Adverse Drug Reactions in Elderly Patients: Alternative Approaches to Postmarket Surveillance," *J. Hlth. Law*, **33**, 383–454 (2000).

Rioux, P., "Clinical Trials in Pharmacogenetics and Pharmacogenomics: Methods and Applications," *Am. J. Health-System Pharm.*, **57**, 887–898 (2000).

Schachtner, J.M., et al., "Prevalence and Cost Savings of Therapeutic Interchange Among U. S. Hospitals," *Am. J. Health-System Pharm.*, **59**, 529–533 (2002).

United States General Accounting Office. (2000). "Adverse Drug Events: The Magnitude of Health Risk is Uncertain Because of Limited Incidence Data." GAO/HEHS-00-21.

THE SOCIAL DIMENSION

Economic Implications of Pharmacogenomics

C.E. Reeder, Ph.D. and W. Michael Dickson, Ph.D.

I. Introduction

Elucidation of the human genome and identification of mutations that lead to disease have opened a new pharmacologic era in medicine. As an emerging research field, pharmacogenomics offers tremendous potential benefits in the areas of drug development and therapeutics. Improved clinical outcomes, decreased drug development costs, improved efficacy, reduced adverse drug events, and improved diagnostic testing, among others, could flow from the application of genomic technology to the study of drug response (Kurth, 2000). Although the potential advantages of pharmacogenomics-based drug development has received considerable discussion, little attention has been devoted to the economic implications of pharmacogenomics.

The purpose of this chapter is to review and discuss the potential economic effects of pharmacogenomics at the individual, health care payer, and producer levels. This chapter uses a broad perspective in discussing the value of pharmacogenomics. To appropriately assess the value of pharmacogenomics, one must include not only the clinical but also the economic and humanistic costs and consequences of treatment. Although no product developed by pharmacogenomics has yet come to market, pharmacogenomics has the potential to change the course and outcomes of many highly

Pharmacogenomics: Social, Ethical, and Clinical Dimensions, Edited by Mark A. Rothstein.
ISBN 0-471-22769-2   Copyright © 2003 Wiley-Liss, Inc.

morbid or even fatal conditions. Such conditions generate enormous economic costs not only in terms of actual health care expenditures, but also in human suffering, lost productivity, and diminished quality of life.

II. EFFECTS ON DRUG DISCOVERY AND DEVELOPMENT

The current costs and time to develop a new drug are substantial. It has been estimated to take 12 to 15 years and $250 to $500 million for a new chemical entity to reach the market (DiMasi et al., 1991; DiMasi, 1994). Pharmacogenomics offers the opportunity to reduce these figures by improving the drug discovery and clinical trial processes. Currently, failure rates during preclinical and clinical phases of drug development are very high (Sander, 1978). Typical statistics from the pharmaceutical industry suggest that only one of every 10,000 molecules synthesized becomes a marketed medicine. This high failure rate becomes even more abysmal when market success is measured in dollar sales. It has been estimated that only about one in 60,000 compounds synthesized by pharmaceutical companies can be regarded as "highly successful" (i.e., global sales in excess of $100 million) (Redwood, 1993). Moreover, only about 12% of the drugs that enter the human testing phases of drug development ever receive marketing approval. The remainder fail because of unacceptable toxicity, lower than expected therapeutic effects, or lack of market viability (when potential market size is too small to justify further drug development) (Hansen, 1979).

Pharmacogenomics can be used to predict drug efficacy, the probability of adverse drug events, as well as pharmacokinetic and pharmacodynamic effects (Emilien et al., 2000). Refinements in inclusion and exclusion criteria and improvements in response assessments based on genetic information should make clinical trials more efficient and effective. Understanding the genetic profiles of patients early in the clinical phases of drug development will enable researchers to conduct trials in subgroups of patients who are most likely to benefit from the drug with minimal if any side effects. By understanding the relationship between genotype and drug response, a pharmaceutical company should be able to design a clinical trial that maximizes both safety and efficacy endpoints and thereby obtain faster FDA approval to market the drug.

All of these benefits of pharmacogenomics translate into economic consequences. By targeting patient subgroups, the number of people needed to conduct clinical trials will decrease, thus reducing trial times and cost (Michelson and Joho, 2000). This reduction in sample size will be achieved without a loss of statistical power because only subjects who are likely to respond to the treatment will be included in the study. Moreover, pharmacogenomics should benefit the drug development process by improving the preclinical research environment. Predictive toxicology, drug metabolism studies, and other nonclinical measures of drug suitability should be enhanced by linking genomic data with other biomolecular information. The economic implications of these potential leaps in drug discovery and development are very significant. According to one estimate, pharmacogenomics could save up to 45% of clinical drug development costs (Aldridge, 2001). Clearly, a significant part of the savings is likely to accrue from the early detection of undesirable drug properties and the identification of subjects who will most likely benefit from the therapy.

Veenstra and colleagues suggested three mechanisms by which genetic variations can affect safety and efficacy: (1) drug targets (receptors), (2) drug transport mechanisms, and (3) drug-metabolizing enzymes (Veenstra et al., 2000). With an understanding of which patients (genotypes) are most likely to respond to an experimental treatment regimen, clinical trials can be designed to assess the efficacy of the drug with minimal losses due to adverse drug events. Clinical trials will require fewer subjects and less time to complete. This will mean earlier access to treatment for patients and decreased development time and cost for the pharmaceutical company. It has been estimated that at its peak, pharmacogenomics will save approximately $17 million or 21% in the drug discovery process. These savings will accrue by improving the quality of drug targets, by decreasing discovery time for new drug targets, and by eliminating poor drug targets earlier in the drug discovery process (Meyer and Thompson, 2001).

One of the economic benefits touted for pharmacogenomics-based drug development is the ability to identify promising compounds earlier in the drug development process. With better information about molecular targets it will be possible to remove from consideration those compounds that lack the necessary specificity for the desired site and to focus efforts on the remaining drug candidates. These less promising compounds will, most likely, never move beyond phase I of the clinical trial process, or the drug will be reevaluated for possible safety and efficacy in a smaller subset of the original target population. Because the most expensive part of drug testing occurs in phases II and III, there can be considerable economic gains from early termination of unpromising compounds and developing only those agents that are likely to succeed in the marketplace.

Pharmacogenomics also offers the potential to "rescue" previous pharmaceutical market failures. Drugs that were never marketed or were withdrawn from the market because of toxicity or side effects may be reevaluated for safety and efficacy. Genetic data on the pool of individuals who exhibited untoward effects of these drugs could be explored for a common genotype. If a strong association exists between a gene variant and the incidence of the adverse drug events, the drug could be reintroduced for patients with the appropriate genetic profile, thus "rescuing" the drug from market failure and offering another treatment alternative for patients. Recent examples of potential "blockbuster" drugs that have been recalled from the market for adverse drug events or toxicity include cerivastatin, cisapride, alosetron, and troglitazone (U.S. FDA). On the basis of estimates of the average research and development costs for pharmaceuticals, these product removals represent $1 to $1.5 billion in lost research investment. The ability to recover R&D expenses from these market failures by reintroducing these agents for a targeted patient group would provide short-term revenues for the industry. Drug products that have failed in late-stage clinical trials may also be reassessed for their effectiveness in particular subsets of patients.

III. EFFECTS ON THE MARKET

There are some concerns that revenue from the sale of pharmacogenomics-based drugs will be less because the market size for any new therapy will be smaller. In general, the objectives of pharmacogenomics are to reduce the attrition rate in drug development and improve the accuracy of drug candidate selection with a resulting decrease in preclinical development time. This is somewhat of a "two-edged" sword. Drugs developed for specific genotypes will not be used in as many individuals as drugs developed through the traditional drug development process. Genetic testing will segment the market for a particular drug into smaller, genetically homogeneous subsets of the population than would be treated with the drug absent the test.

As market size decreases, the total cost of drug development will be spread over a smaller number of users, which will in turn increase the average price of pharmacogenomics-based drugs. Given a smaller probable market size, large drug manufacturers may be reluctant to invest the requisite resources to bring a drug to market or may choose to focus only on the most common and profitable subgroups of the population. This will either increase the price of the drug for the subgroup or require special incentives, similar to existing orphan drug policy, to induce companies to produce these newer drugs with limited market potential. (This is somewhat of a "Catch-22"; pharmacogenomics-based drugs will be highly patient specific, very safe and effective. As a result, they will be used in a smaller number of patients, unit prices will increase, and industry revenues will likely decline.) Moreover, this diminished market size effect may motivate larger pharmaceutical companies to focus their R&D resources on the development of drugs that treat the more prevalent genotypes at the expense of less common genotypes (Rothstein and Epps, 2001).

However, this effect may be countered by an increase in the number of new drug targets that will be disclosed through genomics research. Mapping the human genome has increased our understanding of disease processes at the molecular level. According to Drews, this elucidation could result in 5,000 to 10,000 new drug targets (current drug therapy is based on fewer than 500 molecular targets) (Drews, 2000). Pharmaceutical companies may exploit pharmacogenomics by identifying ways to develop and market new therapies that increase or at least do not reduce market shares. Rather than exclude "nonresponders" (in the clinical trial) from labeled indications, companies may use pharmacogenomics to determine whether such patients may benefit from other doses of the drug or from use of the drug in combination with other agents (so-called "rational polytherapy"). Pharmacogenomics could lead to patients taking multiple medications, each of which has a very specific, safe, and efficacious indication.

In either case, the global impact on drug development will be a function of the total revenue effect of pharmacogenomics. If the increased specificity and smaller target markets of pharmacogenomics-based drugs can be balanced by an increase in the number of drug targets, the total revenue to the pharmaceutical industry may still be sufficient to induce continued research and development. The likely overall economic effect will be an increased use of higher-priced agents that must be balanced against potentially lower expenditures for other health care services.

Pharmacogenomics will likely reshape the structure of the pharmaceutical industry from one focused on drug development, manufacturing, and marketing to one focused on health care outcomes. That is, the pharmaceutical industry will integrate with other sectors so that it can deliver therapeutic and health outcomes and not just produce drugs that offer promise of improvement. Currently, the pharmaceutical industry pursues its objectives in a rather linear fashion, with research leading to drug development that in turn feeds the drug manufacturing process, which ultimately leads to the marketing of new drugs. Pharmacogenomics offers a new paradigm for the structure and process of the pharmaceutical industry in which the industry evolves from a pharmaceutical to a health care focus. This new structure couples the availability of population genotyping with bioinformatics in a way that will permit companies to adapt their research and development to meet future market needs (Richmond et al., 1999).

Although larger firms may avoid some smaller markets, the existence of "niche" therapeutic categories will present an opportunity for smaller, genetics-based biotechnology firms to enter the market and produce the drugs. In addition to meeting demand, this could foster the growth of smaller biotechnology firms that can operate at a scale that is profitable for the market size. In effect, the pharmacogenomics industry may be composed of two sectors: large companies that research, develop, and produce products for high-prevalence polymorphisms and a second "cottage" sector that serves the "orphan drug" market.

IV. EFFECTS ON PATIENT CARE

Pharmacogenomics offers significant promise for improvements in patient care. Genetic differences among individuals account for much of the variation in effectiveness and toxicity of drug therapies. If these genetic differences can be exploited through the principles of pharmacogenomics, the specificity of drug therapy will be enhanced, outcomes will be improved, and drug toxicity will be reduced. Improvements in patient outcomes through safer, more effective therapies also offer potentially significant economic advantages. Clearly, the most cost-effective therapy is the one that works the first time. Knowledge of the patient's genotype can improve outcomes through early intervention and treatment with the correct drug. Consider, for example, drug therapy in Alzheimer disease, where a positive response to some Alzheimer medications depends on the patient's Apo E subtype (Regaldo, 1999). Given the rather narrow therapeutic window for initiation of therapy in Alzheimer disease (about 18 months), prescribing the correct drug during this window is especially critical and there is no time for empirical therapy.

The costs of drug-related morbidity and mortality are substantial. Ernst and Grizzle modeled the cost of managing drug-related problems in a hypothetical cohort of ambulatory patients. Based on the model, the overall cost of drug-related morbidity and mortality exceeded $177 billion in 2000. Hospital admissions accounted for approximately 70% of these costs, whereas long-term care admissions accounted for about 18% (Ernst and Grizzle, 2001). To the extent that pharmacogenomics-based drug therapies can reduce this burden, the consequences to the health care economy could be quite favorable. Cost avoided by a substantial reduction in adverse drug events should be realized as a benefit of pharmacogenomics-based therapies.

Cutler and McClelland have demonstrated this logic in their recent assessment of the value of changes in medical technology (Cutler and McClennan, 2001). In analyzing the costs and benefits of technological advances for the treatment of heart attacks, low-birth-weight infants, depression, and cataracts, the benefits of the technological changes exceeded their costs. For a fifth condition, breast cancer, benefits and costs were about the same. The authors conclude that, in general, technological improvements in medical care are worth the associated increase in cost when improvements in medical care, survival, and quality of life are considered. It seems plausible to extend this same logic to pharmacogenomics-based drug therapies.

An additional cost that must be considered for pharmacogenomics-based drugs is testing cost. To optimize or "tailor" drug therapy to individuals, each person must be genotyped. Traditionally, lab test costs have been paid by the individual or some third party (e.g., insurers or governments). In the case of genotyping for the purpose of specific drug use, testing costs may be bundled with the drug as part of its market price. Most likely, pharmaceutical companies will market their drugs with genetic tests for response and safety. There will also be a derived demand for the services of genetic counselors to help patients and prescribers interpret and act on test results.

The cost associated with genetic counseling may well dwarf the direct expenditure on testing.

The most obvious testing cost is the direct cost of the genotyping, that is, the price of testing a population for the presence of specific alleles. Given the likely economies of scale that will be realized from population-based (or at least large segments of the population) testing, genetic tests should not be very expensive. According to one estimate, the costs of genetic tests are likely to range from about $10 for single mutations to $250 for multiple genetic screens (Flockhart, 2001).

A significant increase in health care expenditure will flow from the increased demand for genetic testing. Screening and diagnostic testing will precede the use of pharmacogenomics-based drugs and will create an additional economic burden on the health care system. When diagnostic and screening tests become available, it is reasonable to expect at least some individuals to demand the tests even if treatments are not yet available. Once pharmacogenomics-based drugs become available, an additional increase in testing cost will be from "derived demand." That is, the demand for testing will grow in concert with the demand for pharmacogenomics-based drugs. The economic impact of the costs of services associated with testing could be substantial. Additional physician or clinic visits will be needed in conjunction with individual testing. Also, genetic counseling will be required for many individuals once the genotype has been determined. The magnitude of the economic effects will be driven by how many tests a person will receive. For example, will there be individual testing for specific genetic variations, or will there be an "omnibus" test at birth, the results of which will be used over the lifetime of the individual? In the latter case, population-based testing could be a type of capital investment for which the benefits and savings will be long-term.

Availability of diagnostic and screening tests also raises questions about who should be tested, when they should be tested, and who should pay for the testing. The possibilities range from testing the entire population and all newborns (or embryos) with all known screens, to selective testing for specific patient groups or diseases, to testing only when symptoms are present. Each of these possibilities presents a different economic burden. Moreover, genetic testing will increase the population at risk for treatment to everyone in society. Because each person has a unique genetic shgnature, the demand for testing!will increase, as many people will desire to know their lifetime risk of developing a disease. This increased demand for testing will be independent of the availability of treatment for the disease. Educating patients and health care providers on appropriate test use will be necessary and will generate an additional cost stream.

Complicating this issue are the imperfections of testing. Most diseases are not caused by a mutation in a single specific gene, but rather by mutations in two or more genes in combination with environmental factors. This situation is confounded further by the concept of variable penetrance. High-penetrance mutations do not cause the disease directly but are strongly associated with development of the disease. Low-penetrance mutations have much lower associations with disease manifestation.

If genotyping is imperfect or the frequency of a particular variant is low, a significant number of false positive results will be generated. For example, if 70% of individuals with a particular genetic mutation actually get the disease, the remaining 30% who test positive but would not develop the disorder will be treated on the basis of positive test results. Similarly, if only 40% of individuals who have a specific mutation actually experience an adverse drug event, 60% of those tested would not receive the drug even though they would not experience the adverse drug event. The extent of false test results has important economic implications. If the rate of false positives is high, many people will be prescribed expensive, unnecessary medicines or, in the case of toxicity screens, will be denied a drug that could result in improved health outcomes. In the case of false negatives, there are costs and consequences associated with a delay in treating a person who should receive therapy. In either situation, the relative cost-effectiveness of the new therapy will be compromised by inadequate test specificity and sensitivity.

Currently, genetic testing for monogenic (e.g., cystic fibrosis) and high-penetrance (e.g., Huntington disease) diseases is directed to specific segments of the population such as individuals with a family history of the disease or individuals in subgroups especially prone to the disease (Richmond et al., 1999). Although genetic testing for high-penetrance diseases is being introduced in the United States, genetic testing for low-penetrance diseases is not yet performed on a wide scale. However, the technology to conduct large-scale screening for low-penetrance diseases is being developed.

In addressing the policy implications of technological change, Schwartz presented two possible scenarios for medicine that are particularly relevant to pharmacogenomics (Schwartz, 1994). One scenario considers the potential for advances in molecular and cell biology to produce cures for illnesses such as cancer, coronary artery disease, and cystic fibrosis. Success in the management of chronic, expensive to treat, highly morbid illnesses could reduce or moderate the growth in overall health care spending. On the other hand, cost increases in health care, fueled by technological innovations, could continue. Historically, rapid technological change in health care has been associated with increased health care spending. A long-term consequence of using pharmacogenomics-based drugs may be the conversion of many rapidly fatal diseases to chronic conditions that generate a stream of long-term utilization and cost. Some diseases may be cured while others have only marginal improvements in outcomes. In either case the individual lives longer and incurs additional medical expense to treat new and existing illnesses. The financial burdens on hospitals, skilled nursing facilities, home care, and others could be enormous.

This situation of escalating costs from improved survival is not new; our society has traditionally spent more for incremental (sometimes minor) improvements in care. From an economic perspective, the question becomes how much more and for how long. Every additional dollar spent on health care is another dollar from the Gross Domestic Product (GDP) that cannot be used for other purposes. This opportunity cost of technological improvement will be one of the primary economic and social issues that must be addressed. Pharmacogenomics-based drug therapies will likely provide a test case. Will these new and expensive therapies be rationed; if so, on what basis? Will our society continue to value improvements in health care above other goods and services such as education, the environment, and Social Security?

V. EFFECTS OF PHARMACOGENOMICS ON HEALTH CARE SYSTEM

In an economic sense, the value of health care interventions resides in their ability to improve physical and psychosocial well-being. Such improvements may translate into lower overall health care expenditure or improved productivity. For 2001, national expenditures for prescription drugs is estimated to be $124.4 billion, and the annual growth in expenditure is expected to continue in the range of 11% to 12% (HCFA, 2001). A recent study evaluated primary drivers for drug expenditure increases over the period 1999–2000. The authors reported that about 42% of the increase in drug expenditure was related to an increase in prescription volume, whereas approximately 36% was attributed to a shift in utilization to newer, higher-priced medicines and the remainder (22%) was explained by drug price increases (NIHCMR, 2001). On the basis of these figures, it is reasonable to expect that the introduction of pharmacogenomics-based drugs will contribute to a continued trend of rising total national expenditure for medicine. As the population continues to grow, prescription volume should increase, *ceterus paribus*. Moreover, as pharmacogenomics-based therapies become available, it is very likely that there will be a shift toward utilization of these newer, more effective, and more expensive treatment modalities.

In light of recent experience with biotechnology-derived products, it is reasonable to expect that pharmacogenomics-based drugs will be expensive relative to traditional modes of treatment (Richmond et al., 1999). If the price of these innovations is viewed out of context from the consumption of other health care goods and services, the response may well be to reduce or deny access to these products. In an era of escalating health care cost, "inputs" to the production of health care, such as pharmaceuticals, physician visits, laboratory/diagnostic services, and hospital care, are often viewed in isolation, as if one sector of the health care economy can be optimized independent of the others (often referred to as the silo mentality). When cost containment is the primary economic objective, a short-term, risk-averse decision framework emerges. The focus shifts from optimizing long-term population health outcomes to minimizing short-run medical expenditure for specific services. Such a short-term cost-minimization strategy may be myopic, because many of the advantages and cost savings of medical innovations are actually realized in the long term.

A recent study provided some evidence of the value of new medicines. Lichtenberg used data from the 1996 Medical Expenditure Panel Survey (MEPS) to evaluate the relationship between the age of drugs and mortality, morbidity, and total medical expenditure (Lichtenberg, 2001). The results suggest that, despite higher prices, newer medicines are associated with lower overall medical care expenditures. An average increase of $18 in drug expenditure was related to a $71 reduction in nondrug spending. Moreover, individuals who used newer drugs were significantly less likely to die or to experience work absenteeism than individuals who used older drugs. Apparently, newer, more effective medicines were efficient treatment substitutes for older medicines.

From an economic perspective, the health care system can be viewed as a "production function" in which many inputs are combined through some process (medical care) to produce certain outputs or "health outcomes" (Fig. 12.1). In an economic sense, many of the inputs to health care production, such as pharmacogenomics-based drugs, are substitutes for or complements to other inputs. That is, they may be used in place of or in addition to the current, standard mode of treatment. Treatment substitution may cause the unit costs to change (increase or decrease), while outcomes are likely to improve. In addition to treatment substitution effects, new technologies also affect utilization and expenditure through treatment expansion. That is, when newer, more effective treatments are introduced, conditions tend to be diagnosed and treated more frequently than in the absence of these newer regimens (Cutler and McClellan, 2001). Pharmacogenomics-based drug therapies and improvements in genetic testing are certain to impact the health care system through their treatment substitution and expansion effects.

To accurately assess the *value* (economic impact) of pharmacogenomics, a long-term health system perspective is especially important. Such a perspective includes not only the direct costs and benefits of health care interventions but also the indirect and intangible costs and benefits such as changes in productivity and health-related quality of life. Although medical advances emanating from the Human Genome Project (HGP) will be highly effective and patient specific, they will also be very expensive by today's standards. From an economic perspective, these new interventions should be evaluated for their marginal benefits and costs. As a society, we will be

"Production" of Health Care

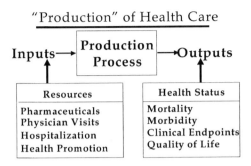

FIGURE 12.1. Health Care as a Production Process.

forced to choose between increased effectiveness and higher cost. At the microeconomic level, sound economic evidence, such as cost-effectiveness studies, will be needed to make efficient and rational use of scarce health care resources. From a societal perspective, we need to appreciate not only the drug costs but also the effects on total health care expenditure and societal well-being of using the newer agents. The value of a pharmaceutical intervention is reflected not only in clinical consequences but also in economic and humanistic outcomes (Kozma et al., 1993). To assess the value of medical innovations such as pharmacogenomics-based drugs, it is necessary to consider their impact on all three types of outcomes simultaneously.

VI. EFFECTS ON PUBLIC AND PRIVATE PAYERS

The economic impact of pharmacogenomics on the behavior of public and private payers will be a major issue. It is reasonable to expect reimbursement or benefit coverage decisions to be made with the same criteria used for other technology assessments. Private companies, such as Rx Intelligence, have emerged to meet the increased demand for information on the relative value of new pharmaceuticals (Rx Intelligence). Payers are incorporating evidence-based, formal technology assessments into their benefit designs so as to achieve more rational and value-based health care delivery. Public and private insurers now require that new technologies actually improve health outcomes. In the United States, professional associations, such as the Academy of Managed Care Pharmacy (AMCP), are publishing guidelines that public and private health care payers can use to formally assess the clinical and economic consequences of new drug therapies (AMCP). In the public sector, the Department of Veterans Affairs and the Department of Defense incorporate evidence-based guidelines and pharmacoeconomic data into their formulary and drug use policy decisions. Internationally, economic assessments of new pharmaceuticals have become a part of the drug approval and reimbursement processes. For example, in 1999. the U.K. established. as part of the National Health Service (NHS), the National Institute for Clinical Excellence (NICE). The purpose of NICE is to provide patients, health professionals, and the public with authoritative, robust, and reliable guidance on current "best practice." This guidance will cover both individual health technologies (including medicines, medical devices, diagnostic techniques, and procedures) and the clinical management of specific conditions (NICE). Given the likelihood that pharmacogenomics-based treatments will be more expensive than currently available therapies, it seems certain that these new technologies will be scrutinized for their value in the health care system.

Currently, physicians and patients determine the demand for pharmaceuticals and employers and insurers assume the risk and cost. As the price of new health care technologies escalates, payers will design and implement strategies to share risk and cost. Defined employer contributions, increased patient cost sharing, and benefit exclusions will be used to help control utilization and cost. In this environment, value-based assessments will be crucial to the adoption of any technological innovation. It is reasonable to expect public and private coverage for new therapies if evidence is provided regarding the costs and consequences of treatment. However, social and ethical dilemmas will certainly arise as therapies whose costs exceed their benefits are debated in the public arena.

VII. PHARMACOECONOMICS

Pharmacoeconomics is the study of the costs and consequences of pharmaceutical products and services (Bootman et al., 1991). The basic question addressed in a pharmacoeconomic evaluation is not whether to use a particular product or service, but rather when and under what circumstances a particular intervention is efficient. Rather than focusing on just product cost, pharmacoeconomics examines the total economic impact of a pharmaceutical product on the health care system. The value of pharmaceuticals is determined by balancing the health system costs and consequences (outcomes) of its use.

As noted by Goldsmith, medical technologies do not inherently increase health care costs. Their effects are broad-based and complex. Advances in medical technology may influence (1) the population at risk, (2) the cost of each treatment, (3) the risk of complications, (4) clinical incomes or hospital revenue, (5) the need for additional treatment, and/or (6) patients' quality of life (Goldsmith, 1994). The pharmacoeconomics of an intervention are particularly important when the intervention is a therapeutic breakthrough (leads to better patient outcomes), is relatively expensive compared

TABLE 12.1. COMPARISON OF PHARMACOECONOMIC EVALUATIONS

Type of Study	Cost Measurement	Consequence Measurement
Cost-consequence analysis	Categorized list of dollars	Categorized list of nonmonetary outcomes
Cost-benefit analysis	Dollars	Dollars
Cost-effectiveness analysis	Dollars	Nonmonetary outcomes
Cost-minimization analysis	Dollars	Assumed equal (not measured)
Cost-utility analysis	Dollars	Quality-adjusted life-years (QALYs)

with traditional modes of treatment, and produces cost savings in the long term.

Table 12.1 summarizes five major types of pharmacoeconomic evaluations: cost-consequence, cost-benefit, cost-effectiveness, cost-minimization, and cost-utility (Drummond et al., 1997; Kielhorn and Graf von der Schulenburg, 2000). In a cost-consequence analysis, a comprehensive list of relevant costs and outcomes (consequences) of alternative therapeutic approaches are presented in tabular form. Costs and outcomes are typically organized according to their relationship to cost (direct and indirect), quality of life, patient preferences, and clinical outcomes (see taxonomy below). No attempt is made to combine the costs and outcomes into an economic ratio, and the interpretation of the analysis is left in large part to the reader.

In a cost-benefit analysis, both costs and consequences are valued in dollars and the ratio of cost to benefit (or more commonly benefit to cost) is computed. Cost-benefit analysis has been used for many years to assess the value of investing in a number of different opportunities, including investments (or expenditure) for health care services. Cost-effectiveness analysis attempts to overcome (or avoid) the difficulties in cost-benefit analysis of valuing health outcomes in dollars by using nonmonetary outcomes such as life-years saved or percentage change in biomarkers like serum cholesterol levels. Cost-minimization analysis is a special case of cost-effectiveness analysis in which the outcomes are considered to be identical or clinically equivalent. In this case, the analysis defaults to selecting the lowest-cost treatment alternative. Cost-utility analysis is another special case of cost-effectiveness analysis in which the value of the outcome is adjusted for differences in patients' preferences (utilities) for the outcomes. Cost-utility analyses are most appropriate when quality of life is a very important consideration in the therapeutic decision.

Identifying and valuing relevant costs and consequences associated with the use of pharmaceutical products is an important first step in any phar-

TABLE 12.2. EXAMPLES OF COSTS AND
CONSEQUENCES INCLUDED IN HEALTH ECONOMIC
EVALUATIONS

Direct Medical
   Physician office visits
   Hospital care
   Pharmaceutical products and services
   Outpatient medical care
   Laboratory tests
   Diagnostic tests and procedures
Direct Nonmedical
   Transportation costs to obtain care
   Housekeeping costs
   Custodial care for dependents
Indirect
   Lost productivity due to death or illness
   Absenteeism
   Changes in productivity due to morbidity
Intangible
   Changes in health-related quality of life
   Social and emotional well-being
   Life Satisfaction

---

macoeconomic evaluation. Costs and consequences (outcomes) may be classified into three major groups: direct, indirect, and intangible (Eisenberg,
1989; Bootman et al., 1991).

Direct costs include both medical and nonmedical expenditures for the
detection, treatment, and prevention of disease. Direct medical costs reflect
resources consumed in the "production" of health care, such as pharmaceutical products and services, physician visits, and hospital care. Direct nonmedical costs reflect expenditures for products and services that are not
directly related to disease treatment but are still related to patient care. Examples of direct nonmedical costs include transportation to a pharmacy or
physician's office and housekeeping during the illness period. Indirect costs
account for changes in productivity of an individual because of illness. The
monetary value of lost or altered productivity is typically used as a measure
of indirect costs. Intangible costs and consequences are nonmonetary in
nature and reflect the impact of disease and its treatment on the individual's
social and emotional functioning and quality of life. Table 12.2 provides
examples of these types of costs and consequences.

Veenstra and colleagues have identified five criteria for evaluating the
cost-effectiveness of pharmacogenomics: severity of the outcome avoided,
drug monitoring, genotype-phenotype association, assay, and polymor-

phism (Veenstra et al., 2000; Flowers and Veenstra, 2000). If the consequences of the outcome avoided are severe or expensive, such as high mortality, significant effects on quality of life, or substantial medical expenditure, then the pharmacogenomics-based therapies will be potentially cost-effective. Likewise if drug monitoring is difficult, unreliable, or not practiced, then the pharmacogenomics-based therapies will be potentially cost-effective. However, if inexpensive, valid methods of drug monitoring are already available, pharmacogenomics may not offer an economic advantage. Also, a strong association should exist between the gene variant and clinically relevant outcomes: The genotype should be highly predictive of the phenotype. Alleles with a high penetrance will be better targets for pharmacogenomics-based therapies and should be more cost-effective. A fast and inexpensive assay for the mutation will be needed. To be cost-effective, the assay must be highly sensitive and specific, and rapid test results will be needed for acute, severe illnesses. Finally, the variant gene should be relatively common or the consequences substantial.

Mullins has identified five conditions under which pharmacogenomics-based drug therapies may be more cost-effective than existing treatment regimens (Mullins, 2001). First, pharmacogenomics-based therapies may be more cost-effective when costly adverse events are associated with genetic polymorphisms that are both predictable and preventable. Second, cost-effectiveness may be realized when common, less severe adverse events that are strongly associated with therapy noncompliance can be predicted and prevented. Third, if the newer therapy is very expensive or if there is a high opportunity cost of not using the most appropriate therapy, economies may be obtained. Fourth, pharmacogenomics-based therapies may be economically superior in cases in which the therapeutic index is narrow and patient variability is high. Finally, pharmacogenomics may be more cost-effective when the cost of events avoided by using the newer regimens is greater than the cost of detecting the events.

In the short-term, increased health care costs will be the primary concern. In the long term, the impact of pharmacogenomics on the total economy must be considered. From a societal perspective, the economic implications of pharmacogenomics can be distilled into a cost versus benefits paradigm. At the margin, do the benefits of pharmacogenomics (health, well-being, productivity) equal or exceed the costs of developing and using the new technology? If not, how much more is our society willing to invest or forego to achieve such benefits? In the allocation of scarce resources, decision makers will be asked to assess the value of new technological advances, such as pharmacogenomics-based drug therapies, to society as a whole and then give priority to those technologies that offer the greatest societal benefit. Pharmacoeconomics provides the tools to assist in these types of resource allocation decisions. Understanding the efficiency of alternative treatment regimens in achieving a therapeutic outcome is necessary for appropriate

TABLE 12.3. Cost of Illness—United States[a]

Diseases/Conditions[b]	Total Cost ($B)	Direct Cost ($B)	Indirect Cost ($B)
Heart disease	183.1	101.8	81.3
Cancer	96.1	27.5	68.7
Cerebrovascular (stroke)	43.3	28.3	15.0
COPD and allied conditions	37.3	21.6	16.2
Pneumonia and influenza	25.6	18.6	7.0
Diabetes	98.2	44.1	54.1
Kidney and urologic diseases	40.3	26.2	14.1
Chronic liver disease and cirrhosis	3.2	1.2	2.1
Septicemia	7.2	4.9	2.3
Alzheimer and other dementia	100.0	15.0	85.0

[a] DHHS, National Institutes of Health, Disease-Specific Estimates of Direct and Indirect Costs of Illness and NIH Support, February 2000.
[b] Ranked by 1998 death rate.

policy decisions. Pharmacoeconomic assessments do not dictate therapeutic decisions but rather elucidate the clinical circumstances under which one therapy is more efficient (achieves desired outcome at a lower cost) than another.

Table 12.3 lists 10 of the top 12 diseases ranked by 1998 mortality rates. Also included are NIH estimates of the cost of illness for each disease (DHHS, 2000). Together these 10 illnesses represent an economic burden in excess of $345 billion. Most of these conditions are thought to have an underlying genetic component and thus to be amenable to treatment with genetics-based treatments. One way to consider the benefits of pharmacogenomics-based therapies is to estimate how much of these illness costs might be avoided or ameliorated by the new therapies. Cost avoidance represents the potential savings from pursuing the development and use of pharmacogenomics-based therapies. If only 50% of the direct medical costs associated with treating these conditions were avoided, over $144 billion in health care costs could be saved.

Understanding the human genome should lead to improved therapies that may have a favorable impact on health care expenditure. In many cases, pharmacogenomics-based therapies will be therapeutic (and economic) substitutes for other more invasive procedures such as surgery or will reduce the need for expensive support care such as long-term assisted living or skilled nursing care.

Pharmacogenomics will also have potential macroeconomic effects, particularly on productivity, a measure of the average output of an individual. As an individual's productivity increases, output increases (more goods and services are produced) and the person has a better chance to improve his or her standard of living. In the aggregate, increased productivity is reflected in the (GDP). The GDP reflects the value of all currently produced goods and services sold on the market during a particular time interval (Gordon, 2000).

Good health and improvements in health contribute to enhanced productivity that can translate into an increase in the GDP. In assessing the economic impact of pharmacogenomics, it is important to include gains in productivity associated with reduced morbidity and mortality because these may more than offset the cost of using the new technologies. That is, if the GDP grows at a faster rate than the increase in health care expenditure, the percentage of GDP devoted to health care will actually decline. Therapeutic improvements will decrease morbidity, which in turn should reduce absenteeism for patients and caregivers, increase productivity while working, and extend productivity in an aging population (Pardes et al., 1999). The GDP should benefit from pharmacogenomics-based therapies that can reduce absenteeism for both patients and caregivers and thus improve productivity. As shown in Table 12.3, the indirect costs (absenteeism and lost productivity) associated with these highly prevalent conditions are estimated to be $345 billion. A reduction of these indirect costs would represent substantial savings to the economy through pharmacogenomics.

In addition to improvements in productivity, growth in new service sectors has been identified as a potential macroeconomic benefit of pharmacogenomics. The likely investment and growth in new services to support genetic research and pharmacogenomics-based drug development should add to the U.S. GDP. Richmond has identified five business areas in which growth and development should have a positive effect on national and world economies (Richmond et al., 1999). First, development of new technologies, such as research equipment, and drug delivery systems for pharmacogenomics-based drugs, will add to the national output. Second, growth in bioinformatics should generate new jobs and capital investments. This new sector will require individuals with expertise in acquiring, storing, and managing genetic information as well as conducting and interpreting genetic tests. Demand for this expertise should create a number of new jobs for the economy. Third, growth should be expected in the diagnostic sector, where firms will be needed to meet the demand for new diagnostics, especially those needed for high-throughput genotyping. Fourth, emergence of centralized diagnostic services will expand the output of the services to meet the need for efficient, low-cost testing. Finally, the medical management sector of the economy should also grow along with the increased use of pharmacogenomics-based therapies. Because pharmacogenomics-based drugs will likely be expensive and targeted to specific genotypes, appropriate utilization monitoring will be an important component of their efficient use.

As a component of the U.S. economy, pharmacogenomics may add to the GDP by inducing capital investment, creating new jobs, and enhancing global competitiveness. The investment and output of the biotechnology industry during the past few years could be a bellwether for the potential impact of pharmacogenomics. In 1997, capital investment in the biotechnology sector was about $5.5 billion. By 2000, this figure increased by a factor of 6.8 times over the 1997 investment. During 2000, the biotechnology sector raised a record $7.21 billion in 91 initial public offerings (IPOs), with an average value of $79.26 million. In 67 follow-on offerings, the industry generated another $16.42 billion, with an average value of $245.12 million per offering. Figuring all sources of fund-raising together, the industry raised $36.9 billion during the year 2000, more than three times the amount in 1999 (Bioworld, 2001). In 1999, there were over 1200 firms in the biotechnology industry with sales in excess of $16 billion. The pharmacogenomics sector can be expected to induce similar growth and investment. Moreover, the rate of mergers and acquisitions among large pharmaceutical companies and smaller biotech and diagnostic firms should escalate. However, given the nature of genetics research and pharmacogenomics, a thriving niche industry is likely to parallel large-firm R&D.

VIII. CONCLUSION

In an era of escalating health care costs, "inputs" to the production of health care, such as pharmaceuticals, physician visits, laboratory/diagnostic services, and hospital care, are often viewed in isolation from each other as if one sector of the health care economy can be optimized independent of the others. When cost containment is the primary economic objective, a short-term, risk-averse decision rubric emerges.

In a microeconomic sense, most of the inputs to health care production are either substitutes for or complements to other inputs. That is, they may be used in place of or in addition to the current or standard mode of treatment. To accurately assess the *value* or economic impact of pharmacogenomics (or any health care technology), a long-term health system perspective is especially important. Such a perspective includes not only the direct costs and benefits of health care interventions but also the indirect and intangible costs and benefits such as changes in productivity and health-related quality of life. Although medical advances emanating from the HGP will be highly effective and patient specific, they will also be very expensive by today's standards. As a society, we may be forced to choose between increased effectiveness and higher cost. At the microeconomic level, sound evidence, such as cost-effectiveness and health outcome research studies, will be needed to make efficient and rational use of scarce health care resources. If public and private payers can establish clear reimbursement criteria and explicit methods of assessing the value of pharmacogenomics, then only cost-effective regimens are likely to be introduced to the market (Danzon and Towse, 2000).

The ability to select specific genotypes for clinical trials should improve drug development success rates, reduce clinical trial time, and perhaps decrease clinical trial cost; however, therapeutic market sizes will likely decrease. It is also possible that pharmacogenomics technology will permit drug companies to reassess the potential benefits of past "drug failures." Many promising drugs have been withdrawn from the market because of unexpected adverse drug events that may have been related to genetic variations rather than general drug toxicity. If so, these agents may still have a valuable place in drug therapy. Finally, the issue of testing and testing costs must be addressed. Will this additional cost be part of the drug price, or will it be an indirect cost of pharmacogenomics-based drug use? More broadly, will testing be universal so that the genotype of a person is known early in life so that treatment regimens can be optimized over a his or her lifetime?

Pharmacogenomics will also have a significant influence at the macroeconomic level. As biotechnology and pharmacogenomic companies grow from small "cottage" industries to meet the demand for new therapies, they will create many new jobs and require a work force with advanced technical skills. There could also be a realignment of the pharmaceutical research and manufacturing sectors. Major manufacturers have already begun to acquire or form alliances with smaller biotechnology companies in anticipation of the shift in drug discovery from the traditional laboratory setting to the genomics platform and targeted drug development. A new equilibrium could occur in which pharmacogenomics/biotechnology companies provide the research and development for new therapies, and traditional pharmaceutical manufacturers adapt their production and marketing skills to supply the new regimens.

Pharmacogenomics represents a true paradigm shift for health care. The potential to effectively treat or cure most diseases will come at a price. These therapeutic advances will create economic and ethical dilemmas at all levels of the economy. At the individual level, there will be economic barriers to access, particularly for disadvantaged or disenfranchised populations. The costs of diagnostic tests and treatments may be prohibitive, and mechanisms (e.g., insurance, income subsidies, rationing) to distribute and pay for care must be devised. With limited health care resources, trade-offs between cost, access, and quality are inevitable.

At the health care provider level, choices will be made among many therapeutic regimens that differ in effectiveness and cost. At the producer level, pharmaceutical companies must decide whether targeted drug development is economically feasible. Will there be a sufficient return on investment to justify the risks involved in pharmacogenomic research? Will companies only pursue research and drug development for diseases for which there is a sufficient market (number of cases) to be profitable? If market imperfections exist in the development, production, and distribution of pharmacogenomics-based innovations, should governments intervene to regulate the markets? These questions will arise as the pharmacogenomics market emerges and can be answered best when the full economic effects are considered.

Pharmacogenomics will yield drug therapies that are safer and more effective than existing treatment regimens. Pharmacoeconomics provides the tools to evaluate the efficiency of these novel therapies and can help guide policy makers in decisions regarding the efficient allocation of scarce health care resources. The final verdict will rest on the cost of the new therapies balanced against the value they bring to the health care system.

REFERENCES

Aldridge, S., "Customizing Drugs to Individual Genetic Profiles," *Genet. Eng. News,* **21 (14)**, 29, 67 (2001).

AMCP Format for Formulary Submissions. *www.amcp.org*/publications

Bioworld. The BioWorld Biotechnology State of the Industry Report, 1998–2001 (2001). *www.bioworld.com/bw/public.htm?path=resources/ann/index.html*

Bootman, J.L., et al., *Principles of Pharmacoeconomics,* Cincinnati, OH: Harvey Whitney Books Company, p. 35 (1991).

Cutler, D.M. and M. McClellan, "Is technology change in medicine worth it?" *Hlth. Affairs,* **20 (5)**, 11–29 (2001).

Danzon, P. and A. Towse, "The Genomic Revolution: Is The Real Risk Under-Investment Rather Than Bankrupt Health Care Systems?" *J. Hlth. Serv. Res. Policy,* **5 (4)**, 253–255 (2000).

Department of Health and Human Services, National Institutes of Health. "Disease-specific Estimates of Direct and Indirect Costs of Illness and NIH Support." (2000). www1.od.nih.gov/osp/ospp/ecostudies/COIreportweb.htm

DiMasi, J., "Risks, Regulations, and Rewards in New Drug Development in the United States," *Regul. Toxicol. Pharmacol.* **19**, 228–235 (1994).

DiMasi, J., et al., "Cost of Innovation in the Pharmaceutical Industry," *J. Hlth. Econ.,* **10**, 107–142 (1991).

Drews, J., "Drug Discovery: A Historical Perspective," *Science,* **287**, 1960–1964 (2000).

Drummond, M.F., et al., *Methods for the Economic Evaluation of Health Care Programs,* Oxford: Oxford University Press, pp. 6–17 (1997).

Eisenberg, J.M., "Clinical Economics: A Guide to the Economic Analysis of Clinical Practices," *JAMA,* **262**, 2879–2886 (1989).

Emilien, G., et al., "Impact of Genomics on Drug Discovery and Clinical Medicine," *Q. J. Med.,* **93**, 391–423 (2000).

Ernst, F.R. and A.J. Grizzle, "Drug-Related Morbidity and Mortality: Updating the Cost-of-Illness Model," *J. Am. Pharmaceut. Assn.,* **41 (2)**, 192–199 (2001).

Flockhart, D., "Genetic Testing for Drug Response," *Pharm. Therapeut.*, **26 (2)**, suppl. 11 (2001).

Flowers, C.R. and D.L. Veenstra, "Will Pharmacogenomics in Oncology Be Cost-Effective?" *Oncol. Econ.*, **1 (11)**, 26–33 (2000).

Goldsmith, J., "The Impact of New Technology on Health Costs," *Hlth. Affairs*, **13 (3)**, 80–81 (1994).

Gordon, R.J., *Macroeconomics*. Reading, MA: Addison-Wesley, pp. 4–6 (2000).

Hansen, R.W., "The Pharmaceutical Development Process: Estimates of Development Costs and Times and the Effects of Proposed Regulatory Changes." In R.I. Chien (ed.), *Issues in Pharmaceutical Economics*. Lexington, MA: Lexington Books (1979).

Health Care Financing Administration, Office of the Actuary (2001).

Kielhorn, A. and J.M. Graf von der Schulenburg, *The Health Economics Handbook*. Chester, UK: Adis International, pp. 15–33 (2000).

Kozma, C.M., et al., "Economic, Clinical, and Humanistic Outcomes: A Planning Model for Pharmacoeconomic Research," *Clin. Therapeut.*, **15**, 1121–1132 (1993).

Kurth, J.H., "Pharmacogenomics: A Genetic Tool for Identifying Patients at Risk," *Drug Inform. J.*, **34**, 223–227 (2000).

Lichtenberg, F.R., "Are the Benefits of Newer Drugs Worth Their Cost: Evidence From the MEPS," *Hlth. Affairs*, **20 (5)**, 241–251 (2001).

Meyer, J. and J. Thompson, "The Real Cost of Pharmacogenomic Research," *Genet. Eng. News*, **21 (13)**, 16, 19, 67 (2001).

Michelson, S. and K. Joho, "Drug Discovery, Drug Development and the Emerging World of Pharmacogenomics: Prospects for Information in a Data Rich Landscape," *Curr. Opin. Mol. Therapeut.*, **2 (6)**, 651–654 (2000).

Mullins, C.D., "Pharmacogenomics: Methodological Considerations for Evaluating Outcomes and Cost-Effectiveness." Presented at the Ninth Annual Invitational Conference on Pharmacoeconomics. University of Arizona College of Pharmacy, Tucson, AZ, January 24, 2001.

National Institute for Clinical Excellence. *www.nice.org.uk*

Pardes, H., et al., "Effects of Medical Research on Health Care and the Economy," *Science*, **283**, 36–37 (1999).

Prescription Drug Expenditure in 2000: The Upward Trend Continues. Washington, DC: National Institute for Health Care Management Research and Educational Foundation. (May 2001).

Redwood, H., "New Drugs in the World Market: Incentives and Impediments to Innovation," *Am. Enterprise*, **4 (4)**, 72–80 (1993).

Regaldo, A., "Inventing the Pharmacogenomics Business," *Am. J. Health-System Pharmacy*, **56**, 40–50 (1999).

Richmond, M.H., et al., "*Human Genomics: Prospects for Health Care and Public Policy*," London, Pharmaceutical Partners for Better Health, pp. 7–9, 48–50, 60–61 (1999).

Rothstein, M.A. and P.G. Epps, "Ethical and Legal Implications of Pharmacogenomics," *Nat. Rev. Genet.*, **2**, 228–231 (2001).

Rx Intelligence. www.rxintelligence.com

Sander, C., "Genomic Medicine and the Future of Health Care," *Science*, **287 (17)**, 1977–1978 (2000).

Schwartz, W.B., "In the Pipeland: A Wave of Valuable Medical Technology," *Hlth. Affairs*, **13 (3)**, 70–79 (1994).

U.S. Food and Drug Administration. Medwatch. *www.fda.gov/medwatch/safety.htm*

Veenstra, D.L., et al., "Assessing the Cost-Effectiveness of Pharmacogenomics," *AAPS PharmSci.*, **2 (3)**, 1–11 (2000) article 29 (http://www.pharmsci.org)

PHARMACOGENOMICS AND THE SOCIAL CONSTRUCTION OF IDENTITY

MORRIS W. FOSTER, PH.D.

I. INTRODUCTION

We are all potentially members of a number of different groups, most of which have not yet (and may not ever) actually been socially constructed as identities that are recognizable to others, or even to ourselves. Once formed, though, these categorizations are loaded with meanings that are often complex and contradictory but that necessarily structure and constrain the possibilities for the social interaction and action of those upon whom they are imposed or who claim them. Differentiation of social identity often leads to differentiations in social status and opportunity, heightening those of some groups and lowering those of others such that the latter may be considered minority populations (that is, minorities in relative social standing rather than minorities in absolute numbers). Those social differentiations can lead to unequal access to health care resources, resulting in health disparities. At the same time, though, social differences also tend to be used as proxies for biological differences that may constitute yet another source for interindividual disparities in health status. It is that potential conflation of social and biological processes affecting health that holds the greatest danger in the use of socially constructed identities like race and ethnicity to conceptualize medical research and practice.

Pharmacogenomics: Social, Ethical, and Clinical Dimensions, Edited by Mark A. Rothstein.
ISBN 0-471-22769-2   Copyright © 2003 Wiley-Liss, Inc.

The practice of associating health status (as manifested by physical signs of well-being and illness) with social identity is at least as old as Greek and Roman travelers' accounts of some peoples as "sickly" and others as "robust" and as current as the ongoing epidemiological practice of categorizing disease risk and incidence by race and ethnicity (Feifer, 1985). Early European explorers, in another example, associated Native American identities with a greater vulnerability to diseases to which Europeans were more resistant, such as measles (Ramenofsky, 1987). In fact, though, it was not Native Americans' social identities that predisposed them to illness (or the moral inferiority that Europeans extrapolated from the epidemics that took so many Native lives), but their lack of previous viral and bacterial exposures.

Particular observable characteristics are often selected to symbolize differences between groups or categories of people. Although usually physical in nature, the relation of those observable characteristics to social divisions is arbitrary. That is to say, there is no necessary relation between having social identity A and physical characteristic B. The use of the latter to symbolize or signal the former is purely a social convention. Once selected, though, those proxy characteristics take on the appearance of being timeless and immutable, but in fact are not. Race (characterized primarily by skin color) is, for example, a European construct that has been a prominent means for differentiating people socially only for the last five centuries, yet it is still frequently treated as though it is a fundamental biological category (Graves, 2001).

Not surprisingly, then, the social implications of new technologies (such as pharmacogenomics) typically are discussed with respect to *existing* social groups, often with special attention to already disadvantaged populations (Rothstein and Epps, 2001a). Biomedical innovations in particular, however, have the potential for making previously covert or insignificant physical characteristics both observable and meaningful and, hence, for forming *new* social groups around those characteristics. Thus attention to how social differentiation and group formation are accomplished with reference to the physical characteristics of disease may tell us much about how pharmacogenomics might contribute to social constructions of identity and group formation.

Indeed, one could reasonably ask whether having a disease is more or less socially significant than being able to treat a disease. There are, of course, many diseases that people have for which effective treatments are available, thus decreasing the social significance both of the illness and of the therapy. At the same time, there are other diseases, such as Huntington disease, for which no treatments are effective and that have significant personal and social implications as a result (Horowitz et al., 2001). In between these two extremes are a large number of diseases for which some treatments are available that are effective in some but not in others. That is, not all those who get a disease have access to an effective therapy for it, are responsive to that therapy, or are diagnosed in time for the therapy to be effective. Cancer, heart disease, and diabetes are good examples of that intermediate health domain in which diagnosis and therapy come together but only in an imperfect fit. By heightening the social significance of differential drug responses, pharmacogenomics may contribute to the ways in which health disparities are reified as (and also reinforce perceptions of differences between) people with differing social identities.

II. "Elementary, my dear Watson"

Extrapolating from prior examples of group formation to future possibilities is a deductive process, and so it is perhaps not so unusual to bring Arthur Conan Doyle's Sherlock Holmes into the discussion. As devoted readers will testify, Conan Doyle's stories are filled with physical details, particularly those relating to the persons and behaviors of his characters. Some of those physical traits are immediately observable to other characters in the stories, whereas other physical traits are apparent only after their logical relation to human actions are made evident by Holmes.

In fact, the story titled "The Red-headed League" provides illustrative examples of two primary ways in which groups are constructed on the basis of physical characteristics (Conan Doyle, 1892). In that story, a "very stout, florid-faced, elderly gentleman with fiery red hair," named Mr. Jabez Wilson, comes to Holmes for advice. He is the owner of a small pawnbroker's shop who has recently learned, through a newspaper advertisement, of a vacancy in the Red-headed League, an organization founded from a bequest that pays a weekly stipend to "red-headed men who are sound in body and mind and above the age of twenty-one years" in exchange for "purely nominal services." The ad drew a very large crowd of men with various shades of red-colored hair, but Mr. Wilson was successful in obtaining the vacancy. As a consequence, he spent four hours a day in an office copying out the *Encyclopedia Britannica*, on condition that he not leave the office during that time. Mr. Wilson was able to hire an assistant at a much lower salary to run his pawn shop, the stipend from the League leaving him with a considerable financial benefit in the difference. Then suddenly, after eight weeks, he arrived at the office to find that the Red-headed League had been dissolved. And it was this mystery that he placed before Holmes. By observing the knees of Mr. Wilson's assistant's trousers, the amount he asked for wages, and his fondness for photography, Holmes was able to deduce that the League had been a ruse to keep Mr. Wilson out of his business each day so that his cellar could be used to dig a tunnel into the basement vault of a next-door bank.

A. EXTENSION AND NARROWING

In this story, as in the entire Holmesian canon, the great detective uses a number of physical features to deduce the identity of one or more criminals. The interesting point about this particular story, though, is that Holmes is not the only character using physical characteristics to define a group. The criminals who organized the Red-headed League also have done so through their advertisement for men with red hair. The difference between the two methods, however, is in how people are evaluated for membership in the Red-headed League as contrasted with Holmes' identification of the members of the criminal conspiracy behind it.

In the case of Holmes' famous method, *multiple* clues (or, for our purposes, observed physical characteristics) are used to *narrow* the range of those who might be members of the identified group of suspects. In the ideal world, which was almost always attainable by Holmes through Conan Doyle's fictional devices, that range could be narrowed down to a single individual. Latter-day mystery writers, though, have preferred to keep intact a larger group of likely suspects until the very last pages of their books to heighten dramatic tension and then narrow the number one by one through logical arguments (usually conducted at a dinner party with all the suspects conveniently present) that refine the necessary and sufficient conditions for group membership until there is only one possible member of the group.

In contrast, in the case of the originators of the Red-headed League, a *single* physical characteristic is used to *extend* the range of those who might be members of the identified group, while still delimiting it in a manner that ensures that the resulting group will include some or all of the individuals they actually are targeting. By overidentifying potential group members, the originators of the League created the impression that their motivation for forming the group was something other than economic gain (indeed, that it was the opposite, altruism). The group defined by men with red hair was used as a convenient proxy for the smaller group defined by men with red hair who owned a pawnshop adjacent to the bank that was to be robbed.

It should be said, of course, that both processes of group formation, narrowing and extension, are appropriate for their respective purposes. Indeed, although Holmes often makes fun of "official" detectives who try to solve crimes by rounding up a large group of "likely" suspects, he observes that that same approach, as used by the originators of the Red-headed League, "was a curious way of managing it, but, really, it would be difficult to suggest a better."

We should keep in mind that Conan Doyle was a physician, and it is said that he modeled Holmes after a teacher who was famed as a great diagnostician (Booth, 2000). Certainly, the narrowing method used by Holmes is reminiscent of a clinical process that progressively excludes possible causes for a set of physical symptoms until the diagnosis is clear (Accardo, 1987). In contrast, the method of extending membership in a group to be sure thereby of including most if not all affected persons is reminiscent of a public health approach that uses self-evident characteristics (such as skin color or ancestry) as proxies for contacting those most likely to be at risk for a particular illness, even though the proxy characteristics are not the proximate causes of the disorder.

B. PHARMACOGENOMICS IN THE MIDDLE GROUND

As a developing technology, pharmacogenomics presently is caught between these two approaches and is likely to remain in that uncomfortable middle ground for the foreseeable future (Foster et al., 2001). On the one hand, individualized treatment (arrived at by progressively narrowing the observable genetic characteristics until they comprise an individual genotype) is the long-term goal of pharmacogenomics. However, neither the technology nor the clinical practice nor the ethical protections are yet sufficiently matured to make individualized medical treatment a reality. Consequently, much of pharmacogenomic research and, presumably, initial clinical applications (once those become available) are and will be phrased in terms of racial, ethnic, and other social groups of people who are the more likely suspects to respond to a particular drug.

Even if the technology were more mature and privacy protections were stronger, however, the promise of individualized treatment may never be fully realized by pharmacogenomics for the very practical reason that it is more cost-effective to research and develop therapies that will be used (and paid for) by the largest groups of people possible (Rothstein and Epps, 2001b). At the same time, critiques of the use (and misuse) of race and ethnicity as diagnostic and therapeutic proxies have become increasingly prominent in biomedicine (Oppenheimer, 2001; Schwartz, 2001). Critiques also have been made of their use in genetics research and genetic medicine (Foster et al., 2001; Graves, 2001). In many ways, dividing study participants and patients by race and ethnicity is akin to using a meat cleaver to prepare a slide for an electron microscope. The resulting extended categories are overly inclusive and do not reflect biological definitions of "population," while they exclude some individuals who may share the genetic feature(s) in question.

Although race and ethnicity may, one day, no longer be used as medical proxies, social constructions of identity nonetheless will continue to affect the ways in which we conceptualize pharmacogenomic findings and applications. Thus it is a useful exercise to ask how new groups formed by the alternative process of narrowing may affect pharmacogenomics. The genomic technologies that we already or soon will have can make a number of previously hidden physical characteristics discoverable—and so available as definitions for membership in a whole range of newly constituted groups. Although narrowing may seem the ideal answer to minimize the inherent risks in using extended proxy groups such as race and ethnicity for genomics, it also could entail risks of its own. As readers may recall, Sherlock Holmes, although almost always a perspicacious practitioner of narrowing, was not the happiest of individuals, which led him to resort to his own personalized drug response as therapy.

III. Disease Groups: Asserted and Imposed

We can begin by asking how disease groups are formed. Sufficient criteria include having the disease in question, being at risk for the disease, or having a compelling relationship with someone who has the disease. Each of these criteria is, in most cases, observable in some manner, both to those involved and to others. Disease susceptibility, for instance, can be physically manifested through a predisposing gene or the evidence of a family history of illness. Thus a common disease experience (either personal or vicarious) creates a basis for social solidarity among group members.

In some cases, disease group membership is an optional matter of *self-assertion*. That is, the social profile of the disease is relatively low such that only those directly affected by its experience construct identities and groups around it. This is particularly so for rare diseases the physical manifestations of which are not recognized by most people, such as Marfan syndrome, and for covert diseases that have no outwardly observable physical symptoms, such as hypertension. In other cases, though, disease group membership is an *imposed* identity to the extent that the physical manifestations are unavoidably evident to others as observable social facts. AIDS is an example of a widely recognized illness that has resulted in the formation of a largely imposed disease group. Of course, persons with such imposed identities also may assert group membership as a way of resisting externally formed characterizations (particularly stigmatizing ones), which many HIV-positive people certainly have done (Bartos and McDonald, 2000). Persons with disabilities, such as members of the deaf community, have been particularly active in asserting their own interpretations of what it is to be deaf and to resist what others perceive as self-evident physical limitations (Stebnicki and Coeling, 1999).

Detectable physical features, however, may be broader than just those identifiable by sight. DNA-based technologies offer the prospect of making discoverable physical characteristics (polymorphisms and haplotypes) that previously could not be known. Thus, although one cannot look at someone and genotype them, it may become increasingly possible to look at someone's lab report or even medical or insurance records and learn their disease or drug-response predispositions.

A. A Question of Control

An important distinction between asserted and imposed disease identities is the extent to which individuals can establish and maintain control, first over the observable physical symptoms and, following from that, over the identity itself. For instance, diabetics who are able to control their symptoms are not necessarily physically recognizable as such. At the same time, though, these are the most likely persons to make use of diabetes educators and to attend diabetes support groups, thereby asserting a diabetic identity that is different from one that might be imposed on them by others (Pooley et al., 2001). The problem of symptom control is similar to that experienced by members of marginalized racial and ethnic groups who attempt to gain control over the outward names and symbols by which they are known to others so as to gain control over how they are treated by others.

Disease groups, though, are much more protean than racial and ethnic groups because of the possibilities for greater control over physical manifestations offered by relatively rapid technological and informational developments. For instance, although people with impaired vision could be treated as a distinct group, the general availability (at least in the U.S.) of eyeglasses, contact lenses, and laser surgery has allowed the large majority of that potential category to function as though they had normal vision with little reason either to assert a distinct identity or for others to impose one on them. Similarly, leprosy, a disease that once resulted in the ostracizing of anyone visibly affected by it in European populations, now can be controlled by various drug therapies (although more than half a million cases still occur each year in third world countries because of a lack of access to health care surveillance and drugs).

These examples underline the point that imposed identities tend to bring with them stigmatization and discrimination, not so much because of the severity of the disease as because of the loss of bodily control that its severity implies (Hatty and Hatty, 1999). Minority group status in general also is about loss of economic, political, or some other kind of control. Often, minority groups organize themselves to regain control over their identities as a way of resisting that disenfranchisement, as in the case of African Americans and the civil rights movement. Similarly, the voluntary self-assertion of a disease identity also is an act of resistance (both individual and collective) in the face of losing control, in effect an effort to publicly demonstrate that control can be regained through social bonding with other affected or potentially affected persons.

Disease identities, once established narrowly on the basis of directly related symptomatic or genetic features, are, however, vulnerable to being extended to associations with proxy identities. Thus "African American" became a proxy for sickle-cell disease, leading to discriminatory laws that targeted African Americans without regard to their individual genetic risks for the illness while ignoring its presence in other socially defined populations (Markel, 1992). Such *extensions by proxy* are almost always imposed rather than self-asserted.

B. DRUG RESPONSE GROUPS

Groups that may form around differential drug response most likely will be smaller subsets of those formed around having or being susceptible to a disease. Indeed, one's drug response status is, arguably, most socially significant in the context of an actual disease or a heightened disease risk. As such, drug response subgroups may endanger the solidarity of larger disease groups—particularly those based on voluntary self-assertion—by creating potentially competing interests in research investment, product development, and product maintenance. They also may promote conflicting public messages as to whether disease-related symptoms are "manageable" and thus as to whether a disease and its susceptibility are socially threatening. For instance, members of a particularly vocal subgroup representing those who suffer from disease A but who are not responsive to drug X that allows others to control its symptoms may threaten the social image of those others by extending the public perception of their lack of drug response to all members of the disease group. This would be a case of *simple extension*, in which characteristics of a drug subgroup are extended to the disease group as a whole, in contrast to *extension by proxy* in which characteristics are extended to nonmedical groups such as race or ethnicity.

In contrast, the positive message of a subgroup that is responsive to a drug that allows control of otherwise disruptive symptoms may discourage public investment in the research and development of alternative drugs for subgroups that are not responsive to that particular drug. This would be a case of *narrowing by special case*, in which public understanding of the disease group is narrowed to the case of just one of its drug subgroups to the exclusion of its other drug subgroups. That process stands in contrast to *narrowing by exclusion* (the true Holmesian method) in which potential members of a class are eliminated by logical criteria rather than special emphasis.

In general, smaller subgroups will be less influential but potentially better focused than larger subgroups. Smaller subgroups also will be more vulnerable to stigmatization and discrimination. Thus extending the drug response problem of a subgroup to the entire disease group or to a larger, more influential social proxy group may be a useful strategy. An example of that strategy for disease groups may be found in making breast cancer a women's issue (although a small number of men also get breast cancer), which has resulted in significant increases of federal and private funding for breast cancer research and drug development. Conversely, narrowing a disease group to a successful example of subgroup drug response may be a useful strategy when a disease has become highly stigmatized. An example of that strategy can be found in the cocktail treatment that retards the development of AIDS in some (but not all) HIV-positive people. Publicizing the possibility of living with HIV for that subset of people who both respond to the cocktail and can afford its component drugs has reduced some of the negative public perceptions associated with that disease (Lert, 2000).

The discussion so far assumes that most drug response groups will be formed through self-asserted identities. Certainly, the drugs or therapies that patients take are mostly hidden from public view, although not from insurance company databases (or from those dinner guests who check out their hosts' medicine cabinets). Moreover, public knowledge of different kinds of drugs and therapies is fairly limited, with the exception of widely prescribed drugs that benefit from expensive advertising campaigns (such as Prozac) or drugs that receive high-profile news coverage (such as Cipro in treating anthrax). Thus most efforts to form drug response groups will be on the initiative of those who know the most about the drugs in question— people with the diseases for which they are prescribed.

Nonetheless, it is possible that some drug response groups will be imposed by outsiders. Persons who are affected by the physical symptoms of a disease but do not respond to a drug that alleviates them in others may stand out as different. That difference may result in imposed limitations. For instance, an individual with epilepsy whose seizures can be controlled by a drug will be able to drive a car and work in a "dangerous" occupation or workplace, whereas an individual with epilepsy who is not responsive or only partially responsive to that drug may not be allowed to drive a car or be employed in potentially hazardous occupations (Ozuna, 1993). These differences also may be observed through genetic testing for drug response. Thus, as it becomes more possible to control everyday risks with pharmacogenomic therapies, it also becomes possible to identify the subgroups of individuals for whom that control is not available—and to minimize their risks (and those of others around them) by nonmedical rules and regulations.

As in the case of disease groups, once drug response groups are imposed, it is likely that some of the persons so categorized will resist their attendant loss of social control (symbolized by their presumed loss of bodily control) by asserting ownership of the drug response identity or by asserting a counter-identity. Indeed, studies of minority-majority group relations show that the process of imposed minority status almost always co-occurs with a process of asserted identity, as those who are subjected to minority status resist being singled out by asserting control over the imposed identity (Scott, 1990) The success of that self-assertion in the face of imposed categories will depend on public perception, public policy, and judicial decisions.

C. THE GROUPNESS OF GROUPS

Not all groups, of course, have the same social dynamic. Some groups are very well organized, with a formal process for determining membership, distributing shared resources, making collective decisions, and selecting leaders. Other groups are essentially informal categories of people who share some defining characteristic(s) but have no organized basis for interacting with one another. And there are many points in between those two poles. Disease and drug response groups will fit into that spectrum at different places, depending on how they are formed.

Self-asserted groups, for instance, will tend to be organized around discrete occasions when members can be physically present to one another, such as support group meetings and other kinds of gatherings for mutual help and public advocacy. Purely asserted groups will be comprised of members who have chosen a disease or drug response identity for themselves and so will be made up of self-selected activists. Other people with that same disease or drug response (or genetic predisposition to those) who do not self-select will not be included in purely asserted groups, which means that the social and cultural characteristics of asserted group membership will tend to be relatively homogeneous or limited because of a similarity in motivation for self-identification among necessarily active members.

In contrast, imposed groups will necessarily include all persons whom outsiders identify as belonging to them by virtue of some observable physical characteristics. Thus imposed groups will have a much more diverse membership because they will not be limited just to those individuals who are motivated to assert membership. As a consequence, purely imposed groups will tend to be organized by external forces (that is, by those who impose the category on others) rather than by a shared, internal social dynamic.

Once imposed groups become established public categories that have some significant consequences for how members interact with others—particularly when that interaction is somehow marginalized or devalued—the development of an internal social dynamic among members is likely. At least some members of the imposed group will resist their categorization by organizing themselves in ways to gain greater control over the identity inflicted on them. Those activists are likely to be the same people who would assert an identity in the absence of its imposition. Nonetheless, the fact that others impose the identity will tend to draw members to the group who would not normally be pure activists. The greater the limitations imposed on social interaction, the greater the social solidarity among those so grouped.

D. TYPES OF GROUPS

Although individualized medicine may not emerge through drugs customized for each person's genotype, it may appear in the form of the entire collection of drug response subgroups of which an individual is a member. Each of us will be a member of a multitude of different drug response subgroups and, except for identical twins, no two people will be members of precisely the same set of subgroups. The greatest similarities in these personal universes of drug response predispositions will be among close biological relatives. In contrast, more heterogeneous racial and ethnic categories will be comprised of overlapping but nonetheless often widely divergent individual universes of drug response predispositions.

Still, the historical human preference has been to focus on the physical evidence of just a few characteristics (such as skin color, last name, dialect, etc.) to group large numbers of people. Despite the best intentions of pharmacogenomics, that historical inclination suggests that a handful of drug response predispositions may become publicly salient in grouping people in the future. Because of what we know about ethnic and racial group formation, drug response groups may take one of three primary forms: (1) well-organized groups of self-asserted activists who are engaged in mutual support and public advocacy; (2) socially diverse, weakly organized collections of people who are identified by others as having a similar drug response based on observable evidence; (3) better-organized groups of people who are socially constrained by others as a consequence of their identifiably similar drug responses and who respond to this by constructing a heightened social solidarity with one another. In some instances, a fourth form may emerge: (4) well-organized groups of people primarily identified (and socially constrained) by others as a consequence of their similar drug response, and relegated to minority status as a consequence, who resist that minority status by trying to take control of their imposed identity.

Type (1) groups differ from types (3) and (4) in the motivation for members of (1) to become socially known because of their drug response status, whereas members of (3) and (4) are reacting to that already-extant public knowledge and differ in type based on the degree to which they are organized to resist it. Members of type (2) groups lack that internal organization.

Disease groups based on chronic conditions are most likely to produce drug response subgroups that have a long enough duration to develop into better- or well-organized social categories, imposed, asserted, or both. For drug response (or lack thereof) to become a significant social fact leading to

the formation of groups (1), (3), and (4), the physical manifestations probably must be of an everyday nature over an indeterminate period of time. In
medical terms, the group is more likely to form over chronic than acute conditions. Thus lack of response to a diabetes drug will be of greater social
moment than lack of response to an influenza drug. At the same time, being
known to be more susceptible to influenza because of a lack of drug response
(based on discoverable genotypic information) may become a social con-
straint on certain kinds of employment in which one is exposed to a large
number of people on an everyday basis (such as teachers or health care
workers). Employees who can control their susceptibility to influenza will
cost less in missed workdays and health benefits than those who cannot, who
may end up as members of a type (2) group. Systematic exclusion from
employment on the basis of drug response predispositions also may trigger
self-assertion and resistance to those barriers.

Although lack of response and only partial response to a drug are more
likely to be of social moment than effective response, it is possible that positive response also may be socially noteworthy, particularly where drug
responders are relatively few in number and the effects of the response are
particularly evident. Continued availability of so-called "orphan" drugs, for
instance, often is promoted by advocacy groups comprised of people whose
symptoms they control effectively. Pharmacogenomic therapies to which
only relatively small numbers of people respond may also become orphaned
(Scheuerle, 2001). Similarly, the affordability of effective pharmacogenomic
drugs may become a basis for group formation. Participants in managed care
plans and Medicare participants are already raising the problem of being
priced out of an effective drug (Pollock, 1988), a situation that easily could
arise with new pharmacogenomic therapies and the patents that protect their
developers' profits. Other pharmacogenomic therapies may allow some
groups of people to undertake activities that generally are prohibited for
people with their underlying health status, such as individuals with epilepsy
who are not normally allowed to drive a car or commercial airline pilots who
are forced to retire at a particular age. Thus rare responders may come to
constitute minority groups, members of which are activists for changes in
the imposed social perceptions or rules that limit their lives.

IV. CONCLUSION: THE RETURN OF SHERLOCK HOLMES

The Sherlock Holmes stories often are read as the triumph of the scientific
method of deduction. There is, though, another, darker interpretation to be
advanced. That is the frustration of science and technology in changing
everyday social practice. Throughout the canon, Holmes is consistently
thwarted by the standard operating procedures of the Scotland Yard "regu-
lars." He also is annoyed—and even disturbed—by the inability of Victorian

society to value and conform to a scientific way of organizing social interaction, leading to his famous eccentricities (Jann, 1995).

Pharmacogenomics is in much the same position as Holmes, offering a scientific approach for identifying individual drug response—if only society can keep itself from constructing social groups based on those genetic predispositions. But, of course, selectively differentiating one person from another is an essential element of human social interaction, one that pharmacogenomics is not going to alter. At best, we can try to manage the conditions that support the formation of groups, reducing the motivations that may lead individuals and others to treat drug response as a socially identifying characteristic. As in the example of poor vision, this means having corrective or therapeutic technologies that are everyday, universally available, and relatively inexpensive.

We also know, however, that pharmacogenomics will not develop in an everyday, universal, or inexpensive manner—and probably will not reach those goals any time soon. Pharmacogenomic therapies will be available for some diseases before others. and some members of disease groups will not respond to those therapies. Pharmacogenomic research and development already is expensive, and the costs will be reflected in the drugs that come to market. Thus social differentiation in the long middle term, as pharmacogenomics strives to reach the point of truly individualized medicine, is inevitable, as are inequities linked to those different identities.

In the end, neither scientists nor laws and regulations will be able to control the ways in which pharmacogenomics contributes to the construction of social identities and group formation. Nor will there be some evil Professor Moriarty who organizes a grand conspiracy to use pharmacogenomic information to take advantage of groups of people with particular drug response predispositions. Instead, out of potentially thousands of polymorphic interindividual differences, a relatively small number of perceived variations in drug response will, for a variety of historically grounded reasons that have little to do with genomic or scientific rationales, become publicly salient bases for marking some groups of people as different from others. Human social organization depends on such selective differentiations between categories of people (Durkheim, 1933; Weber, 1946). Given that unavoidable practice, the question becomes how we treat (both socially and medically) those drug response identities and groups that end up being constructed.

ACKNOWLEDGMENTS

This publication was made possible by Grant ES-11174 from the National Institute of Environmental Health Sciences (NIEHS) and the National Human Genome Research Institute (NHGRI), NIH. Its contents are solely the responsibility of the author and do not necessarily represent the official views of the NIEHS and/or NHGRI, or NIH. The author is indebted to Richard Sharp and Mark Rothstein for their comments on a previous version.

REFERENCES

Accardo, P.J., *Diagnosis and Detection: The Medical Iconography of Sherlock Holmes*, Rutherford, NJ: Fairleigh Dickinson University Press (1987).

Bartos, M., and K.K. McDonald, "HIV as Identity, Experience or Career," *AIDS Care*, **12**, 299–306 (2000).

Booth, M., *The Doctor and the Detective: A Biography of Sir Arthur Conan Doyle*. New York: St. Martin's (2000).

Conan Doyle, A., *The Adventures of Sherlock Holmes*. New York: Harper and Brothers (1892).

Durkheim, E., *The Division of Labor in Society*. Glencoe, IL: Free Press (1933).

Feifer, M., *Tourism in History: From Imperial Rome to the Present*. New York: Stein and Day (1985).

Foster, M.W., et al., "Pharmacogenetics, Race, and Ethnicity: Social Identities and Individualized Medical Care," *Ther. Drug Monit.*, **23**, 232–238 (2001).

Graves, J.L., *The Emperor's New Clothes: Biological Theories of Race at the Millennium*. New Brunswick, NJ: Rutgers University Press (2001).

Hatty, S.E., and J. Hatty, *The Disordered Body: Epidemic Disease and Cultural Transformation*. Albany, NY: State University of New York Press (1999).

Horowitz, M.J., et al., "Psychological Impact of News of Genetic Risk for Huntington Disease," *Am. J. Med. Genet.*, **103**, 188–192 (2001).

Jann, R., *The Adventures of Sherlock Holmes: Detecting Social Order*. New York: Twayne Publishers (1995).

Lert, F., "Advances in HIV Treatment and Prevention: Should Treatment Optimism Lead to Prevention Pessimism?" *AIDS Care*, **12**, 745–755 (2000).

Markel, H., "The Stigma of Disease: Implications of Genetic Screening," *Am. J. Med.*, **93**, 209–215 (1992).

Oppenheimer, G.M., "Paradigm Lost: Race, Ethnicity, and the Search for a New Population Taxonomy," *Am. J. Publ. Hlth.*, **91**, 1049–1055 (2001).

Ozuna, J., "Ethical Dilemmas in Epilepsy and Driving," *J. Epilepsy*, **6**, 185–188 (1993).

Pollock, A., "High Cost of High-Tech Drugs is Protested," *NY Times*, Feb 9: A1, D23 (1988).

Pooley, C.G., et al., " 'Oh It's A Wonderful Practice. . . . You Can Talk To Them': A Qualitative Study of Patients' and Health Professionals' Views on the Management of Type 2 Diabetes," *Hlth. Soc. Care Community*, **9**, 318–326 (2001).

Ramenofsky, A.F., *Vectors of Death: The Archaeology of European Contact*. Albuquerque, NM: University of New Mexico Press (1987).

Rothstein, M.A., and P.G. Epps, "Pharmacogenomics and the (Ir)Relevance of Race," *Pharmacogenomics J.*, **1**, 104–108 (2001a).

Rothstein, M.A., and P.G. Epps, "Ethical and Legal Implications of Pharmaco-genomics," *Nat. Rev. Genet.*, **2**, 228–231 (2001b).

Scott, J.C., *Domination and the Arts of Resistance*, New Haven, CT: Yale University Press (1990).

Scheuerle, A., "Limits of the Genetic Revolution," *Arch. Pediatr. Adolesc. Med.*, **155**, 1204–1209 (2001).

Schwartz, R.S., "Racial Profiling in Medical Research," *N. Engl. J. Med.*, **344**, 1392–1393 (2001).

Stebnicki, J.A., and H.V. Coeling, "The Culture of the Deaf," *J. Transcult. Nurs.*, **10**, 350–357 (1999).

Weber, M., *From Max Weber: Essays in Sociology.* New York: Oxford University Press (1946).

Pharmacogenomics: Considerations for Communities of Color

Laurie Nsiah-Jefferson, M.P.H.

I. Introduction

According to Francis Collins, by 2020, pharmacogenomics may be the standard approach for disease management. Pharmaceutical manufacturers will market new designer drugs for diabetes mellitus, hypertension, mental illness, cancer, and many other conditions. By 2029, it is conceivable that every tumor will possess a precise molecular fingerprint that catalogs the genes that have gone awry, and therapy will be targeted individually to maximize safety and efficacy (Collins, 2001).

When new technologies are not available to all members of a society, they can exacerbate existing inequalities in the system of medical care. Currently the structure of the U.S. health care system is fundamentally flawed. Racial and ethnic disparities in health status and health care, more than 40 million uninsured, spiraling health care costs, and inadequate levels of culturally competent medical care are all evidence of this problem.

What are the ethical, legal, social, and scientific implications of pharmacogenomics? Given the current disparities and the role—however illegitimate—of skin color and ethnicity in the distribution of medical services, how might new approaches to therapy affect socially defined groups of African, Asian, and Hispanic descent? These questions are complex individually, and they become more so when intersected.

Pharmacogenomics: Social, Ethical, and Clinical Dimensions, Edited by Mark A. Rothstein.
ISBN 0-471-22769-2   Copyright © 2003 Wiley-Liss, Inc.

This chapter utilizes a bioethics framework derived from the work of Jorge Garcia (Garcia, 1992) and Annette Dula (Garcia, 1992; Dula and Goering, 1994). They observe that racism and insensitivity by society and the medical profession toward communities of color have resulted in unequal access to care, disparate quality of care, and a lack of respect for cultural differences. These experiences have brought about mistrust, suspicion, and hostility by some members of minority communities (Garcia, 1992). As a result, communities of color have a different perspective on the health care system and new technologies may not be nearly as effective in practice as in theory in these communities. It is less the technology and more the social behavior, including bias in medical decision making and health policy, that will determine how pharmacogenomics will affect communities of color (Thiel, 1999).

A few words on terminology are appropriate here. "Communities of color" is defined as communities of African, Asian, American Indian, and Hispanic descent. This is not a monolithic group, as each has its own history and cultures. However, they often share a marginalized place in society because of past and present economic, political, and social oppression. Other terms include African American, a person having origins in any of the African American racial groups of Africa who was born in America; American Indian or Alaskan Native, defined as a person having origins in any of the original peoples of North America and maintaining cultural identification through tribal affiliations or community recognition; Hispanic, a person of Mexican, Puerto Rican, Cuban, Central and South American, or other Spanish culture or origin, regardless of race; Asian or Pacific Islander, a person having origins in any of the original peoples of the Far East, Southeast Asia, the Indian subcontinent, or the Pacific Islands including China, India, Japan, Korea, the Philippines, and Samoa; and white, a person having origins in any of the original peoples of Europe, North Africa, or the Middle East. There are also people in this country who are "multiracial" (Byrd and Clayton, 2002).

The terms race, ethnicity, and minority are also used in this chapter. Race refers to a group that is socially defined on the basis of physical characteristics. Ethnicity refers to a shared culture and lifestyle, especially as reflected in language, folkways, religious and other institutional forms, material culture such as clothing and food, and cultural products such as music, literature, and art. Finally, the term minority group refers to a part of a population differing from others in some characteristics, which often subject them to differential treatment (Byrd and Clayton, 2002; Van de Berghe, 1967).

Scholars have noted that members of communities of color tend to be antimajoritarian—against the belief that those constituting a simple majority should make all of the rules for all members of the group. They are also antiutilitarian—against the belief that actions derive their usefulness as means to an end, especially good for the greatest number of people, because communities of color have suffered under majoritarian theory. They may be antisituationist—against the belief that behavior or decisions should be chiefly context-specific instead of categorical principles, because many individuals from communities of color do not trust subjective decision making by most individuals and entities in the dominant culture (Garcia, 1992).

Members of racial and ethnic minorities express greater levels of mistrust of health care providers and the medical establishment than white Americans because of prior breaches of trust that have occurred between minorities and the scientific and medical communities (Smedley et al., 2002). In addition, ethnic minority patients perceive higher levels of discrimination in health care settings than nonminorities. For example, in a survey of 781 African American and 1003 white cardiac patients, African American patients were four times more likely than white patients to mistrust the health care system (LaVeist et al., 2000). Another study found that 30% of Hispanics and 35% of African Americans believe that racism is a major problem in health care compared with 16% of whites. Well over half of the minorities in this sample (58% of Hispanics and 65% of African Americans) were very or somewhat concerned that they or a family member could be treated unfairly when seeking medical care because of their race or ethnic background; less than 25% of whites endorsed this view. Finally, nearly three times as many African Americans as whites think they receive a lesser quality of care (Lillie-Blanton et al., 2000). A study by the Henry J. Kaiser Foundation also indicated these same views. Hispanics, African Americans, Asians, and Native American consumers believe that there are several forms of discrimination in the health care system and that there are treatment disparities because of insurance status, race, and gender. African Americans were found to be the most distrustful of all racial/ethnic groups (Henry J. Kaiser Family Foundation, 1999). These perceptions are based on many factors, including the factors that health care is based on profit, that many providers hold negative views toward minority groups, and that health professionals lack concern and care for them (Smedley et al., 2002). In other words, communities of color share a perspective on health care that may differ significantly from prevailing paradigms. Therefore, qualities often taken for granted by the dominant culture, such as efficiency, advancing regulation, and autonomy may not be as important as other considerations to members of these communities (Dula and Goering, 1994).

II. BIOLOGICAL CONCEPTS OF RACE

Driven since ancient times by folk beliefs, religious teachings, and social customs based on differences in physical appearance of various geographic populations, by the eighteenth century race became the subject of formal theoretical speculation and scientific investigation (Levi-Strauss, 1951; Smedley,

1999; Byrd and Clayton, 2000). As an extension of Western culture's intellectual preoccupation with human inequality, race became the focus of empirical and scientific inquiry for the next three centuries, codifying a color-coded, racial hierarchy of humans (Banton, 1986; Smedley, 1999; Byrd and Clayton, 2000). By the middle third of the twentieth century traditional biology- and anthropology-based ideas of race and "races of man" that had become dominant during the nineteenth century began losing influence as more objective anthropological, genetic, paleontological, archeological, linguistic, biogeographical, and biological studies proved the unity of the human species, the common African origins of all racial groups, and the biological insignificance of the old parameters of racial classification such as skin and eye color, hair texture, physical features, and skull size and shape (Banton and Banton, 1975; Banton, 1986; Cavalli-Sforza, 1994; Smedley, 1999; Cavalli-Sforza, 2000). Yielding to a deluge of scientific evidence, race is now known to be a sociocultural concept. Despite the scientific dismissal of any significance to race, groups of people sharing certain physical characteristics are often treated differently based on stereotypical thinking, discriminatory institutions and social structures, shared world view, and social myths (Diamond, 1999; Feagin and Feagin, 1999; Smedley, 1999; Byrd and Clayton, 2000; Feagin, 2000; Byrd and Clayton, 2002).

Eighty-five percent of all human genetic variation occurs between two persons of the same ethnic group, 8% occurs between tribes and nations, and 7% between the so-called major races (Rose et al., 1984), and only 0.012% of the variation between humans in total genetic material can be attributed to differences in race (King, 1995). The utility of racial categories is further limited by the intermingling of human populations, and clinically important mutations such as hemoglobin arise in populations that straddle racial categories (Foster et al., 2001).

Joseph Graves argues that racism has persisted in U.S. society because adequate scientific reasoning has not entered into the equation when people think about genetic diversity. He uses a scientific model to explain how to ask the proper questions about the nature of population differences and how, if at all, that diversity can be correlated with differences with a whole host of factors, including responses to medications. Graves shows that when doctors and researchers look for race as an explanation, it is social class position and environment, diet, stress and other environmental factors that cause the majority of health disparities, and it is therefore useless to lump disparate individuals into a racial group for purposes of medical treatment (Graves, 2001).

As general knowledge of the role of the gene in biology and in the practice of medicine increases, the definitions of race and ethnicity are likely to evolve in social discourse to reflect a more scientific understanding of biological diversity. The value of race as a predictor of disease and response to, or compliance with, drug therapy is an area of considerable debate. The varying frequency of certain genetic variations relevant to drug response among ethnic groups offers a perspective that may inform the debate. If nothing else, the uneven distributions of certain polymorphisms should elevate the level of discourse by emphasizing genetic differences over racial differences (Epps and Rothstein, 2002).

III. Racial/Ethnic Disparities in Health and Health Care

Despite notable progress in the overall health status of Americans, African Americans, Hispanics, American Indians, Alaskan Natives, and some Asian Pacific Islander subgroups experience continuing disparities in the burden of illness and death compared with the population as a whole (Henry J. Kaiser Family Foundation, 2002). At no time in the history of the United States has the health status of communities of color equaled or even approximated that of white Americans. For instance, the decreased life expectancy of African Americans has been noted over centuries and has proved stubbornly resistant to change. Two periods of health reform specifically addressed the correction of race-based health disparities, the first period, 1865–1872, was linked to the Freedman's Bureau Legislation, and the second, 1965–1975, included the inception of Medicare and Medicaid and the community health center movement. Both periods had a dramatic and positive impact on the health status of minority populations. Since 1980, however, African American health status has generally stagnated and diminished compared with that of the white population. Moreover, overall progress during the 1980–1995 expansion period was much slower than during the 1965–1980 period. Differential outcomes may explain the racial inequalities in health care at both the system and individual provider-patient level (Byrd and Clayton, 2001).

The federal government identifies six priority areas in which racial and ethnic minorities experience serious disparities in access to medical care and health outcomes: infant mortality, cancer screening and management, cardiovascular disease, diabetes, HIV/AIDS, and immunizations. These six areas reflect areas of disparity known to affect multiple racial and ethnic minority groups at all life stages. The national goals for improvement in these six areas were originally set forth in Healthy People 2000, the nation's health objectives agenda, which has as a major goal the elimination of racial disparities in health (U.S. Department of Health and Human Services, 2000). They were reiterated in Healthy People 2010 (U.S. Department of Health and Human Services, 2000).

Several factors affect racial disparities in health. Traditional explanations point to social, economic, and environmental factors, including income, educational and employment status, lifestyle choices, occupational and environmental exposures, housing, nutrition, and cultural beliefs (Williams, 1998). Another explanation relates to access to health care, particularly concerning discrepancies in private or public insurance or public health services (Blendon, 1989; Weinick, 2000). Other explanations include the lower reimbursement payments and administrative burdens that discourage provider participation in medical assistance programs that many minorities rely on for their care. Still others highlight a lack of trust in, and familiarity with, resources, which inhibits minorities from accessing early intervention and treatment (Doescher, 2000). More recently, disparities have been explained by racial/ethnic discrimination in health care revealed through (1) the structure of the health system, which provides disparate levels of resources and services, and (2) provider bias (conscious or unconscious), manifested at the national, state, and local levels and facilitated at the policy, practice, and service levels (Mayberry, 1999; Geiger, 2001; Smedley et al., 2002).

A. HEALTH STATUS OF COMMUNITIES OF COLOR

On average, African Americans, Hispanics, and Native Americans have higher mortality rates than whites. Only Asians have lower mortality rates than whites. However, data aggregating diverse ethnic groups mask the higher mortality rates of particular Asian subpopulations, such as the Vietnamese. When overall mortality is examined by a measure of socioeconomic conditions, differences between African Americans and whites are reduced, but not eliminated. Mortality is a crude indicator of health status, but it demonstrates how critical the disparities are for racial and ethnic minorities. For some groups, these disparities begin early in life and are sustained throughout life. African American infant mortality rates are more than double those of whites (14% vs. 6%), and Native American infant mortality rates (10%) are more than one and a half times higher than those of whites. Among adults, African American death rates are approximately 55% higher than those of whites. African Americans succumb to diabetes at three times the rate of whites, American Indian/Alaskan Natives at more than twice the rate of whites, and Hispanics at more than one and a half times the rate of whites. African American men have the highest rate of prostate cancer incidence and mortality (Henry J. Kaiser Foundation, 2002).

Despite higher morbidity and mortality rates for cardiovascular disease, African Americans and Hispanics are less likely to receive high-technology cardiac procedures, such as cardiac catheterization and coronary revascularization. African American women present with late-stage breast cancer more frequently than white women, possibly related to barriers to life-saving diagnostic services and treatment. Higher rates of late-stage diagnosis may account for 14% of the difference in breast cancer survival rates. Despite having the highest incidence of cervical cancer, Asian American and Pacific

Islander women have the lowest screening rates for cervical cancer; cervical cancer rates for Vietnamese women are the highest among all groups, nearly five times the rates of white women (Mayberry, 1999).

B. Access to Health Care

Access to health care can be defined in many ways, including by insurance status, number of physician visits in the last year, and treatment interventions once an individual is in the health care system. Furthermore, access can relate to whether the patient's health care is satisfactory, comprehensive, and meeting the overall needs of the patient from a social, cultural, spiritual, or other perspective.

Individuals from communities of color are more likely to be uninsured than non-Hispanic whites. Over one-third of Hispanics (37%) are uninsured. They are three times more likely to be uninsured than non-Hispanic whites. Nearly one-fourth of African Americans and one-fifth of Asian Americans and Pacific Islanders (AAPI) (Veenstra and Higashi, 2000) and American Indian/Alaskan Natives (AI/AN) are uninsured (Henry J. Kaiser Family Foundation, 2002).

Among preschool age children, 8% of AAPI and Hispanic children did not have a physician visit in the past year compared to 5% of whites and African American children. For school age children (6–17 years), Hispanics AI/ANs, and AAPIs are one-and-a-half to two times more likely not to have visited a physician in the last two years compared with African Americans and whites. One of the reasons that African American children are covered at a higher rate than AAPI and Hispanic children is because of the safety net of Medicaid, for which some AAPI and Hispanic children are not eligible because of their immigrant status and their parents' fear of applying for their native-born children. Other families are uninsured because of working in small businesses or being self-employed. Hispanic, AI/AN, and AAPI children are two to three times as likely as white and African American children to lack a usual place of care (Henry J. Kaiser Foundation, 2002).

The picture is even grimmer for adults. One-third of Hispanic and AAPI men in fair and poor health have not visited a doctor in the past year. Hispanic, African American, and AAPI women in fair or poor health are also less likely to have visited a physician than non-Hispanic white men (Henry J. Kaiser Foundation, 1999).

The level of care also may be inadequate. Many uninsured individuals receive their care through "safety net" providers such as community health centers and public hospitals, which usually have limited budgets and are facing competition from Medicaid managed care providers. Many who are on Medicaid today will be ineligible in the short term and will become uninsured (Lewin and Altman, 2000).

1. ACCESS TO PHARMACEUTICALS.   Lack of access to drugs often leads
to declines in health status (Lurie et al., 1984). This decline is explained by
increased ambulatory care visits to obtain medications and the shifting of
care to provider settings, such as hospitals and nursing homes, where prescription drug reimbursement is available (McCombs et al., 1994). The cost
of prescription drugs and drug utilization vary considerably among individuals (Doonan, 2001). The poor, uninsured, and elderly have reduced
access to drugs and rely on safety net providers for pharmaceuticals (Committee on the Changing Market Managed Care and the Future Viability of
Safety Net Providers, 1999). With few exceptions, Medicare does not cover
outpatient prescription costs. Consequently, older Americans spend three
times as much of their income on drugs as the general population. Thirty-
five percent of seniors do not have prescription drug coverage, and many
with coverage have high deductibles and caps on total dollar coverage.
The three out of ten standard Medigap policies that offer prescription drug
coverage are very expensive (Doonan, 2001).

Medicaid has become the largest single purchaser of prescription drugs.
In 1995, Medicaid provided coverage to 15.8% of Americans, including
women, children, the elderly, and individuals with disabilities. A disproportionate percentage of minority populations were on Medicaid. For
example, in 1997, 19% of African Americans were on Medicaid (Mayberry,
1999). To control utilization, several states limit the number of prescriptions
that a beneficiary can have filled per month or per year. Although many of
these programs allow for exceptions (e.g., life-threatening illnesses) or
provide some mechanism to allow for review and considerations, some do
not. Federal law allows states to impose limits on all such drugs in a therapeutic class, the maximum or minimum quantities per prescription, and the
number of refills, if such limitations are necessary to discourage waste
(Scholesberg and Jerath, 1999). Many states impose copayments for prescription drugs, ranging from fifty cents to five dollars. Certain categories
of individuals, such as pregnant women, children, and the elderly, cannot
be charged a copayment. In addition, states may require prior approval of
prescription drugs before they are dispensed for any medically accepted
indication (Scholesberg and Jerath, 1999).

A few studies have examined the impacts of Medicaid and non-
Medicaid copayments on drug utilization and health care costs. In a 1993
study, Reeder et al. noted an 11% decrease in prescription use after South
Carolina established a 50-cent per prescription copayment. This increase was
significantly greater than in Tennessee, a comparison state with no copayments (Reeder et al., 1993). Another study using survey data from the 1992
Medicare Beneficiary Survey found that elderly and disabled Medicaid
beneficiaries who live in states with prescription drug copayments have
lower prescription drug utilization than their counterparts in states without
copayment, and three-fourths of the difference was directly attributed to
copayment policies. The study predicted that Medicaid copayments

reduce annual prescription drug utilization by 15.5% (Stuart and Zacker, 1999).

These findings are also supported by a study of non-Medicaid HMO members as a whole; a $1.50 copayment was associated with an 11% decrease in drug utilization (Harris and Stergachis, 1990). Similarly, persons in high coinsurance plans in the RAND Health Insurance Experiment reduced their use of most medications by 50–70% compared with plans with copayments. Lower income was related to greater reductions in prescription drug use (Lohr et al., 1986). Some state Medicaid plans also use prescription caps that limit the number of prescriptions that may be written for a person. Prescription caps are associated with a reduction in the utilization of prescription drugs (Soumerai et al., 1994; Martin and McMillan, 1996; Soumerai et al., 1997), especially by Medicaid beneficiaries who used multiple prescription drugs. Prescription limits are also associated with an increase in nursing home admissions for Medicaid beneficiaries with chronic illness (Soumerai et al., 1994). In a subsequent study of disabled Medicaid beneficiaries with schizophrenia, prescription caps led to increased community health center visits, increased use of emergency mental health services, and increased days of hospitalization (Scholesberg and Jerath, 1999).

C. DISPARITIES IN TREATMENT

Numerous studies demonstrate that even for those able to gain access to the health care system, disparities in treatment based on race and ethnicity are profound, even controlling for insurance coverage, education, and income. For example, Moore found that race was the strongest predictor for the receipt of drug therapy, with African Americans 41–73% less likely to receive drug agents (Moore et al., 1994). In a study of over 4000 elderly nursing home residents with cancer who experienced daily pain, African Americans were 63% more likely than whites to receive no pain medication (Bernabei et al., 1998). In diabetes care, African Americans were more likely to be treated with insulin but less likely to receive daily injections or to self-monitor their glucose levels (Cowie and Harris, 1997) A study of African American and white Medicaid-insured children in Detroit found that African American children were much more likely than their white counterparts to receive inadequate therapy, which included obsolete fixed-combination medications rather than recommended single-entity prescriptions, and were less likely to receive steroids for adrenergic inhalers, despite higher rates of health care and higher rates of hospitalizations (Bosco et al., 1993).

A study on racial differences in receipt of analgesics found that nearly three-fourths (74%) of white patients compared with 57% of African American patients received analgesics for lower extremity fractures in emergency departments (Todd et al., 2001). An assessment of racial/ethnic differences in physicians' prescriptions of patient-controlled analgesia for postoperative pain found that after adjustment for age, gender, preoperative use of narcotics, health insurance, and pain site, ethnicity persisted as an independent predictor of the amount of pain medication prescribed (Ng and Dimsdale, 1996; Jenkins, 2000). In regard to HIV disease, a study evaluating the use of anti-retroviral drugs and prophylactic therapy to treat *Pneumocystis carinii* pneumonia (PCP) in an urban population infected with HIV found that 63% of eligible whites and 48% of eligible African Americans received anti-retroviral therapy, and PCP prophylaxis was received by 82% of eligible whites but only 58% of eligible African Americans (Moore et al., 1994).

A study of the differences in white and nonwhite neighborhoods in pharmacy stocking of opioid analgesics found that over two-thirds of the pharmacies that did not carry any opiates were in predominantly nonwhite neighborhoods. After adjustments, pharmacies in predominantly nonwhite neighborhoods were significantly less likely to have adequate opiate supplies than pharmacies in white neighborhoods. Reasons for inadequate stock included little demand, concern for disposal, fear of fraud and illicit drug use, fear of robbery, and other problems (Morrision and Wallenstein, 2000).

In 2002, the Institute of Medicine reported that racial/ethnic minorities tend to receive lower-quality health care than whites do, even when insurance status, income, age, and severity of conditions are comparable. The report emphasized that differences in treating heart disease, cancer, and HIV infection partly contribute to higher death rates for minorities. According to Alan Nelson, chairperson of the committee, "Disparities in the health care delivered to racial and ethnic minorities are real and associated with worse outcomes in many cases, which is unacceptable" (Institute of Medicine, 2002). The study noted that there are many possible reasons for ethnic disparities in health care. Unequal treatment occurs in the context of persistent discrimination in many sectors of American life. Some evidence suggests that bias, prejudice, and stereotyping on the part of health care providers may contribute to differences in care. The report further noted that although it is reasonable to assume that the vast majority of health care providers find prejudice morally abhorrent, several studies show that even well-meaning people who are not overtly biased or prejudiced typically demonstrate unconscious negative racial attitudes and stereotypes. In addition, the time pressures that characterize clinical encounters, as well as the complex thinking and decision making they require, and uncertainty about a patient's condition may increase the likelihood that stereotyping will occur (Smedley et al., 2002).

1. ASSUMPTIONS AND STEREOTYPES.   One explanation for disparities in care relates to the assumptions held by health providers about racial and ethnic groups. Race-based health research and practice reveal a series of dubious assumptions and negative profiles of African American and other patients of color that lead to inferior medical treatment. Medicine sometimes attaches characteristics to particular racial and ethnic groups, and physicians

and other health care providers, for a variety of reasons, resort to these profiles in making individual treatment decisions. For example, African American patients are often assumed to be less likely to survive invasive medical procedures and less likely to respond to the standard course of treatment because of biological differences (Bowser, 2001). Race-based research links to a broader, institutionalized pattern of racial profiling of African Americans in clinical decision making. From a social context, they are also believed to be less compliant, more likely to engage in risky behaviors, to have less social support, and to be angrier and less intelligent, and to have difficulty understanding their medical condition and regimens (Schulman et al., 1999; Van Ryn and Burke, 2000).

2. CULTURALLY AND LINGUISTICALLY APPROPRIATE HEALTH CARE. Health care organizations nationwide vary widely in their ability to provide culturally and linguistically competent care. Some organizations are meeting the needs of their diverse populations; others are struggling to find the tools and resources; still others are not even aware of the importance of culturally competent health care. Lack of culturally competent care has been shown to cause miscommunication, mistrust, and poor outcomes. Provision of culturally and linguistically appropriate care, supported by administrative and policy supports and procedures, facilitates higher-quality health care (Cross et al., 1989; Campinha-Bacote, 1998 and 1999).

A few studies have highlighted cultural differences and pharmaceutical use. For example, some Asian cultural groups believe that multiple drugs are more effective for therapy than a single agent. This is based on the fact that multiple herbal ingredients are prescribed in traditional medicine. Many traditional Puerto Ricans and Mexicans feel that Western medicine is too strong and potent. In addition, some members of both of these groups often expect quick relief from symptoms; despite this fact they are cautious about the detrimental side effects of potent American medicine. In many Southeast Asian countries medicines are only prescribed on a short-term basis. Individuals, particularly recent immigrants, may have different preferences regarding the type of drug prescribed, tolerance of side effects, dosage form preference, and other aspects of drug therapy (Levy, 1999). How drugs are marketed and discussed in clinical encounters will affect acceptance by specific population groups.

3. ENVIRONMENTAL FACTORS.   The environment plays a large role in the disproportionate susceptibility of people of color to disease. There is a higher prevalence of harmful environmental exposures, including air pollution, lead, asbestos, and other toxins in minority and low-income neighborhoods. Minorities of color also disproportionately work in jobs with higher physical and psychosocial health risks (migrant workers, fast food workers, garment industry workers, and factory workers) and encounter stressors from living

in this environment (Weintraub, 2002). In addition, minority communities are frequently the targets of alcohol and tobacco marketing.

4. THE IMPACT OF DISPARITIES IN CARE ON TREATMENT WITH PHAR-MACOGENOMICS-BASED DRUGS. There are five possible scenarios involving pharmacogenomics that could be manifested because of unequal treatment based on race/ethnicity: (1) The patient will not be prescribed any drugs at all; (2) the patient will be offered traditional medications, based on the assumption that the pharmacogenomic drugs will be too expensive, Medicaid will not pay for them, or the process for prescribing and obtaining approval to utilize pharmacogenomic drugs is simply too complicated for the health care system or the patient; (3) the patient will receive a pharmacogenomic drug, but it will be the drug on the market for the "black male," and therefore it will be prescribed without a genetic test, which could lead to an adverse reaction; (4) the patient will be given the correct pharmacogenomics drug without adequate education or counseling about the drug; or (5) "a black male" will be prescribed the pharmacogenomics-based drug for a white male because the "white" drug is most convenient or affordable or available. Most of these possibilities stem from biased clinical decision making and a lack of equity in access to genetics services.

To reduce bias in clinical decision making, promulgation of detailed decision rules for prescribing medications should be considered by major providers and payers. In addition to rules based on practice and clinical protocols for decision making and utilization of genetics in medicine, including pharmacogenomics-based drugs, the development of prescribing report cards based on race/ethnicity and/or genotype and possibly tied to financial incentives may be helpful. Provider education against bias also will be necessary. Legal reforms, such as monitoring Medicaid program practices and finding mechanisms to facilitate sustained relationships between providers and patients, would also be helpful (Bloche, 2000).

IV. PHARMACOGENOMICS AND HEALTH DISPARITIES

What impact will pharmacogenomics have on promoting health, preventing and curing disease, and delivering appropriate care in communities of color? What are the implications for differential health status, access to care, and disparate treatment for pharmacogenomic research and the delivery of pharmacogenomics-based medicine to communities of color?

A. GENETIC VARIABILITY

Genetic variability unquestionably affects how a drug acts on different people, but genes for drug activity and response are not localized in socially defined races (Graves, 2001). Pharmacogenetic studies have demonstrated

that race and ethnicity are not always correlated with polymorphic variation, and thus race and ethnicity may not be a good indicator of drug metabolism. Whereas pharmacogenetic studies have revealed some important
genetic differences between Africans and non-Africans, a traditional social
category such as "African American" remains problematic, because the term
includes persons who have genetic markers from Africans and non-Africans.
This weakens the case for using race- or ethnicity-specific research categories
to gauge drug response (Foster et al., 2001). Graves illustrates this point by
noting that if sub-Saharan Africans only mated with sub-Saharan Africans
and Europeans only mated with Europeans there might be unique lineages,
but this has not occurred, particularly in America, because of the history of
chattel slavery (Graves, 2001). Other explanations of variation in allele frequency, such as geographic locality, further complicate the interpretation of
associations between drug response and social categories and further limit
the usefulness of social labels in predicting drug response (Foster et al., 2001).
An experimental design to find out what genes are responsible for a drug
response of an individual is preferable to an inquiry limited by using a
person's socially constructed race (Villarosa, 2002).

It is also important to understand the source and significance of genetic
variations. The Pima Indians have the highest rates of diabetes in the world;
Tay–Sachs disease is primarily found in Ashkenazi Jews. Contemporary literature indicates that these differences stem from reproductive isolation, not
race. Genetic traits common to persons with sickle-cell disease are related to
malaria frequency and not our social view of race. This is why the disease
can be found in high frequency in Yemen, West Africa, Greece, and Saudi
Arabia.

Some researchers have speculated that biologically based racial differences in clinical presentation or response to treatment may justify racial differences in the type and intensity of treatment provided. Several studies find
important differences among racial/ethnic groups in physiological response
to drugs and consider genetics as the major underlying determinant. These
differences relate to metabolism, clinical effectiveness, and side effects profiles. Studies have shown differences in response to beta-blockers, diuretics,
calcium channel blockers, converting enzyme inhibitors, central nervous
system agents, antihistamines, alcohol, and analgesics. Studies also show
that plasma concentrations and responses to antidepressants vary considerably between patients treated with similar doses. About 5% of Caucasians
are poor metabolizers of some antidepressants, and some may develop
adverse reactions to these drugs (Bertilsson, 1997). Genes of the major histocompatibility groups have been associated with susceptibility to drug-
induced adverse reactions in Ashkenazi Jews and, to a slightly lesser degree,
in Eskimos (Corzo et al., 1995). Northern Canadian Indians are extremely
rapid acetylators of isoniazid. In addition, racial and ethnic group differences
are found in response to drug therapies such as Enalapril, an angiotensin-
converting enzyme inhibitor used to reduce heart failure (Exner et al., 2001).

These differences in response to drug therapy, however, are not due to "race" but rather to the distribution of polymorphic traits among population groups (Wood, 2001).

B. DRUG TRIALS

Will minorities be involved in clinical trials for pharmacogenomic drugs, especially trials that are segmented by genotype? Will safeguards be in place to guard against discrimination based on genetic information?

1. INVOLVEMENT IN DRUG TRIALS. A convincing argument could be made that people of color should be involved in clinical trials because if there are drugs that work better for certain genotypes correlated with racial/ ethnic groups, the members of these groups need to be involved in that research. Nevertheless, recruitment within racial/ethnic groups may be difficult. Using broad social categories, such as race/ethnicity, to identify populations for pharmacogenetic research may cause both investigators and pharmaceutical companies to understate genetic variability within those populations. Members who were not part of the research studies may, as a result, have limited access to diagnostic tests and pharmaceutical products developed from the studies, either because they have different polymorphisms or because the prevalence of the polymorphism is not known for these populations (Foster et al., 2001). This highlights the point that the people who benefit the most from the research are usually the ethnic groups who have been included in the studies (Lucino, 2000). Differences in drug response are explained by shared environments and pathogenesis of diseases rather than absolute biological differences (Epps, 2000).

Currently, NIH mandates that all racial/ethnic groups be included in federally funded studies unless there is a compelling scientific basis for exclusion. NIH considers the recruitment of sufficient numbers of ethnic minorities into studies crucial to minority health by permitting the analysis of data for that specific group. The analysis, however, should not just consider race as a broad category, but it is essential to probe the variables behind the race category. These variables might include diet, stress, racism, urban setting, family dynamics, and other factors. If race is deconstructed, we will uncover the more subtle variables that determine different responses to drugs. This will also widen the applicability of results, not only based on race, but on life experiences and environment for different groups of people (Epps and Rothstein, 2002).

It has also been noted by some researchers that pharmacogenomics could reverse the current regulatory push to include more diverse grouping of people in clinical trials. Under the rubric of pharmacogenomics, reviewers might deem it unethical to test a new drug on people whose genes suggest they will not respond well to the drug (Times, 2000). Laws and FDA regulations may have to be changed to accommodate the need for targeting patients with rare diseases or with subtypes of otherwise common diseases. This approach will set the stage for testing whether targeting small populations with select drugs is superior to treating many patients with the best drug for a given disease. The outcome may vary from one case to another (Mancinelli et al., 2000).

Others believe that, within the context of pharmacogenomics, clinical trials constructed around a particular polymorphism should not conflict with inclusiveness guidelines. And even though subpopulation frequencies of any polymorphism vary across different ethnic groups, the frequency rarely drops to zero percent for any ethnic group. For a given polymorphism, at least someone will possess the variation in each group. The end result may be greater difficulty in recruiting sufficient participants who not only share a particular SNP but who also represent the ethnic diversity of the national population in sufficient numbers. The effect of higher standards may add to the cost and time needed to get a drug through the approval process (Epps and Rothstein, 2002).

Given this analysis, the question is, What is the relevance of the FDA's 1998 rule requiring sponsors of new investigational drugs to submit effectiveness data by race/ethnicity, age, and gender (Scott, 2001)? Should race be considered at all? It appears that there is a contradiction between what we now know—that race is a social, not a biological, concept—and recruiting subjects for genetic testing and research, including pharmacogenomics, by race. Researchers must be careful that utilization of racial/ethnic distinctions do not serve to reinforce racial/social categories.

The case of BiDil highlights the debate concerning the utility and ethics of using social labels of race to determine the effectiveness of a drug for a group of people. NitroMed, Inc., is working in partnership with the Association of Black Cardiologists to conduct final testing of a drug designed to treat African American patients with heart failure. The drug, BiDil, is believed to be the first government-approved medication that will be developed for and marketed specifically to a racial group. The clinical trial includes only African Americans. The drug was developed because there is evidence that African Americans did not respond well to widely successful heart failure drugs such as beta-blockers and ACE inhibitors. One theory is that African Americans may produce less nitric oxide in their bodies than Caucasians. Because the effectiveness of ACE inhibitors hinges on the production of nitric oxide, BiDil, a pill combining two long-established heart drugs, works by restoring depleted nitric oxide levels and protecting nitric oxide formed naturally in the body (Sealey, 2001).

The trial will last a year and include 800 patients. All participants will continue to use their regular medications. Although preliminary data suggest that African Americans respond well to the new drugs, some question marketing a drug to a particular race, because it is not scientifically sound. It is possible that if NitroMed starts aggressively advertising to African Americans, these patients and their doctors could ignore other potentially effective drugs. Because patients do not always respond to the same drug, marketing to a race can be misleading, unethical, and even dangerous (Sealey, 2001).

BiDil, although not an example of a medicine based on genotype, illustrates concerns over racial profiling for drug treatment recommendations. Similarly, where the elimination of racial disparities in health status may be linked with the availability of therapies best suited for persons with a genotypic characteristic that occurs with greater frequency in a particular ethnic group, the question is whether the benefit of designing drugs tailored to common genotypes outweighs the danger of perpetuating partially misleading perceptions of biological heterogeneity between racial classifications and homogeneity within racial classifications (Epps and Rothstein, 2002).

2. PHARMACOGENOMIC DRUGS UNDER MEDICAID. Because a disproportionately large number of minorities in the U.S. utilize Medicaid as their insurer, it is important to think about pharmacogenomics and minority Medicaid populations. Who would be eligible to access expensive pharmacogenomic drugs? What classes of drugs would be covered? If a person has a genotype for a drug that may be out of the Medicaid formulary, will Medicaid make an exception for that person? And, if so, how much paperwork and time would be needed for approval? What will be the cost of genetic testing? Will the demand for genetic counselors increase? How will the costs of new drugs affect Medicaid budgets?

Most of the implications for low-income people under Medicaid will probably be negative. In light of current practices under Medicaid, the following may be potential implications of pharmacogenomics.

PRESCRIPTION LIMITS. Several states under Medicaid control utilization and costs by limiting the number of prescriptions that a beneficiary can have filled per month or per year. It is possible that, because pharmacogenomic drugs will decrease or eliminate trials of different drugs or dosages, there will be less need for multiple prescriptions to calibrate the drug dosage or choose the correct drug (Scholesberg and Jerath, 1999), thereby reducing the cost of care. On the other hand, it is possible that individuals may require several different types of drugs, which may be all or partially pharmacogenomic drugs. This leads to the possibility that patients would be limited to a smaller number of pharmacogenomic drugs or a smaller number of drugs overall because of the higher costs of these drugs.

COST SHARING. Most states impose cost sharing on patients in the form of copayments ranging from fifty-cents to five dollars. In 1996, 29 states and the District of Columbia had adopted prescription drug copayments for all groups except for children, pregnant women, institutionalized individuals, and the categorically needy in Medicaid HMOs (Scholesberg and Jerath, 1999). It is possible that the potential higher costs of pharmacogenomic drugs will increase cost sharing for all drugs paid for by Medicaid patients, therefore further decreasing drug utilization and access. As noted above, this would increase illness, health care visits, and admissions to hospital and nursing homes. Therefore, Medicaid patients who are already in a system that requires prior authorization, prescription caps, and other limits may experience an increase in differential access to pharmaceuticals compared with the general population.

RESTRICTING ACCESS TO SPECIFIED DRUGS THROUGH DRUG FORMU-LARIES.   Medicaid managed care organizations or those that serve Medicaid clients save money by buying pharmaceuticals in bulk and limiting coverage to only those drugs listed on the formulary. If pharmacogenomics causes a segmentation of patients by genotype, it will make it harder to buy drugs in bulk.

When pharmacogenomic drugs begin to fill the market, formularies may be overly restrictive. Historically, drugs to treat HIV/AIDS, mental illness, and other chronic conditions have often been difficult to obtain, especially for people of color. Reports from both the commercial market and Medicaid managed care indicate that Medicaid managed care plans will require providers to first use older, less expensive drugs in their formularies before filling prescriptions for newer drugs like Clozaril, to treat mental illness. This scenario does not bode well for pharmacogenomic drugs (Scholesberg and Jerath, 1999).

PRIOR AUTHORIZATION.   Authorization can be a hindrance to both the physician and the patient. If the physician thinks that the approval of a drug is in question, rather than wasting time, he or she may just prescribe the more conventional pharmaceutical.

DRUGS TO BE INCLUDED IN THE FORMULARY.   Currently, a covered out-patient drug may be excluded from the formulary with respect to a specific disease or condition for an identified population only if, based on the drug's labeling, the excluded drug does not have a significant, clinically meaningful therapeutic advantage in terms of safety or efficacy over other drugs in the formulary and there is a written explanation (available to the public) for the basis for exclusion. This regulation should promote pharmacogenomic drugs because they are tailored to particular genotypes to maximize safety, efficacy, and clinical outcome compared with other drugs for individuals with a specific genotype. Therefore, at least in theory, under this regulation pharmacogenomic drugs should be able to prove better safety and efficacy over a traditional drug.

V. POTENTIAL BENEFITS AND COSTS WITH THE USE OF PHARMACOGENOMIC MEDICINES IN COMMUNITIES OF COLOR

A. PUBLIC HEALTH APPROACHES

Pharmacogenomics may be beneficial to people of color because of their high rates of morbidity and mortality from certain cancers, hypertension, cardiovascular disease, asthma, HIV/AIDS, Alzheimer disease, clinical depression, and other diseases. Thus more effective therapies help those individuals most in need of treatment.

Pharmacogenomics has the potential to provide a vast array of interventions to prevent and treat diseases. For example, prevastin may be more effective in lowering blood lipid levels in people with a variant of the CETP gene compared with other drugs. Prevastin could help reduce the risks of cardiovascular disease, which is disproportionately prevalent in African American and Hispanic populations. Pharmacogenomics also offers the promise of targeted primary chemoprevention.

For pharmacogenomics to fulfill the promise of targeted interventions, clinical and epidemiological studies are urgently needed to (1) assess how drug responses vary among individuals within different genotypes; (2) determine the prevalence of relevant genotypes in the population or subpopulations; and (3) consider whether and to what degree other environmental factors, such as drugs, diet, and environmental toxins or pollutants, interact with genetic factors to explain drug response. Clinical trials and observational epidemiological studies are crucial for providing the population-level data to utilize pharmacogenomics in public health (Khoury and Morris, 2001).

B. GENE-BASED MEDICAL CARE

The potential impact of pharmacogenomics on individualized treatment in medicine is touted widely. Tailored medicines have been introduced into the treatment of HIV disease, which affects African Americans and Hispanics at a rate more than twice that of other populations in the U.S. In treating AIDS, doctors prescribe combinations of three or four drugs at a time to hold the virus at bay. Although there are now 17 drugs from which to choose combinations, there are also at least 120 HIV gene variants that render one or more of the drugs ineffective. To learn which medicines would be useless to prescribe, doctors have increasingly been having their patients' viruses analyzed for drug-resistant gene variants. Ineffective treatment with the wrong drug and the associated cost and suffering can be thus avoided. Other current and upcoming drugs on the market include drugs to be utilized by poor metabolizers of CYP2D6 drug-metabolizing enzyme and those with long-QT syndrome in cardiology (Hollan, 2001).

Although the incidence of breast cancer is lower for African American and Hispanic women, their mortality rates are higher than for Caucasian women. Most traditional cancer drugs help less than half of patients, exposing them to the toxic side effects without the prospect of improving their health. Herceptin is a drug designed to treat metastatic breast cancer whose tumors are shown to express abnormally high amounts of the protein called HER2. For these patients, up to 30% of women with breast cancer, Herceptin can bind to HER2, slowing tumor growth. Herceptin is one of four drugs in various developmental stages that attack tumors with overactive cancer genes. Intended for a variety of cancers, brain, melanoma, prostate, breast, and lung, they will be prescribed only after first verifying cancer gene overactivity. For women of color, the use of this drug could be very beneficial, especially because their cancers are detected later and they therefore experience higher mortality rates. Thus pharmacogenomics could be particularly important for minorities who seek treatment late. They, of all patients, cannot afford to use a drug that is ineffective, or even dangerous (Hollan, 2001).

C. PRODUCTION OF DRUGS FOR SMALL POPULATION GROUPS

It has been postulated that profit-conscious companies will use pharmacogenomics to aim their drug development efforts toward genetic subgroups of people who can best afford them, further marginalizing already underserved populations (Times, 2000). Some drug responses are linked to ethnicity; some drugs may have only a very small potential market. This could have negative implications for testing and marketing. Despite this possibility, there is some evidence that even if big pharmaceutical companies do not produce these products small biotechnology companies will produce drugs for this market.

If pharmacogenomic screening serves primarily to shrink a drug's market size, companies are unlikely to push testing forward. The FDA may need to mandate pharmacogenomic testing of certain kinds of pharmacogenomic drugs. An example of a small market is the 14% of Alzheimer disease patients with the APOE4 allele; no companies report having drugs in development for this particular subpopulation. Dr. Robert Green, senior director of product development at Genetech, states, " No intelligent pharmaceutical company wants to play for a small fraction, even in a very large market" (Thiel, 1999). Developing a product for 5% of the population means that you have to screen out 95%, "for one thing, that's very expensive." He continues, "Even a small number of false positive diagnostic results would make drugs for tiny sub-populations impractical. If biotech and pharmaceutical companies are unwilling to research and develop niche market drugs, governmental subsidies, akin to those under the Orphan Drug Act, may be needed."

VI. Conclusion

The promise of personalized medicine heralds a new era of medical care with potential benefits for America in general and communities of color in particular, because of the disproportionate impact of disease for this group. However, we must not lose sight of the fact that the fundamental structure of the U.S. health care system is flawed. Unequal access to treatment and services and divergent public perceptions and trust in the system are only some of the ways in which this is manifested. New technologies do not eliminate, but normally exacerbate, existing inequalities in the system of care. For pharmacogenomics to have a positive effect on health disparities, we must address the fundamental problems that lead to disparities in health status and care. We will have a far greater impact on health status outcomes for communities of color if we look to pharmacogenomics-based drugs as an enhancement of the health care system and not a panacea that ignores the social, political, and structural causes of health disparities.

Acknowledgment

I would like to thank Phyllis Griffin Epps, J.D., from the University of Houston Health Law and Policy Institute, for her thoughtful and insightful comments.

References

Banton, M., *Racial Theories*. Cambridge: Cambridge University Press (1986).

Banton, M. and H.J. Banton, *The Race Concept*. New York: Praeger (1975).

Bernabei, R., et al., "Management of Pain in Elderly Patients with Cancer," *JAMA*, **279**, 1877–1882 (1998).

Bertilsson, L., "Pharmacogenetics of Anti-depressants: Clinical Aspects," *Acta. Psychiat. Scand. Suppl.*, **391**, 14–21 (1997).

Blendon, R., "Access to Medical Care for African American and White Americans: A Matter of Continuing Concern," *JAMA*, **261**, 278–281 (1989).

Bloche, G., "Race and Discretion in American Medicine," *Yale J. Hlth. Policy L. Ethics*, **1(1)**, 95–132 (2000).

Bosco, L.A., et al., "Variations in the Use of Medications for the Treatment of Childhood Asthma in the Michigan Medicaid Population," *Chest* **104**, 1727–1732 (1993).

Bowser, R., "Racial Profiling in Medicine," *Mich. J. L. Race*, Vernellia R. Randall, **2002**, 79–133 (2001).

Byrd, M. and L. Clayton, *An American Health Dilemma: A History of African Americans in the Health System: Beginning to 1900*. New York: Routledge (2000).

Byrd, M. and L. Clayton, *An American Health Dilemma. Volume 2. Race, Medicine and Health Care in the United States 1900–2000*. New York: Routledge (2002a).

Byrd, M.W. and L. Clayton, "Understanding and Eliminating Racial and Ethnic Disparities in Health Care: A Background and History," In: *Unequal Treatment: Confronting Ethnic and Racial Disparities in Care*, Institute of Medicine, National Academy Press, Washington, DC (2002b).

Campinha-Bacote, J., *The Process of Cultural Competence in the Delivery of Health Care Services*, Transcultural C.A.R.E. Associates, Ohio (1998 and 1999).

Cavalli-Sforza, L.L., *The History and Geography of Human Genes*. Princeton, NJ: Princeton University Press (1994).

Cavalli-Sforza, L.L., *Genes, People and Language*. New York: North Point Press/Farrar, Strauss and Giroux (2000).

Collins, F.S., "Implications of the Human Genome Project," *JAMA*, **285**, 540–544 (2001).

Corzo, D., et al., "The Major Histocompatibility Complex Region Marked by HSP and HSP70-2 Variant is Associated with Clozapine-induced Agranulocytosis in Two Different Ethnic Groups," *Blood*, **86**, 3835–3840 (1995).

Cowie, C. and M.I. Harris, "Ambulatory Medical Care for Non-Hispanic Whites, African-Americans, and Mexican Americans with MIDDM in the U.S.," *Diabetes Care* **20**, 142–147 (1997).

Cross, T.L., et al., "Toward a Culturally Competent System of Care," Vol. 1: Monograph on Effective Services for Minority Children Who Are Severely Emotionally Disturbed. Washington, DC: CASSP Technical Assistance Center, Georgetown University Child Development Center (1989).

Diamond, J., *Guns, Germs and Steel: The Fates of Societies*. New York: W.W. Norton & Company (1999).

Doescher, M.P., "Racial and Ethnic Disparities in Perceptions of Physicians' Style and Trust," *Arch. Fam. Med.*, **9**, 1115–1163 (2000).

Doonan, M., "The Economics of Prescription Drug Pricing," Schneider Institute for Health Policy, Heller School for Social Policy and Management, Brandeis University, Waltham, MA, pp. 5–6 (2001).

Dula, A. and S. Goering, *It Just Ain't Fair: The Ethics of Health Care for African-Americans*, Wesport: Praeger, pp. Introduction, 5–6 (2001).

Epps, P.G., *White Pill, Yellow Pill, Red Pill, Brown Pill: Pharmacoeconomics and the Changing Face of Medicine: The Challenges and the Impact of Humane Genome Research in Minority Communities*. Zeta Phi Beta Society, Inc., National Educational Foundation, July 7–8, 2000, Philadelphia PA (2000).

Epps, P.G. and M.A. Rothstein, "Pharmacogenomics: Ensuring Equity Regarding Drugs Based on Genetic Difference." In: J. Licinio and M.-L. Wong (eds.), *Pharmacogenomics: The Search for Individualized Therapies*. Weinheim, Germany: Wiley-VCH (2002).

Exner, D., et al., "Lesser Response to Angiotensin-Converting-Enzyme Inhibitor Therapy in African Americans as Compared to White Patients with Left Ventricular Dysfunction," *N. Eng.. J. Med.*, **344**, 1351–1357 (2001).

Feagin, J., *Racist America: Roots, Current Realities, and Future Reparations*. New York: Routledge (2000).

Feagin, J. and C. Feagin, *Racial and Ethnic Relations*. Upper Saddle River, NJ: Prentice Hall (1999).

Foster, M., et al., "Pharmacogenetics, Race and Ethnicity: Social Identities and Individualizes Medical Care," *Therapeut. Drug Monitoring*, **23**, 232–237 (2001).

Garcia, J., "African-American Perspectives, Cultural Relativism and Normative Issues: Some Conceptual Questions.," In: H.F. Flack and E. Pellegrino (eds.), *African-American Perspectives on Biomedical Ethics*. Washington, DC: Georgetown University Press, pp. 11–66 (1992).

Geiger, J., "Racial and Ethnic Disparities in Diagnosis and Treatment: A Review of the Evidence and A Consideration of Causes," Unequal Treatment: Confronting Ethnic and Racial Disparities in Care. Washington, DC: National Academy Press (2001).

Graves, J., *The Emperor's New Clothes*. New Brunswick, NJ: Rutgers University Press (2001).

Harris, B. and A. Stergachis, "The Effect of Drug Co-payments on Utilization and Cost of Pharmaceuticals in a Health Maintenance Organization," *Med. Care*, **28**, 907–917 (1990).

Henry J. Kaiser Family Foundation, "Race, Ethnicity and Medical Care—A Survey of Public Perceptions and Experiences," 1–29 (1999).

Henry J. Kaiser Family Foundation, "Racial and Ethnic Disparities in Access to Health Insurance and Health Care," *UCLA Center for Health Policy Research*, **XI**, 3 (2002).

Hollan, T., "The Making of the Pharmacogenomic Prescription," *Gene Letter, GeneSage* (Jan. 2001).

Institute of Medicine, "Minorities More Likely to Receive Lower-Quality Health Care, Regardless of Income and Insurance Coverage," Press Release (March 20, 2002), *www.omhrc.gov/omhrc/pressreleases/2002press0320.htm* (last visited July 6, 2002).

Jenkins, I.L., *The Human Genome Project: Ethical, Legal, and Social Implications for the Minority Communities: The Challenges and Impact of Humane Genome Research for Minority Communities*. Zeta Phi Beta Society, Inc., National Educational Foundation, July 7–8, 2000, Philadelphia, PA (2000).

Khoury, M., "Challenges in Communicating Genetics: A Public Health Approach," Centers for Disease Control and Prevention (2000).

Khoury, M., "Genetics and Public Health: A Framework for the Integration of Human Genetics into Public Health Practice," *Genetics and Public Health in the 21st Century*. Centers for Disease Control and Prevention (2000).

Khoury, M. and J. Morris, "Pharmacogenomics and Public Health: The Promise of Targeted Disease Prevention," Centers for Disease Control and Prevention (2001).

King, R., "Race: An Outdated Concept," *Am. Med. Women's Assoc. J.*, **10**, 55–58 (1995).

LaVeist, T.A., et al., "Attitudes about Racism, Medical Mistrust and Satisfaction with Care among African Americans and White Cardiac Patients," *Med. Care Res. Rev.*, **Suppl 1**, 146–161 (2000).

Levi-Strauss, C., *Race and History*. Paris: United Nations Educational, Scientific and Cultural Organization, pp. 123–163 (1951).

Levy, R., "Cultural Diversity and Pharmaceutical Care," National Pharmaceutical Council, Virginia (1999).

Lewin, M.E. and S. Altman, *America's Health Care Safety Net: Intact But Endangered*. Washington, DC: National Academy Press (2000).

Lillie-Blanton, M., et al., "Race, Ethnicity and the Health Care System: Public Perceptions and Experiences," *Med. Care Res. Rev.*, **57**, 218–235 (2000).

Lohr, K., et al., "Use of Medical Care in the Rand Health Insurance Experiment: Diagnosis and Service Specific Analyses in a Randomized Control Fill," *Medicare*, **24**, S729–S801 (1986).

Lucino, J., "Pharmacogenomics and Ethnic Minorities." *Psychiat. Times* **XVII(11)**, 1–6 (2000).

Lurie, N., et al., "Special Report: Termination of Medi-Cal Does It Affect Health?" *N. Engl. J. Med.*, **3**, 480–484 (1984).

Mancinelli, L., et al., "Pharmacogenomics: The Promise of Personalized Medicine," *AAPS Pharmsci 2002*, **2**, 1–11 (2000).

Martin, B. and J. McMillan, "The Impact of Implementing A More Restrictive Prescription Limit on Medicaid Recipients—Effects on Cost, Therapy, and Out of Pocket Expenditures," *Med. Care*, **34(7)**, 686–701 (1996).

Mayberry, R., "Racial and Ethnic Differences in Access to Medical Care," *Henry J. Kaiser Family Foundation*, 1–3 (1999).

McCombs, J., et al., "The Costs of Interrupting Anti-Hypertensive Drug Therapy in A Medicaid Population," *Med. Care*, **32**, 214–226 (1994).

Moore, R., et al., "Racial Differences in the Use of Drug Therapy for HIV Disease in an Urban Community," *N. Engl. J. Med.*, **330**, 763–768 (1994).

Morrision, R. and S. Wallenstein, " 'We Don't Carry That'—Failure of Pharmacies in Predominately Non-White Neighborhoods to Stock Opioid Analgesics," *N. Engl. J. Med.*, **342**, 1023–1026 (2000).

National Medical Association, *National Colloquium on African American Health: Racism in Medicine and Health Parity for African Americans: The Slave Health Deficit*. Washington, DC: National Medical Association, 11 (2002).

Ng, B. and J. Dimsdale, "The Effect of Ethnicity on Prescriptions for Patient Controlled Analgesia for Post-Operative Pain," *Pain* **66(1)**, 9–12 (1996).

Reeder, C., et al., "Economic Impact of Cost Containment Strategies in Third Party Programmes in the U.S. (Part 1)," *Pharmaeconomics*, **4**, 92–103 (1993).

Rose, S., et al., *Not In Our Genes: Biology, Ideology, and Human Nature*. New York: Penguin (1984).

Rothstein, M.A. and C.A. Hornung, "Public Attitudes About Pharmacogenomics," this volume.

Ryn, V., "The Implications of Social Cognition and Theory for the Provider: Contribution to Race/Ethnicity Disparities in Health Care," Commissioned Paper Prepared for Physicians for Human Rights Committee on Racial and Ethnic Disparities in Diagnosis and Treatment in the United States Health System, Washington, DC (2001).

Scholesberg, C. and S. Jerath, "Fact Sheet: Prescription Drug Coverage Under Medicaid," *National Health Law Project, 2002* (1999).

Schulman, K., et al., "The Effect of Race and Sex on Physicians' Recommendations for Cardiac Catheterization," *N. Engl. J. Med.*, **340**, 618–626 (1999).

Scott, R., "FDA Requires New Drug Applications Present Effectiveness and Safety Data for Gender, Racial and Age Sub-Groups" (2001), *www.law.uh.edu/healthlawperspectives/food/980213FDARequires.html* (last visited July 6, 2002).

Sealey, G., "Race and the Heart, 1st Drug Developed for African American Heart Failure Patients," *ABCNews.com* (2001).

Smedley, A., *Race in North America: Origin and Evolution of a World View*. Boulder, CO: Westview Press (1999).

Smedley, B., et al., *Unequal Treatment: Confronting Racial Disparities in Health Care*, Washington, DC: National Academy Press (2002).

Smith, D., "Addressing Racial Inequalities in Health: Civil Rights Monitoring and Report Cards," *Hlth. Politics Policy L.*, **23**, 75–105 (1998).

Soumerai, S., et al., "Effects of Limiting Medicaid Drug Reimbursement Benefits on the Use of Psychotropic Agents and Acute Mental Health Services by Patients with Schizophrenia," *N. Engl. J. Med.*, **331**, 650–655 (1994).

Soumerai, S., et al., "Determinants of Change in Medicaid Pharmaceutical Cost Sharing: Does the Evidence Affect Policy?" *Milbank Q.*, **75**, 11–34 (1997).

Stuart, B. and C. Zacker, "Who Bears the Burden of Medicaid Drug Co-Payment Policies?" *Hlth. Affairs*, **18**, 201–212 (1999).

Thiel, K.A., "Pharmacogenomic Medicine-Technology Outpacing the Health Care System," *BioSpace.com*. *2002* (1999).

Times, S.P., "Which Medicine to Take May Depend on Genes," *St. Petersburg Times*, St. Petersburg, FL (2000).

Todd, K., et al., "Ethnicity and Analgesic Practice," *Ann. Emerg. Med.*, **35**, 11–16 (2001).

U.S. Department of Health and Human Services, "Eliminating Racial and Ethnic Disparities in Health," Office of Minority Health (1999).

Van de Berghe, P.L. Race and Racism: A Comparative Perspective. New York: John Wiley and Sons, Inc. (1967).

Van Ryn, M. and J. Burke, "The Effect of Patient Race and Socio-Economic Status on Physicians' Perceptions of Patients," *Social Sci. Med.*, **50**, 813–828 (2000).

Veenstra, D. and M. Higashi, "Assessing the Cost-Effectiveness of Pharmacogenomics," *AAPI PharmSci. 2001* (2000).

Villarosa, L., "Beyond African American and White in Biology and Medicine," *N.Y. Times*, Sect. F, p. 5, col. 2 (Jan. 1, 2002).

Weinick, R., "Racial and Ethnic Differences in Access to and Use of Health Services," *Med. Care Res. Rev.*, **57**, 36–54 (2000).

Weintraub, I., "Fighting Environmental Racism: A Selected Annotated Bibliography," *http:llegj.lib.uidaho.edu/egj01/weintol.html* (last visited July 6, 2002).

Weisse, P., et al., "Do Gender and Race Affect Decisions About Pain Management?" *J. Gen. Intern. Med.*, **4**, 211–217 (2001).

Williams, D., "African American Health and the Role of the Social Environment." *J. Urban Hlth.: Bull. N.Y. Acad. Med.*, **75**, 300–321 (1998).

Wood, A., "Racial Differences in Response to Drugs-Pointers to Genetic Differences," *N. Engl. J. Med.*, **344**, 1393–1395 (2001).

CONSTITUTIONAL ISSUES IN THE USE OF PHARMACOGENOMIC VARIATIONS ASSOCIATED WITH RACE

JOHN A. ROBERTSON, J.D.

I. INTRODUCTION

Progress in genetic knowledge and gene sequencing has renewed attention to the use of racial and ethnic classifications in health care. Most scientists now believe that the categories of race and ethnicity are predominantly social and cultural constructs, rather than meaningful indicators of actual genetic difference. Little time has elapsed, on an evolutionary scale, since the ancestors common to all modern humans dispersed from Africa (Owens and King, 1999; Olson, 2001). It is markedly improbable that in that brief time geographically dispersed humans encountered selection pressures at once so strong and so materially different that group-specific traits have evolved beyond those that influence such superficial traits as skin coloration. Genetic research, however, has now shown that some allelic variations in disease susceptibility and in drug responsiveness correlate with race and ethnicity as commonly understood. Although race matters biologically hardly at all, in some instances it may be significant for the provision of good medical care to all population groups. Before exploring the ethical and legal standards for taking such differences into account, I first describe the genomic developments that have raised this issue.

Pharmacogenomics: Social, Ethical, and Clinical Dimensions, Edited by Mark A. Rothstein.
ISBN 0-471-22769-2   Copyright © 2003 Wiley-Liss, Inc.

II. THE EMERGENCE OF PHARMACOGENOMICS

Growing knowledge of the human genome has made medical care based on genetics a likely future possibility. Pharmacogenomics—the study of genotypic variation in drug response—may come to play an important role in such a future, with physicians routinely using genetic testing to determine a person's drug response profile before prescribing a drug. Developed and used on a wide scale, pharmacogenomics medicine has enormous potential to improve health care.

The surge of interest in pharmacogenomics stems from growing knowledge that individual responses to drugs vary considerably. In some cases, these variations reflect genetic differences. For example, because of a genetic polymorphism, some persons may not produce the receptors needed for drugs to dock to a cell. Other polymorphisms could affect the development of proteins/enzymes that enable a drug to pass through the cell wall, to be distributed throughout the cell, or to be excreted. The clinical relevance of drug response polymorphisms has already been established for genetic variations affecting a person's response to drugs for leukemia, asthma, and other conditions. It is now known that cytochrome P450 genes play a major role in producing enzymes responsible for metabolizing one-third to one-half of all drugs now in use. It is likely that polymorphisms in those genes will affect responsiveness to cholesterol-lowering statins, antibiotics, analgesics, antipsychotics, and selective serotonin reuptake inhibitors like Prozac and other antidepressants. Variations in other genes might affect the responsiveness of heart patients to such standard therapies as beta-blockers and angiotensin-converting enzyme (ACE) inhibitors.

Many other examples of pharmacogenomic effectiveness are likely to emerge as government agencies sponsor and drug companies conduct research in this field. As knowledge about specific drug responses and drugs accumulates, that knowledge will be introduced into medical practice. Genetic testing for drug response, like genetic testing for susceptibility and carrier states, may become an essential and established part of medical practice, with many patients not receiving drug prescriptions unless they have first taken a genetic test confirming the safety and efficacy of the drug for them. Although pharmacogenomic testing will not be relevant to all drugs or illnesses, it will be significant enough to change drastically much of clinical medicine.

III. RACE AND PHARMACOGENOMIC VARIATION

Incorporation of pharmacogenomic testing and prescription into medical practice, however, may also require that attention be given to race and ethnicity. Investigation of the pharmacogenomic basis of drug safety and efficacy has uncovered some differences in drug response profiles correlated with race and ethnicity, just as there are variations in genes that increase risks to health (e.g., sickle-cell anemia, colon cancer) correlated with race and ethnicity (Weber et al., 1999; Graves, 2001).

Consider, for example, the fact that in the United States, African Americans are more likely than whites to develop congestive heart failure and are roughly twice as likely to die of the disease. They do not, however, benefit as much from ACE inhibitors, the front-line drug therapy for heart failure, as do whites. Many African Americans have polymorphisms in the MSD720 gene, which causes them to produce lower amounts of nitrous oxide in response to ACE inhibitors than whites. Because the bioactivity of endogenous nitrous oxide is known to be lower in African Americans than in whites, standard ACE inhibitors will generally have little effect in African Americans with that variation. Unless identified and given another drug, African Americans with that variation will receive less effective treatment for heart failure and will die at a higher rate, as now occurs (Noah, 1998).

Although not all African Americans have these variations, enough do to justify testing them for a different set of genetic polymorphisms than whites are tested for. These variants affect cell receptors and drug pathways, both of which are necessary for drugs to have therapeutic effect. Genetic testing for those variations before selection of a drug could provide African Americans with more effective therapy for heart disease and other illnesses than they now receive (Exner et al., 2001; Yancey et al., 2001).

Another example of racial differences in drug responsiveness is variations in the CYP 2D6 gene, which codes for the enzyme responsible for the metabolism of beta-blockers and tricyclic antidepressants. The gene is functionally absent in 8% of whites and 4% of Asians, but it is missing in 25% of African Americans because of an allele generally not found in those other groups. Those lacking the enzyme would be at a risk of excess α-adrenergic blockade and severe toxicity from inability to metabolize the drug that those without the variation avoid. In some cases, lowering the dosage will alleviate the toxicity and preserve some therapeutic effect. In other cases, the drug will be contraindicated altogether, and some other drug will be needed to treat the condition. Whether the variations interfere with receptors, enzymes, or particular cellular pathways, they affect the ability to metabolize drugs (Wood, 2001). Identifying them in advance would greatly aid physicians in deciding the best course of treatment.

Racial and ethnic differences in drug response are important clinically, but they also show the importance of the underlying genetic determinants of drug response. Once scientists identify the genetic determinants of differential response to drugs for heart disease, such as enalapril and carvedilol, the genetic variations can be tested for directly. In the future, DNA microarrays will be able to test for many drug response genotypes simultaneously, thus avoiding the need to rely on appearance and self-reports of race and ethnicity to determine what tests to order. This will improve accuracy, for not all persons with the same racial and ethnic background will have the same drug response polymorphisms. For the near and mid-future, however, race, gender, and ethnicity will remain indicators of genetic polymorphisms affecting health status and drug responsiveness.

This brief discussion shows that racial distinctions reflecting underlying genotypic variation may play or come to play an important role in future health care. Given the short time in which humans have had to diverge genetically, it is likely that there will be a limited number of pharmacogenomic differences between population groups defined by race or ethnicity (Nebert and Menon, 2001).

Yet the differences that exist could have great significance for the health of the individuals involved. A large differential now exists between the health care that minorities and other groups receive, because of social, cultural, and economic factors, and in some cases, racial bias and prejudice (Institute of Medicine, 2002). However, some of the differences in health outcomes for African Americans and other groups may be due to genetic polymorphisms that cause a higher frequency of some diseases or that reduce the person's ability to take in or metabolize drugs. Although the minority community has many pressing health care needs that require attention, ensuring minority access to pharmacogenomic medicine is also important and must be attended to as pharmacogenomic research proceeds.

Attention to the racial and ethnic implications of pharmacogenomic medicine may require both public and private actors to use race and ethnicity in determining what research to fund, what genetic tests to offer or require, or what drugs to prescribe. Before addressing the various situations in which gene-based racial and ethnic categories might be relevant, I discuss the constitutional and legal criteria that must be met if explicit racial and ethnic categorization is to be acceptable. After clarifying the legal issues, I address the legitimacy under those criteria of public policies and private actions that use gene-based racial and ethnic categories in research, diagnosis, and therapy.

IV. CONSTITUTIONAL STATUS OF RACIAL CLASSIFICATIONS

How should society take account of racial and ethnic differences in disease and illness, and of the genetic differences that may underlie some of them? May it ever legitimately take those differences into account? If so, what limits are required? Are there ever duties to use race and ethnicity in making health policy and health care decisions? In what cases would racial and ethnic categories be prohibited?

The constitutionality of racial classifications in medical care will turn on the strength and significance of the genetic and medical data, the specific ways in which they are used, and whether persons are harmed or disadvantaged by those categorizations. To overcome the suspect status of race, proponents of racial and ethnic categorization in medicine must show that these categories are necessary to serve compelling health interests that cannot be as effectively achieved by nonracial categories. An important factor in that assessment will be whether racial and ethnic categories unfairly advantage or disadvantage particular racial groups or individuals.

A. Treating Differently to Treat Similarly: The Acceptability of Race and Ethnicity in Medical Decision Making

The key issue in determining the constitutionality of racial and ethnic classifications in pharmacogenomic medicine is whether racial categories are being used for valid nonracial medical goals that cannot be as effectively achieved by nonracial means. If race and ethnicity strongly correlate with particular kinds of health care problems and with remedies for treating them, it is both more efficient and good medicine to offer certain tests to members of one racial or ethnic group and not to others, and to diagnose or prescribe therapy accordingly. In those circumstances, use of race and ethnicity in many biomedical research, diagnostic, and medical treatment situations should be constitutionally acceptable, if not obligatory.

Meeting the health care needs of all citizens is a compelling governmental interest. If race and ethnicity correlate with certain drug response genetic markers, using them to decide which tests, diagnoses, and therapies to recommend may serve the health care needs of persons in those groups more effectively than ignoring race and ethnicity would. Indeed, not taking account of race and ethnicity could mean that tests, diagnoses, and therapies are being used inappropriately, because they may have been developed on the basis of information derived from particular racial or ethnic groups, for example, Caucasians, that does not uniformly apply to non-Caucasians.

A second reason for the acceptability of race and ethnicity in medicine is that using race to diagnose or treat the health problems of people does not denigrate them or deny others equal protection of the law. Indeed, if individuals are to be treated equally with regard to their health care needs, in some circumstances their race or ethnicity must be taken into account. The racial classifications at issue would be based on established genetic, physical, and medical facts and not on contested facts about the impact of social and cultural history or stereotypes about the inferiority or superiority of a particular race. The claim is not that racial and ethnic groups are owed differential treatment because of past discrimination, but that genetic variation or other physical difference makes one type of genetic test or drug appropriate for one group but not for other groups.

A third reason for the acceptability of race and ethnicity in medicine is that no individual is being denied something that he or she might otherwise have received on the basis of race, as occurs in controversies over affirmative action, transracial adoption, and electoral redistricting. At issue is not a competition for a scarce resource, such as a job or an educational slot. Race functions to allocate the precise type of treatment received, and in some cases whether one is treated at all, but in doing so it does not deny or prevent other racial and ethnic groups from being treated. Without attention to racial differences in genotype and phenotype in research and therapy, the health care needs of all individuals will not be met. For example, if race were not considered, African American and white patients would both get enalapril for their heart failure, but only whites would benefit, and the advantages of carvedilol and Bidil might have gone unnoticed.

To be sure, funds may go to some projects defined on racial grounds, but the purpose is not to favor those groups as such but to use that category as the best means to serve the compelling purpose of meeting the health care needs of all citizens. Because some health care differences are race based, race must be taken into account to meet the health care needs of those persons.

But this does not mean that people are being treated unequally on the basis of race. Instead, race is necessary if government programs directed at improving the health of all citizens or the safety and efficacy of drugs for all patients are to fulfill their goals. Treating people differently in polymorphisms tested is necessary to treat people equally in health outcomes. A useful analogy may be drawn from bankruptcy and immigration law. The Constitution requires that Congress enact "a uniform rule" for bankruptcy and for immigration. Yet in each case the courts have recognized that uniformity may include differences in how the states, in the case of bankruptcy, define homestead exemptions, and, in the case of immigration, "legal separation," which affects the immigration status of certain individuals (*Stellwagin v. Clum*, 1918, *Nehme v. INS*, 2001). Similarly, equal treatment of the health needs of individuals may require that they be treated differently as their genotype and phenotype demand. Equality exists even if people with the same disease receive different treatments as determined by race, because race may be a scientifically valid surrogate for health risks and common genotypes. Race is thus necessary to ensure that government health programs serve the needs of all people.

None of these points holds for racial classifications in affirmative action in employment and education, in transracial adoptions, and in electoral redistricting. In each case other individuals or groups are directly impacted or are denied something on the basis of race. In most cases, the preferences are not based on objective, scientific physical differences but on claims about the fairness of selection criteria, about compensation for past wrongs, about diversity for educational purposes, about distributive claims based on the proportion of the group in society, or about the benefits of being raised by parents of the same race.

In the past, racial classifications have been used to deny African Americans status or services based on negative stereotypes of social and moral inferiority based on skin color. As Justice O'Connor has noted, "racial classifications of any sort pose the risk of lasting harm to our society" (*Shaw v. Reno*, 1993). In contrast, the use of race in health care settings would occur to make sure that health care needs of African Americans are met, not to deny them or other groups needed benefits. Race is not an end in itself, but a surrogate or means of ensuring that everyone's health care needs are met. Because such uses of race are focused on protecting health, they are not likely to further entrench racialist thinking, as some opponents of using racial and ethnic categories in medicine fear (Graves, 2001; Schwartz, 2001).

This purpose, however, does not justify all health-related uses of race. As shown below, some explicit uses of race in the health area would be impermissible because the close connection asserted to exist between health care and race and ethnicity does not exist or exacts an unfair toll on the legitimate interests of other claimants to the resources or services at issue.

In the end, claims about constitutionality are predictions about what the United States Supreme Court would do if faced with a case involving the question at hand. Because of the differences between racial and ethnic categories in medical care and racial and ethnic categories in affirmative action and other contexts, the argument is strong that the medical uses of race, when the facts support it, would be constitutionally acceptable. In a different but related context, the Court recently upheld the use of race as a surrogate marker for achieving legitimate nonracial goals of drawing electoral district lines. A closer look at that case is instructive for how the Court might view medically based uses of race and ethnicity.

In 2001, the Supreme Court found race to be an acceptable factor in drawing congressional district lines to create a Democratic seat (*Hunt v. Cromartie*, 2001). The case involved a dispute about the acceptability of congressional redistricting in North Carolina to ensure election of an African American member of Congress. The Court, in striking down districting schemes designed to ensure election of an African American representative (*Shaw v. Reno*, 1993), had earlier held that the 1968 Voting Rights Act was designed to overcome the voting discrimination in a segregated South and did not permit carving out districts on racial bases alone. Many commentators took this as a sign that the Court had grown unsympathetic toward measures favoring minorities, confirming a pattern seen in other affirmative action settings.

Yet in *Hunt v. Cromartie* the Court approved the lines drawn, although they too were likely to result in the election of an African American member of Congress. The state redistricters were able to convince five justices that their purpose was to ensure, permissibly, the election of a Democratic

Representative, not an African American representative, as the earlier invalidated efforts had. Because the race of voters was more strongly correlated with voting Democratic than was registration as a Democrat, drawing district lines to ensure the election of a Democrat could rationally use African American precincts as indicators of likely Democratic voting. As long as the state's purpose was to ensure election of a Democrat, the fact that the election of an African American also occurred was acceptable. Remarkably, the Court allowed the use of race as a marker of likelihood to vote Democratic, which had the same result as if the goal were to elect an African American, which previous cases had prohibited.

Hunt v. Cromartie supports the conclusion that race may be acceptable to achieve legitimate nonracial goals that cannot be as effectively achieved by race-neutral methods. If the Court in *Hunt* permitted the use of race when white candidates would be at a disadvantage, it is all the more likely to approve it when strong connections to the medical welfare of racial and ethnic groups can be shown and no whites or other groups are harmed as a result. In both cases, the use of race is not to favor or disfavor African Americans or whites in themselves. Rather, race is a closely linked proxy for nonracial goals. Just as race could appropriately be used in *Hunt* as an indicator of partisan voting, so race in the health care context can function as an indicator of the particular health care risks and needs that groups of the population experience differentially.

A legislative goal to avoid the disparate impact on African Americans of a race-neutral medicine, which would disproportionately disadvantage African Americans, should not be impermissible just because racial categories are used. It neither stigmatizes, denigrates, nor harms African Americans or whites. It is a constitutionally acceptable way of meeting the health care needs of the population as a whole. Indeed, failure to take race into account could mean that the health needs of some members of the community will not be met.

The analysis reflects one of the paradoxes or ongoing dilemmas of equal protection law in trying to determine the relevance of sameness and difference among groups and individuals. The obligation to treat equally sometimes requires us to treat differently. Surely treating people of different racial and ethnic groups differently in medical research, diagnosis, and therapy is constitutional if that different treatment directly affects health outcomes. If not, the legitimate health care needs of those people would not be met.

B. IS TESTING ALL A LESS RESTRICTIVE ALTERNATIVE?

A key issue in determining whether a racial or ethnic category is constitutionally acceptable is whether there are nonracial means of achieving the same result. In constitutional law terms, is the use of those suspect categories "necessary" because there is no racially neutral way to as effectively meet the legitimate state goal at issue (*Loving v. Virginia*, 1967)? Opponents of racial and ethnic classifications in health care might argue that the use of race is unnecessary because the underlying genetic variations could be tested for directly. Instead of testing differentially according to race, everyone should be tested for all the known mutations in the gene of concern. Indeed, such direct testing would avoid the imperfections of the use of race as a surrogate marker for underlying genes due to interbreeding among population groups, which has occurred throughout human history: Any identified differences on the basis of race will not apply to all members of each stratified group (Wood, 2001). But testing everyone for every polymorphism may add significant costs and delays. At present, the cost of sequencing the entire genome of every patient is prohibitive. Despite considerable progress, DNA microarrays are not yet sufficiently reliable and inexpensive to be used routinely to test for thousands of mutations at once. A test of all mutations would require first identifying them and putting DNA containing them on a chip, which would then be used for patients with this disease.

As a result, it now appears to be more advantageous to test for different mutations in African Americans and whites than to test everyone for everything. Starkly put, just as men with stomach pains need not be given pregnancy tests or women over 50 need not have a prostate examination, so whites need not be tested for the same CYP 2D6 mutations as African Americans, unless there were cheap and easy ways to test all groups for all mutations of interest. Similarly, haplotype maps for identifying genotypic variations in drug response might appropriately use racial markers, as long as that can be shown to be a cost-efficient way of obtaining information material to clinical decisions.

In the future, the increased efficiency of DNA microarrays may cause the need to use racial and ethnic categories to protect the health of all citizens to wither away. A Moore's law of DNA microchips might come into play, with the amount of DNA contained on a microarray doubling every eighteen months, while the cost of the chip halves. Indeed, at some point it may be possible to have one's entire genome cheaply encoded on a chip or disk, with ready access for thousands of genetic tests. Such chips or disks could be made with blood taken for newborn screening or, as is more likely, with the consent of adult patients deciding which approach best serves their medical and privacy interests.

As the cost of pharmacogenomic testing drops and its efficiency increases, markers such as race, ethnicity, or gender may play a lesser role in selecting genes or polymorphisms within them for testing. When only a few variations can be tested at a time, race or ethnicity may sometimes be the best indicator for determining which mutations to test. When the medical data support that judgment, racial and ethnic categorization would be the least restrictive means of meeting the health care needs of those persons.

V. ASSESSING PUBLIC AND PRIVATE USES OF RACE IN PHARMACOGENOMIC MEDICINE

Having established that genotypic differences scientifically correlated with race or ethnicity *may* be used in biomedical research and treatment situations, I now turn to programs or efforts that have explicitly used racial categories in public or private decision making or might in the future, looking first at the use of racial and ethnic categories in public research programs and then their use in private clinical decision making. In addition to pointing out when the use of race and ethnic markers are *obligatory*, I also discuss some proposed uses that would be inappropriate.

A. USE OF RACIAL CATEGORIES IN BIOMEDICAL RESEARCH

The health care needs of racial and ethnic groups are similar in virtually all respects, but in some cases diseases or drug responsiveness based on genotypic variation may correlate with race or ethnicity. Knowing that those variations exist, when to test for them, and when to take them into account in clinical decisions will require research directed at uncovering genotypic differences that correlate with race or ethnicity.

In the past, inattention to racial differences among subjects in clinical trials has led to most trials being done with white males. As a result, evidence of the safety and efficacy of drugs and other therapies was established for white males only and was assumed to apply in equivalent or weight-adjusted terms to women and minorities (and/or children). Considerable research now shows that that assumption was incorrect. Genetic or other differences between genders and racial and ethnic groups may lead to differential response to drugs, just as those differences may lead to variation in the rate of exposure or contracting of disease and illness. Unless drugs are assessed with all potential users in mind, assessments of safety and efficacy may not apply to the entire pool of patients who will receive the drug.

Congress attempted to deal with this problem in 1993 by requiring that NIH-funded research projects include sufficient women and minorities in research and clinical trials to tell whether drugs or treatments being tested work for them as well (NIH Revitalization Act, 1993). In 1997, Congress directed the FDA to consult with the NIH to ensure that women and minorities were adequately represented in clinical trials of new drugs and devices (Food, Drug and Cosmetic Act Amendments, 1997). Although both pieces of legislation speak generally of "women and minorities," that category clearly involves racial, ethnic, and gender classifications and thus requires heightened scrutiny. The case for race- and gender-specific inclusion in research is not a difficult one. Nonetheless, it is worth analyzing in detail because of the light it throws on other acceptable uses of racial and ethnic categories in medicine.

B. NIH Funding of Research

The 1993 NIH Reauthorization Act sought to deal with the problem of unrepresentative subjects in clinical trials by requiring the Director of NIH to ensure that "in conducting or supporting clinical research . . . (A) women are included as subjects in each project of such research; and (B) members of minority groups are included as subjects in such research" (NIH Reauthorization Act, 1993). In designing clinical trials in which women and minorities are included as subjects under this directive, the Director "shall ensure that the trial is designed and carried out in a manner sufficient to provide for a valid analysis of whether the variables being studied in the trial affect women or members of the minority group, as the case may be, differently than other subjects in the trial." The requirement is inapplicable if the inclusion of women and members of minority groups is "inappropriate with respect to the health of subjects . . . to the purpose of the research . . . or under other designated circumstances." The cost of inclusion is not a permissible factor in determining whether inclusion is "inappropriate."

A sponsor or conductor of research, who would have selected all the subjects on some other basis, now must pay attention to race and gender and include enough women and minorities to make statistically meaningful assessments of drug safety and efficacy for those groups as well. Although no specific race is singled out or numbers specified, the law does impose a requirement to include subjects on the basis of race and gender when subjects would have been chosen on the basis of inclusion criteria insensitive to the needs of persons in those groups. Without a mandatory inclusion requirement, too few representatives of each group might be included to provide meaningful information about them.

The racial and gender classifications at issue here, however, are justified to achieve the government's goal of spending research funds in a way that will yield useful information about all citizens, including women and minorities, not simply about whites or males, as has been the case in the past. Meeting the health care needs of all its citizens, including women and minorities, is a compelling state interest. Because biological and genetic variation in women and minorities may differ significantly from that in white males, requiring their inclusion in funded research is essential if valid information about their health is also to be produced. Alternatives such as private investment by drug companies are not as likely to achieve that goal, for the minority subgroups involved may be too small to justify investment for that purpose.

Requiring inclusion of women and minorities in research is not based on any stereotype of the inferiority or inappropriateness of women and racial minorities, nor does it deny males or whites anything that they are due based on their merit or need. Although motivated in part by the past exclusion of women and minorities from much biomedical research, the NIH program does not purport to compensate for past discrimination, for there is no showing that any individual subject who will now be included was the victim of past discrimination in being selected for clinical trials. To be sure, fewer funds will be available for whites and males, and fewer of them will be included in research than would have been. But there is no entitlement to be a research subject or to be selected for research independent of race or ethnicity, so no individual is denied anything that he would otherwise deserve or receive. Nor is he being denied something on the basis of race or gender because of a judgment of the relative worth or merit of persons in those groups.

C. FDA Requirements

The FDA now encourages inclusion of data on the effect of studied drugs on women and minorities in clinical data submitted on new drug applications. Although an explicit racial and gender classification, the requirement does not act like a subsidy to a racial group in the same way that the NIH funding requirement does. Given the FDA's duty to ensure that drugs are safe and effective, it is reasonable to require that a drug is safe and effective for all the persons likely to be given the drug, including women and minorities. Without such a requirement the FDA might approve the marketing of a drug that is safe and effective only for persons of the race and gender of the subjects who participated in clinical trials, which traditionally might have excluded women or African Americans. Yet the drug could still be prescribed for African Americans or women, who might respond less well or with more adverse reactions. To avoid that result, Congress might reasonably require that the FDA require new drug and device applicants to submit data on minorities and women as well. No whites or males are hurt by this policy. Indeed, a failure to do so would impose a disproportionately greater health care burden on women and minorities. Such a burden would not be unconstitutional, if it is not the FDA's or applicant's intent to exclude those groups from the study, but rather an effect of pursuing other goals in the design and conduct of the study. Yet it is clearly better policy—and certainly constitutional—to require that data on women and minorities also be included in new drug applications.

D. Fund/Conduct African American-Only Research

Another way to obtain the information necessary for racial and ethnic groups to benefit from pharmacogenomic testing would be government grants or programs that directly study racial and ethnic associations with drug responsiveness. If government may mandate inclusion of women and minorities in clinical research to ensure that their health care needs are met, then it may also fund research into the drug response profiles of African Americans or how they respond to particular drugs.

Congress took that approach in the 1993 NIH Reauthorization Act when it also created an Office of Women's Health and an Office of Minority Health Care. The Women's Health Initiative and Office of Women's Health resulted from a recognition that different factors may affect the health of women and men, that women have been excluded from much past clinical research, and that if their health needs are to be met, special attention to research on women is needed, for example, of the effects of estrogen replacement therapy on osteoporosis. This program does not constitute impermissible gender discrimination because men are not being denied research into diseases that they uniquely have, for example, prostate cancer. The purpose is to meet the health needs of a large group of the population that would otherwise go unmet.

Similarly, the Office of Minority Health Care should also be within the power of government. Past disparities in research and clinical funding have contributed to the great differentials that now exist between the health and health care received by whites and minorities. An Office of Minority Health Care attempts to address those problems, to serve the health needs of a major group of the populace that otherwise might go unserved. Neither creating such an office nor funding African American-only studies denigrates or hurts African Americans or denies anything to whites. The amount of money the Office of Minority Health Care has received is quite small, in fact, smaller than that received by the Office of Women's Health Care. Without such an office or research grants that focus on racial differences in disease or drug responsiveness, the health care needs of minorities may not be adequately served. The use of race as a surrogate for the unique health care needs of large numbers of the population thus serves the goal of equality in meeting the health needs of the population, which could not be met as effectively by other means. Given wide legislative discretion in spending for the general welfare, focusing some funding and research on the special needs of minorities should be constitutionally acceptable.

If so, it would also be acceptable to give a grant for study of a drug that is expected to work in a particular racial or ethnic group only. As noted above, African Americans with heart disease respond less well than whites to beta-blockers and ACE inhibitors, the main agents now used for heart conditions. An earlier clinical trial of Bidil, a drug that combines two ACE inhibitors with a nitrous oxide source, showed little beneficial effect on mortality for patients generally, including whites, but a subset of patients, all African American, seemed to have been helped. On the basis of this information, the FDA has agreed to a 2-year trial that will test Bidil and other drugs on 600 African American patients at 100 sites (*Science* Editors, 2001).

Such race-based research should be constitutionally acceptable. The situation is a variation on the issues underlying mandatory inclusion of women and minorities in clinical research. The government would be funding a study that only African Americans could enter, so a racial classification is involved. In this case whites are excluded because there is evidence that the drug will not work for them but may for African Americans. Because most studies for heart disease have included white patients, indeed, have disproportionately benefited them in the past, whites are not being disadvantaged by this instance of government funding of a study that focuses exclusively on a drug that might work in African American heart patients. Without the study, many African Americans may be left with few effective treatments for heart disease, when most whites can be treated with beta-blockers and ACE inhibitors. Nor are whites denied research funding, for there has been ample study of therapies that benefit whites disproportionately, if not exclusively.

Whether there will be sufficient funding for such studies depends on the NIH's and the public's commitment to improving the health care of minority populations, for example, its willingness to ask for and get budgetary support, and then awarding grants to researchers studying these questions. After a long disregard of the health needs of African Americans, there is now a willingness to fund more race-related research. With the large budget increases given to the NIH in recent years and the past neglect of minority health, even larger increases for minority health care should be acceptable. With wide discretion to spend resources to improve the general welfare, Congress and federal agencies may use race and ethnicity in research and treatment funding decisions when medical evidence shows that racial and ethnic differences affect health care.

E. FUND RESEARCH INTO DISEASES THAT DISPROPORTIONATELY AFFECT MINORITIES

If mandatory inclusion laws and African American-only studies are acceptable, then *a fortiori* there should be few doubts about the permissibility of funding research or treatment of diseases that have an especially high incidence among African Americans. Because not all racial groups are evenly distributed in disease categories, choices to fund some diseases more than others could end up favoring some races and genders over others.

This issue is not likely to arise with funding for research into leading causes of death. Although African Americans may be at higher risk than whites for certain cancers or diabetes, such large numbers of African Americans and whites have these illnesses that differences in the degree of risk among population groups are not significant. Rather, the issue would arise only with funding diseases that have a disproportionate impact on African Americans.

Congress thus may constitutionally fund sickle-cell anemia treatment centers across the country, even though 97% of persons with the disease are African American (Lee, 2001). Because this is a disease-, not race-, specific program, no racial classification is involved. Disparate impact on a particular race or ethnic group is not subject to strict scrutiny unless the classifying trait with that impact was chosen to cause that disparate racial impact. Whites, who have usually benefited from laws that have had a disparate impact on minorities, have no grounds to complain if in this instance the disparate impact favors groups that have traditionally been disfavored.

Yet even if one could show that the legislative motivation or purpose here is to help African Americans, this would be an acceptable motivation. As we have seen, the state may explicitly target the health of African Americans because their health care is as important as that of other persons and, because of genetic and other differences, is likely not to be met equally unless explicit attention is paid to race and ethnicity. As long as the health needs of others are being met, singling out African American health needs as an efficient way to meet the health care needs of all the people should be acceptable, whether it is done explicitly or by funding disease categories in which African Americans are disproportionately represented.

Indeed, there are social benefits from using disease-specific rather than racial and ethnic categories in research allocations that might make them preferable to more direct subsidies. Few racial or ethnic groups will have a monopoly on particular diseases, so all groups are likely to benefit to some degree from research funding for specific diseases. Focusing on disease rather than racial or ethnic identity also draws attention to the underlying physical facts and away from racial correlations and the social baggage that they might carry. Surely such choices are constitutionally acceptable.

F. ACCEPTABILITY OF RACE IN HAPLOTYPE MAPPING

The criteria used in the above assessments also suggest that it would be constitutionally acceptable to construct a haplotype map with markers for race. Pharmacogenomics researchers are currently relying on a genome-wide map of single-nucleotide polymorphisms (SNPs) to match genotype with phenotypic drug responses. Because of the large number of SNPs spread throughout the genome, it has become essential to find easier ways to identify and handle them. Recently, scientists have discovered that haplotypes—chunks of DNA perhaps 10,000 base pairs long, which may contain many SNPs— due to linkage disequilibrium tend to be inherited together and thus might be as good indicators of genetic variation among persons and populations as SNPs and be easier to find. The NIH's National Human Genome Research Institute is now funding the development of a publicly available haplotype map. An issue before it is whether haplotypes should be identified by race (Wade, 2001). Doing so would then enable race to be determined in medical or other applications by testing directly for that haplotype.

Although some have objected to identifying racial haplotypes, if haplotypes do vary among racial and ethnic groups, it may be easier to correlate genotype and phenotype in various groups by taking those haplotypes into account. If haplotype mapping by race will more efficiently show drug and disease correlations with race and ethnicity, their use in those contexts should be permitted. Failure to do so might deprive racial and ethnic minorities of good health care.

VI. Taking Account of Pharmacogenomic Differences Among Minorities in Clinical Practice

A second important set of issues arises once precise and reliable drug responsiveness in minorities is known. Here the problem is to determine whether or when public and private decision makers *may* take that race-based information into account, and if so, whether there is also an *obligation* to take it into account.

It seems clear that government as a funding agency and physicians as clinical providers may take race into account when racial correlations with health status or health care are directly relevant. In many clinical situations moral and legal duties to take race into account in order to respond properly to the patient's health needs may also exist, either because of state action or because of state laws banning racial and ethnic discrimination in public accommodations, which include hospitals and physician offices. Three situations in which a physician may have a duty to recognize racial correlations and order a different test or therapy as a result are discussed below.

A. Susceptibility Testing for Preventive Purposes

Good clinical care in many situations will require a physician to know what health risks members of racial and ethnic groups are more likely to face, and to take appropriate action based on that knowledge. Although not strictly speaking pharmacogenomics, there may be genetic variations more prevalent in particular racial or ethnic groups that make them more likely to have particular diseases. Physicians treating patients in those groups will need to be aware of those conditions and the tests that can determine their presence so that appropriate preventive action can be taken. Just as doctors should know that the ACE inhibitor enalapril generally does not work for African Americans in heart failure but that the beta-blocker carvedilol does, they should also be aware of racial differences in genetic susceptibility to other diseases so that preventive measures can be taken.

Although African Americans and whites share many of the same genetic susceptibilities, some racial differences in the incidence or significance of certain disease predictions may exist. In some cases, African Americans may have a risk that whites do not. For example, researchers have recently found that a mutation in the gene coding for a protein essential for blood clotting is more likely in African Americans than whites, thus greatly increasing the risk of heart disease among African Americans (Ackerman, 2001). The mutation was associated with a sixfold increase in serious heart disease and was frequent enough in African Americans (7% vs. nearly 0% in whites) to warrant the attention of physicians. If further research confirms this difference, physicians will have an ethical and legal obligation to consider testing African Americans for this mutation and then advising them of the preventive steps of diet, exercise, or drug therapy that could reduce that risk. Overall, because African American men die of cardiovascular disease 50% more than white men and African American women 65% higher than white women do, knowledge of this race-based genetic variation could be useful in attacking that problem. Highly publicized instances of sports stars such as Hakeem Olajuwon having this condition might not then have occurred.

B. CARRIER AND PRENATAL TESTING

Racial and ethnic groups may also vary in the extent to which they are carriers of autosomal recessive or dominant diseases that could be inherited by offspring. African Americans will be subject to many of the same diseases as whites, for example, cystic fibrosis, but the incidence of the disease may vary. In some cases, however, the incidence of disease may be higher among African Americans, just as it is for some white subgroups, for example, the higher incidence of Tay–Sachs disease and BRCA1 and -2 mutations in Ashkenazi Jews.

Doctors need to inform their patients of carrier and prenatal tests for sickle-cell disease so that they can determine, if they choose, whether they are carriers and, if so, what their options for having a child free of that disease are. Married carriers could avoid producing offspring with sickle-cell disease by remaining childless, adopting, using donor gametes, or undergoing some form of prenatal or preimplantation genetic diagnosis (Xu et al., 1999). A doctor who did not inform an African American couple of the availability of carrier or prenatal testing for sickle cell would probably have committed malpractice and be liable in damage to the parents for wrongful birth, just as he or she might by neglecting to inform white couples of the availability of cystic fibrosis screening, or Ashkenazi Jewish couples of Tay–Sachs testing. Race and ethnicity is relevant here because it will enable a patient to make an informed choice about his or her reproductive future.

In the early 1970s when sickle-cell carrier screening first became available, some African American groups called for legislation to ensure that all African Americans were screened to see whether they were carriers of the sickle-cell trait, either before entry to school or on applying for a marriage license. Although never enacted, would laws requiring all African Americans to be tested for sickle-cell carrier status before marriage (or school) be permissible?

Here, mandatory testing purports to serve the interests of African Americans by informing marriage partners of their sickle-cell carrier status before reproduction, thus enabling them, if both partners are carriers, to avoid having a child with this disease. Of course, testing does have its burdens. People who have the trait but not the disease have suffered unjust discrimination in some insurance and employment settings, usually as a result of misunderstanding of the significance of being a carrier. Persons who test positive will also face decisions about reproductive planning and choices that they might have preferred to avoid.

Although carrier testing is ostensibly a benefit for African Americans, the key point here is that African Americans alone were being singled out for mandatory testing. If testing is a benefit, why were African Americans alone given it, and not other persons at risk for sickle cell-type diseases, such as persons of Mediterranean origin? Why was mandatory carrier testing not required of Caucasians or Ashkenazi Jews who are at comparable risk for cystic fibrosis or Tay–Sachs disease? Because there were some risks from testing, and no other groups were required to be tested, a mandatory testing program for one group could be seen as a harm to them. Alternatively, because many African Americans supported the program, it could be viewed as an important benefit that should also be extended to other groups at comparable risk for genetically transmitted diseases. The same issues would arise if the test were purely voluntary, but the law required that it be offered to African Americans at marriage, and no tests were required to be offered to others.

C. CLINICAL USES OF PHARMACOGENOMIC DIFFERENCES AMONG RACES

As pharmacogenomic knowledge develops, much more will be known about how drug responsiveness varies among racial and ethnic groups. If medically relevant connections are established, it will be the physician's duty in some circumstances to offer race-specific genetic tests of drug responsiveness before prescribing a drug.

If there are racial or ethnic differences in drug responsiveness, then race or ethnicity may be relevant in meeting the health needs of people. In this case race matters because it is relevant to what tests are performed and what drugs are prescribed and thus whether racial and ethnic minorities receive safe and effective therapy for their conditions. Physicians should be familiar with these tests, inform patients about them, and respect the patients' privacy just as with other tests that have a higher applicability to a particular racial or ethnic group. Insurers should also cover those tests and drugs just as they cover other pharmacogenomic tests and drugs.

The most difficult questions here will arise when a pharmacogenomic test for a variation known to be more frequent in African Americans is positive, indicating that the drug is contraindicated on safety or efficacy grounds. If the patient has no better therapeutic alternative, is it permissible to prescribe the drug anyway, despite the risk to health or waste that it poses? As with other pharmacogenomic tests, the question of how automatically exclusionary a test should be is a difficult one. Patients will need to be fully informed of the trade-offs at stake, in light of therapeutic alternatives. In some cases using the drug contraindicated on pharmacogenomic grounds might be the best alternative for the patient, but such cases are likely to be few, and it may be difficult to get insurance coverage for that drug, or even find a physician who is willing to prescribe it (Robertson et al., 2002).

As a result of racial correlations with drug efficacy, white patients would get effective beta-blockers and ACE inhibitors for their heart disease, whereas African Americans get a more limited range of drugs and die at higher rates. But that difference is medically justified by the data and physical or genotypic differences among white and African American patients, which is directly and objectively relevant to the case at hand. The important policy issue is to make sure, if such exclusionary decisions are going to be made, that they are well founded and reliable. Shoddy science and testing procedures will not justify differential treatment, but good science and testing will.

VII. Impermissible Uses of Racial Classifications to Improve Minority Health

Although it may often be permissible for health reasons to classify patients by race or ethnicity, some uses of race in the health area would not be acceptable, for example, unfounded negative stereotypes about the need of African Americans for pain relief or other medical procedures. (Institute of Medicine, 2002). The impermissibility of some uses of race, however, does not negate their permissibility in other areas. As we will see, the impermissible uses are more likely to involve health delivery issues, are less closely tied to the health effects of biological or genotypic differences, and are only indirectly connected with minority health care. In addition, another person is often excluded or hurt by the racial classification at issue.

A. Racial Preferences in Medical School Admissions

In *Regents of University of California v. Bakke*, the famous Supreme Court decision on affirmative action in education, the state gave as one justification for racial preferences for minorities in medical school the need to improve minority health care (*Regents of University of California v. Bakke*, 1978). No justice of the United States Supreme Court bought that argument, for it was clear that there were many more effective ways to improve minority health care than admission preferences, for example, better public health insurance coverage. Nor did the preference require its recipients to practice for a set number of years in minority communities. Minority preferences alone would not serve that goal, because some nonminorities would also practice in minority communities in return for medical school admission.

An NIH policy that set aside a percentage of federal research grants for African American applicants, or which gave them extra points in the evaluation process, would also be constitutionally dubious. The use of racial preferences in public contracting, employment, and professional school admissions have now been severely limited (*Richmond v. J.A. Croson Co.*, 1988; *Adarand Constructors, Inc. v. Pena*, 1995). It would be very difficult to show that these preferences were necessary to achieve a legitimate health purpose that could not be equally well achieved by nonracial means. The race of a researcher or medical school applicant is no indicator that his research or practice will address minority health concerns, much less improve them.

B. RACIAL PRIORITIES FOR KIDNEY TRANSPLANTS

Some persons have suggested that the current system of allocating kidneys for transplant is biased against African Americans and should be changed to include race as a factor in determining allocation of organs (Ayres et al., 1993). Although African Americans disproportionately end up on dialysis with end-stage kidney disease, they receive disproportionately fewer of the kidneys available for transplant. A number of factors explain this difference, including the requirement of a close enough match between donor and recipient to ensure that the graft will take (Epstein et al., 2000). Because African Americans share fewer HLA antigens with whites, who are also the most frequent organ donors, they are often the poorest match and the most infrequently transplanted.

As a result, some persons have suggested that the matching requirement be altered to enable more African Americans to obtain kidney transplants, or that African Americans be allocated a share of kidneys regardless of matching. An explicit preference for African American kidney transplant recipients, however, would be difficult to justify. African Americans currently on the transplant list are not claiming previous illegal discrimination. Nor, if they would not do as well with donated organs because of HLA mismatch, is it tenable to give them a share of donated kidneys proportionate to their numbers on the list. Such a preference would reduce the likelihood that donated kidneys functioned properly in recipients and could reduce the willingness of people to donate.

However, other actions to improve minority access to kidney transplantation in other ways would be acceptable. Education programs directed to

African Americans about their high risk for kidney disease and steps to better inform them of transplant alternatives are clearly in order (Ayanian et al., 1999). They are directly health related, and no individual is denied a needed therapy as a result, as is the case with racial preferences in access to donated kidneys.

C. HEALTH CARE VOUCHERS FOR PARTICULAR RACIAL OR ETHNIC GROUPS

Suppose the government gave each African American person a $5000-per-year voucher to buy health insurance, but only African Americans got this benefit. Although intended to overcome years of medical underservice for the African American community, this program has the appearance of a naked preference for African Americans rather than a program that is trying to improve African American health care. No whites might be denied something that they otherwise would have received, but because they are being excluded on the basis of race, strict scrutiny of the program's racial criteria would be required.

Its supporters would have to show that making a voucher program accessible to all would be too expensive and that African American health was so much worse than the health of whites that no other way to meet their needs would exist, a highly implausible claim. A compensatory argument would face the same difficulty that arises in justifying affirmative action in education or employment on this basis—that present beneficiaries of the racial preference were victims of past illegal discrimination. Although the voucher program discussed here is unlikely to be enacted, such a program helpfully shows that meaningful limits on the use of race in health care do exist.

VIII. DOES ATTENTION TO THE GENETICS OF RACE HURT MINORITIES?

The uses of race discussed above are justified by the clear and direct relationship between some genetic polymorphisms, health and disease outcomes, and conventionally understood racial and ethnic groups. Yet some people are greatly concerned that geneticizing some racial differences in health care may provide a genetic basis for racism in society (Schwartz, 2001; Lee et al., 2001). They think, for example, that attention to race-based variations in drug response genes may support racialist thinking about the genetic underpinnings of complex behaviors and ultimately hurt the interests of racial and ethnic minorities that race-based pharmacogenomics testing purports to help.

A closer look at biological history shows how few genetic differences are likely to be correlated with race, and thus how unlikely it is that recognizing genotypic differences in drug responsiveness or disease risk will promote, rather than eliminate, racism among informed persons. The close genetic similarity of all humans follows from the relatively brief evolutionary history of *Homo sapiens*. Several different species of hominids evolved out of primates some 7.8 million years ago, competed with themselves, and then were replaced by *Homo sapiens*. Much of this evolutionary activity occurred in Africa, with early hominids eventually migrating out of Africa and populating the Middle East and parts of Asia and Europe. About 100,000 years ago a genetic bottleneck occurred in Africa, which led to about 10,000–20,000 *Homo sapiens* migrating out and then moving both east and west, populating the world over time. This group of humans competed with earlier hominids that had populated those areas and, in the last 20,000–40,000 years, eventually replaced them (Paabo, 2001; Olson, 2001; Owens and King, 1999).

If this history is correct, then all humans in the world evolved from a relatively small gene pool that has not had much time or selection pressure to evolve greatly since emerging from Africa roughly 100,000 years ago. As the genetic archeologist Savante Paabo notes, "from a genetic perspective, all humans are therefore Africans, either residing in Africa or in recent exile" (Paabo, 2001). Because their exile has been at most only 100,000 years or so, there has been relatively little time for genetic variation to work itself out, for example, for more favorable mutations to be selected for over time because of greatly different selection pressures. Unlike primates, which have had millions of years more than *Homo sapiens* to evolve, post-Africa humans have simply not been around long enough for greater evolutionary divergence to occur (Olson, 2001). High rates of migration and interbreeding—the "continued sharing of genetic materials"—have also helped maintain all of humankind as a single species (American Anthropological Association, 1998).

The Human Genome Project has confirmed that there are relatively few genetic differences among people, as Francis Collins and Craig Venter stressed when they announced publication of the human genome in June 2000. Geneticists generally agree that "more than 80% of genetic variation is between individuals of the same population, even in small or isolated populations" (Owens and King, 1999). As a result, the idea of "genetically relatively homogeneous groups (races), distinguished by major biological differences is not consistent with the genetic evidence" (Owens and King, 1999). As Jared Diamond has masterfully shown, differences among population groups in the development of technology, social organization, and other aspects of culture have arisen because of the vagaries of geography and food production, not genes (Diamond, 1999).

At the same time, however, one can also recognize that there has been enough time for selection pressure to work on some visible traits, namely, skin color, hair, and facial structure. It is striking that these differences affect those areas of the phenotype that interact with the external world, thus reflecting differential selection by climate in various parts of the world, as analysis of the melanocortin-stimulating hormone receptor gene that underlies much of observed human variation in skin and hair color shows (Owens and King, 1999). The rise of pharmacogenomics is now uncovering genetic variations that correlate with drug responsiveness and ethnic phenotypes. They too may reflect selection pressure from climate and environment, for example, the reproductive fitness effects of sickle-cell heterozygosity in climates with malaria, or genetic drift from particular founder populations.

Despite the striking similarity of human DNA, societies have built elaborate systems of privilege and control on genetic differences in superficial external characteristics, mistaking the results of culture and environment for those of genetic essence. Studying the genetic basis for medical differences among races should help to eliminate, rather than entrench, such confusion. Genetics, in fact, exposes the lie of the racialist claim of the uniqueness or specialness of one race. Differences in external phenotypic traits and drug responsiveness cannot negate the overwhelming genetic similarity among all human populations.

Diffusion of these facts should help educate people about the biological unimportance of race in nearly all contexts, so that race may cease to be the powerful marker it now is. Using race or ethnicity in medically valid research and clinical settings is not likely to undermine reaching that goal, although the dangers of misunderstanding are great enough that special attention to minimizing them is needed. It is unclear, however, that avoidance of direct uses of race by circumlocutions such as ". . . a phenotypic characteristic that occurs with greater frequency in a group of persons roughly approximating a historic racial category " will make a significant difference, much less enter common discourse (Rothstein and Epps, 2001). Instead of race, the preferred term now is ethnicity, "which should be distinguished from race, which has grown increasingly obsolete" (Rothstein and Epps, 2001). It is difficult, however, to see how this usage avoids the danger if there is some genetic variation in drug responsiveness in groups of individuals with similar external phenotypes. Variations in drug metabolism enzymes, however, have no connection with socially constructed attributes of certain population groups.

In sum, biologically. race and ethnicity matter very little in explaining differences among humans. But in a few instances racial and ethnic differences may correlate with genetic polymorphisms connected with disease or drug responsiveness. Recognizing those differences poses little risk of essentializing perceived social differences among races and ethnic groups. Indeed, failing to recognize medically relevant differences could harm members of racial and ethnic groups that the denial of those differences aims to protect.

IX. CONCLUSION

Pharmacogenomic medicine promises more effective health care by more precisely distinguishing persons and their responsiveness to drugs. In some instances racial and ethnic correlations between disease and drug response genotypes may exist, thus making race a useful indicator for when certain genetic tests or treatments should be provided. Although race is a social construct with little biological reality, genetic polymorphisms that associate with race and ethnicity can be taken into account without re-inforcing essentialist notions about social differences among racial and ethnic groups. Constitutional norms against racial and ethnic discrimination should present no obstacle to scientifically based uses of race and ethnicity in biomedical research or in the prevention, diagnosis, and treatment of disease.

ACKNOWLEDGMENTS

The author gratefully acknowledges the comments of Mitchell Berman, Lino Graglia, Owen Jones, and Patrick Woolley on an earlier draft.

REFERENCES

American Anthropological Association Statement on Race (Updated 9/15/00) (1999).

Ackerman, T., "Genetic Mutation Linked to Heart Disease: Researchers Puzzled Over Ethnic Difference," *Houston Chronicle*, April 24, 2001, A-17 (2001).

Adarand Constructors, Inc. v. Pena, 518 U.S. 634 (1995)

Ayanian, J.Z., et al., "Effect of Patient's Preferences on Racial Differences in Access to Renal Transplantation," *N. Engl. J. Med.*, **341**, 1661–1667 (1999).

Ayres, I., et al., "Racial Equity in Renal Transplantation: The Disparate Impact of HLA-Based Allocation," *JAMA*, **270**, 1352–1356 (1993).

Brown v. Board of Education, 347 U.S. 483 (1954).

Civil Rights Act of 1964 (Title II, VII), 42 U.S.C. § 2000a and e.

Craig v. Boren, 429 U.S. 190 (1976).

Diamond, J., *Guns, Germs, and Steel*, New York: Norton Co. (1999).

Editors, "Trial for Ethnic Drug," *Science*, **291**, 2547 (2001).

Epstein, A.M., et al., "Racial Disparities in Access to Renal Transplantation," *N. Engl. J. Med.*, **343**, 1537–1544 (2000).

Exner, D.V., et al., "Lesser Response to Angiotensin-Converting-Enzyme Inhibitor Therapy in African American as Compared with White Patients with Left Ventricular Dysfunction," *N. Engl. J. Med.*, **344**, 1351–1357 (2001).

Food, Drug, and Cosmetic Act, Amendments 21 U.S.C. § 355 (b)(1) (1997).

Graves, J.L., Jr., *The Emperor's New Clothes: Biological Theories and Race at the Millennium.* New Brunswick, NJ: Rutgers University Press, pp. 176–192 (2001).

Gunter, G. and K.M. Sullivan, *Constitutional Law,* 13th ed. Mineola, NY: Foundation Press (1997).

Harris, A. "Equality Trouble: Sameness and Difference in Twentieth-Century Race Law," *Calif. L. Rev.,* **88**, 1923–2014 (2001).

Hunt v. Cromartie, 532 U.S. 234 (2001).

Institute of Medicine, *Unequal Treatment: Confronting Racial and Ethnic Disparities in Health Care* (B.D. Smedley et al., eds.) Washington, DC: National Academy Press (2002).

Lee, R., "Dallas Researchers to Study Treatments, Cure for Sickle Cell," *Austin American Statesman,* June 30, 2001, B7 (2001).

Lee, S.S.-J., et al., "Meaning of 'Race' in the New Genomics: Implications for Health Disparities Research," *Yale J. Hlth. Policy L. Ethics,* **1**, 33–75 (2001).

Loving v. Virginia, 388 U.S. 1 (1967).

National Institutes of Health Revitalization Act, Public Law 103-43, §§ 131–133 (1993).

Nebert, D.W. and A.G. Menon, "Pharmacogenomics, Ethnicity, and Susceptibility Genes," *Pharmacogenomics J.,* **1**, 19–22 (2001).

Nehme v. Immigration and Naturalization Service, 252 F.3d 415 (5th Cir. 2001).

Noah, B., "Racial Disparities in the Delivery of Health Care," *San Diego L. Rev.,* **35**, 135–178 (1998).

Olson, S., "Genetic Archeology of Race," *Atlantic Monthly,* (April, 2001) **287**, 69–80 (2001).

Owens, K. and M.C. King, "Genomic Views of Human History," *Science,* **286**, 451–453 (1999).

Paabo, S., "Human Genome and Our View of Ourselves," *Science,* **291**, 1219–1220 (2001).

Regents of Univ. of California v. Bakke, 438 U.S. 265 (1978).

Reilly, P.R., *Genetics, Law, and Social Policy.* Cambridge, MA: Harvard University Press (1977).

Richmond v. J.A. Croson Co., 488 U.S. 469 (1988).

Robertson, J.A., et al., "Pharmacogenomics Challenges for the Health Care System," *Hlth. Affairs* (July/August 2002).

Rothstein, M.A. and P.G. Epps, "Pharmacogenomics and The (Ir)Relevance of Race," *Pharmacogenomics J.,* **1**, 104–107 (2001).

Schwartz, R.S., "Racial Profiling in Medical Research," *N. Engl. J. Med.,* **344**, 1392–1393 (2001).

Shaw v. Hunt, 517 U.S. 899 (1996).

Shaw v. Reno, 509 U.S. 630 (1993).

Stellwagen v. Clum, 245 U.S. 605 (1918).

Wade, N., "Genome Mappers Navigate the Tricky Terrain of Race," *New York Times,* July 7, 2001, p. A17.

Washington v. Davis, 426 U.S. 429 (1976).

Weber, T.K., et al., "Novel hMLH1 and hMSH2 Germline Mutations in African Americans with Colorectal Cancer," *JAMA*, **281**, 2316–2320 (1999).

Wood, A.J.J., "Racial Differences in the Response to Drugs—Pointers to Genetic Differences," *N. Engl. J. Med.*, **344**, 1393–1395 (2001).

Xu, K., et al., "First Unaffected Pregnancy Using Preimplantation Genetic Diagnosis for Sickle Cell Anemia," *JAMA*, **281**, 1701–1705 (1999).

Yancy, C. W., et al., "Race and the Response to Adrenergic Blockade with Carvedilol Patients with Chronic Heart Failure," *New Engl. J. Med.*, **344**, 1358–1365 (2001).

Epilogue: Policy Prescriptions

Mark A. Rothstein, J.D.

I. Introduction

The year 2003 marks the fiftieth anniversary of the discovery of the double helix structure of DNA by Crick and Watson. It also marks the completion date for the sequencing phase of the Human Genome Project (HGP), officially begun in 1990. In a mere thirteen years the public consortium of researchers has mapped and sequenced the entire human genome of over three billion base pairs (International Human Genome Sequencing Consortium, 2001). A private company, Celera Genomics, sequenced substantially all of the human genome in an even shorter period of time (Venter et al., 2001). In the postgenome stage of research, scientists are constructing haplotype maps, sequencing the genomes of other model organisms, and applying genome-wide research techniques to the study of proteins and the development of pharmaceutical products (see chapter by Mohrenweiser, this volume).

The principal reason for publicly funding genome research is the prospect of preventing, diagnosing, and treating diseases. Because of the speed at which various diseases have been linked with genetic loci and the publicity surrounding each new discovery, it is understandable that the public would believe that treatments and cures for dreaded diseases would begin flowing shortly after each new gene discovery. Unfortunately, although genomic knowledge has been increasing exponentially, clinical

Pharmacogenomics: Social, Ethical, and Clinical Dimensions, Edited by Mark A. Rothstein.
ISBN 0-471-22769-2   Copyright © 2003 Wiley-Liss, Inc.

applications have been increasing only incrementally. Although some experts continue to predict that the genomic age in medicine is imminent (Collins and McKusick, 2001), others are more cautious (Holtzman, 2001). The same range of views applies to pharmacogenomics (see chapters by Omenn and Motulsky and Holtzman, this volume).

The measured pace of progress in pharmacogenomics does not detract from its potential to produce safer and more efficacious pharmaceutical products. Each year, approximately 100,000 deaths in the U.S. are attributed to adverse drug reactions, making it between the fourth and sixth leading cause of death (Lazarou et al., 1998). In addition, an unacceptably low percentage of commonly prescribed medications are effective for any particular individual, ranging from 25% for oncology drugs to 80% for analgesics, with the median in the 50–60% range (see chapter by Manasco, this volume). If pharmacogenomics-based medications could improve safety and efficacy for currently prescribed pharmaceutical products—even without developing new ones—it would be a major medical advance (Phillips et al., 2001). In addition, scientists may be able to perform pharmacogenomics-based research with smaller and more targeted clinical trials, which would be faster and cheaper (Cardon et al., 2000). Furthermore, pharmacogenomic analysis may permit the FDA to approve medications for use only by individuals with certain genotypes, thereby "rescuing" pharmaceutical products that could not obtain prior approval because of adverse reactions in a previously unpredictable subset of the population.

In the understandable emphasis on scientific issues in pharmacogenomics, there has been relatively little discussion in the literature or policy debate about the ethical, legal, and social implications of pharmacogenomics (Moldrup, 2002). Some scholars, however, have begun to address these issues (Buchanan et al., 2002; Robertson et al., 2002). This volume continues and expands on that work by providing the technical analyses necessary for policy development. Among the topics included are public opinion about pharmacogenomics, the scientific state of the art, clinical issues, and ethical, social science, and legal considerations. The contributors have presented a variety of perspectives on pharmacogenomics. They discussed the time frame for developing pharmacogenomics-based drugs, observed that the differential response to drugs is more likely to be along a continuum than an all-or-nothing model, and emphasized that genotype is only one element in the differential response to xenobiotics, thereby underscoring the continuing need to study environmental factors. They considered the ethical issues surrounding participation in research, drug development, and regulation, the application of pharmacogenomics in the clinical setting including the training and roles of various health professionals, and the economic, societal, and legal implications of pharmacogenomics. The following policy analysis is based on the discussions in the chapters of this book as well as other sources.

II. Research

A. Biobanks

One of the profound changes brought about by the HGP is the scale of research and its applications. For example, the sequencing of the human genome was of a scale unprecedented in biology. It involved an international consortium of hundreds of researchers at numerous institutions spending hundreds of millions of dollars to generate an extraordinary amount of genetic data. Another example of the principle of scale to emerge from genome research is the use of oligonucleotide chips to perform thousands of genetic tests simultaneously. Pharmacogenomics also involves a new scale of analysis and application. Indeed, one way of distinguishing pharmacogenomics from pharmacogenetics is the scale of the analysis. Pharmacogenomics uses genome-wide analytical technologies to study the role of genes in differential response to pharmaceutical agents.

The research phase of pharmacogenomics also is being conducted on a large scale. To perform genome-wide research, scientists need access to large banks of biological samples or the genetic information derived from such samples. Often, clinical records must be linked with the samples. Large-scale research of this kind raises important ethical and policy issues.

As indicated in the public opinion survey reported in Chapter 1, the public strongly supports genetic research in general and pharmacogenomic research in particular. Moreover, the public generally is willing to participate in research involving sharing data with researchers nationwide and to permit the analysis of their medical records in addition to genetic samples. These large-scale genetic research projects, however, must be conducted in a manner that satisfies ethical principles of informed consent, confidentiality, and respect for persons. The good will of the public would be easily dissipated if there were a perception that individuals were being taken advantage of for the scientific or financial gain of the researchers.

Biobanks are repositories of biological specimens that can be used for research (Knoppers, 2001). Although millions of samples have been stored in hospitals and academic medical institutions for decades, new forms of biobanks have emerged in the last few years. These include government-sponsored biobanks (typically involving limited, homogeneous populations, such as in Iceland), commercial biobanks (involving for-profit intermediaries collecting and compiling samples for sale to researchers, including pharmaceutical companies), and Internet-based biobanks (a variation of commercial biobanking in which the advertising and arrangements for sample collection are done on-line).

If individual identifiers are removed from the samples and medical records, these de-identified records and materials are not considered "protected health information" under the Privacy Regulations of the Health

Insurance Portability and Accountability Act (HIPAA). Therefore, it is not necessary, under HIPAA, to obtain authorization for their use and disclosure. Nevertheless, institutional review boards (IRBs) may well require additional measures, especially for the prospective collection of new samples and information. In addition, even without individual identifiers, if there are group identifiers, such as race and ethnicity, there is the potential for group-based stigma and other harms (see chapters by Foster and Greely, this volume).

Although there are several ethical concerns with biobanks and other large-scale research repositories (Rothstein, 2002), the most important issue is informed consent. Participants in research involving human subjects must be advised of the intended research to be performed with their specimens. With biobanks, however, the future research uses of the samples are unknown at the time of collection. It would be infeasible to contact each of the donors to obtain consent every time a new research use is contemplated, yet IRBs are reluctant to approve the use of blanket consent for unspecified uses. One way to avoid this problem is for prospective sample donors to be given a menu of possible uses of their samples. Such a list might include research in mental health, HIV/AIDS, cancer, cardiovascular disease, or other areas. The donors would then have the opportunity to approve the use of their specimens for all or some of the listed uses. Other disclosures necessary to obtain informed consent include the financial interests of the biobank and researchers, whether individuals may elect to be notified of research findings, and whether it is possible for an individual to withdraw his or her sample from the bank (Rothstein, 2002).

B. INCLUSION OF WOMEN AND MINORITIES IN RESEARCH

In the 1990s, the NIH, FDA, and CDC separately promulgated regulations requiring the inclusion of women and minorities in federally funded research involving human subjects, unless there is a compelling scientific reason for noninclusion (NIH, 1994; FDA, 1997; CDC, 1995). The purpose of these regulatory initiatives was to eliminate the entrenched practice of including predominantly white males in clinical trials, thereby limiting the information about safety, efficacy, and dosing in the population at large (Rothenberg, 1996; DHHS, 2002).

Although the use of genotype-based trials has the potential for enrolling study participants with unequal distribution along ethnic lines, the need for genotypic homogeneity would appear to be a compelling scientific justification under the regulations for such inclusion criteria. Nevertheless, where feasible, researchers should recruit individuals with a particular genotype from all ethnic groups rather than using a convenience sample composed of individuals exclusively from the ethnic group with the highest frequency of the genotype of interest.

C. Genetic and Environmental Factors

Pharmacogenomics may help to explain much of the variation in response to medications, but it will not explain all of the difference. Drug interactions, age, nutrition, renal and liver function, and a variety of environmental factors influence the essential pharmocodynamics and pharmacokinetics of drug response (Evans and Relling, 1999; Omenn, 2001). Thus, to understand the role of genetic factors, it will also be necessary to study the effects of environmental factors, both individually and in combination. Only with greater knowledge about the role of both genetic and environmental factors will it be possible to develop drug response profiles for individual patients as well as patient-specific prevention strategies.

The National Institute of General Medical Sciences (NIGMS) of the NIH has initiated the Pharmacogenetics Research Network focusing on the study of "how genes affect the way people respond to medicines." NIGMS defines the ultimate goal of pharmacogenetics research as helping physicians tailor doses of medicines to a person's unique genetic makeup, thereby making medicines safer and more effective for everyone. Information regarding pharmacogenetic research is available at
http://www.nigms.nih.gov/pharmacogenetics/index.html.

The National Institute of Environmental Health Sciences (NIEHS) within the NIH is the lead institute for the Environmental Genome Project (EGP). Relying on technology developed by the HGP, the goal of the EGP is to identify polymorphisms of environmental disease susceptibility, develop a central database of these polymorphisms, and foster population-based studies of gene-environment interaction in disease etiology (NIEHS, 2001). Although the EGP is more concerned with toxicology than pharmacology, broad scientific understanding of gene-environment interaction will have widespread applicability, including in pharmacogenomics.

Additional information regarding progress of the EGP is available at
http://www.niehs.nih.gov/envgenom/home.htm.

D. Translational Research

Research on individually tailored medications must be considered in light of the lack of adequate regulation of genetic testing in general as well as the unsatisfactory current state of pharmaceutical efficacy. In considering whether pharmacogenomic medications are appropriate to move from the research setting to the clinical setting, it will be necessary to conduct translational research (research designed to assess the appropriateness of incorporating newly emerging technologies into the clinical setting) and develop clinical guidelines. Such research, of course, should be undertaken for all pharmaceutical products, including those considered to be standard therapy. By demonstrating its clinical utility through translational research, pharmacogenomic research could serve as a catalyst for developing a new paradigm of evidence-based prescribing of medications.

Some NIH-sponsored research deals with applying basic science insights to clinical disease and therefore could be considered translational. Nevertheless, translational research is supported primarily by the Agency for Healthcare Research and Quality (AHRQ), located in the Department of Health and Human Services. Unfortunately, the level of funding for translational research is woefully inadequate to meet current needs, and the negative trend in appropriations makes it clear that there will not be adequate funds to support future research related to the integration of pharmacogenomic medications into clinical care. In fiscal year 1999, the AHRQ established an initiative called Translating Research Into Practice (TRIP). TRIP was designed "to generate new knowledge about approaches that promote the utilization of rigorously derived evidence to improve patient care" (AHRQ, 2002a). Among its initiatives are efforts to reduce disparities in health care and to use information technology to translate research findings into sustainable health improvements. Unfortunately, for fiscal year 2003, President Bush requested an appropriation of $251,700,000 for AHRQ, a decrease of 16.2% from fiscal 2002. Non-patient safety research and training grants (which includes the TRIP program) were cut by $66,404,000, including all of the funds for new grants and all but $7 million for existing grants (AHRQ, 2002b).

The White House and Congress have heralded the fact that the NIH budget was doubled to $27.3 billion in the five-year period ending in fiscal 2003. Nevertheless, the benefits of basic and clinical research supported by the NIH will not be realized if there is little or no grant support for translational research. Neither the pharmaceutical industry nor health care practitioners are well positioned to perform this research. A national commitment to improving health outcomes cannot end at the clinical research phase. Substantial new resources must be committed to translational research.

E. Postmarketing Surveillance

The use of smaller, genotype-matched clinical trials increases the importance of postmarketing (phase IV) surveillance. In 1998, the FDA's Center for Drug Evaluation and Research established a new Office of Post-Marketing Drug Risk Assessment (CEDR, 2002). The office is charged with conducting epidemiological studies involving possible adverse outcomes caused by medications. Adverse event reporting is the key to identifying possible problems with an approved medication. The FDA's Adverse Events Reporting System consists of an electronic and paper system for pharmacovigilence. In the past, however, reporting has sometimes been incomplete, untimely, or poorly regulated.

Pharmacogenomics is going to make postmarketing surveillance a more important part of the drug approval process because of smaller and more targeted clinical trials. Therefore, an improved, comprehensive system of postmarketing surveillance is essential. As part of this effort, increased attention must be given to educating health care providers in identifying and reporting adverse events.

When some pharmaceutical products are approved, the FDA will ask the manufacturer to commit to postmarketing phase IV studies. These studies are mandatory for accelerated approval of products to treat serious and life-threatening illnesses ("fast-track" products) and deferred pediatric studies where safe use in children needs to be clarified. Other studies may be requested by the FDA before or after granting marketing approval if the FDA concludes that additional information is important to the prescribing and use of the product.

According to a report submitted to Congress by the FDA in 2002, pharmaceutical companies had completed only 882 out of 2400 reporting commitments (FDA, 2002). Currently, for drugs with mandated postmarketing studies, the FDA can withdraw the drug from the market, modify its label, or deem it misbranded. Nevertheless, the FDA has never pulled a drug from the market for failure to complete a postmarketing study in the absence of reports of adverse events. New legislation to provide the FDA with subpoena power and the authority to impose civil monetary penalties for failure to complete a postmarketing study is long overdue.

III. CLINICAL IMPLICATIONS

A. ROLES OF HEALTH PROFESSIONALS

Physicians and pharmacists must take steps to prevent the two most common errors in prescribing and dispensing medications: (1) giving an incorrect medication or dosage inappropriate for the patient's diagnosis; and (2) giving a medication that will adversely interact with another medication the patient is taking. Although avoiding the first of these errors is primarily the responsibility of the physician, both physicians and pharmacists have been assigned the duty to avoid the second type of error. Discoveries in pharmacogenomics may give rise to a third potentially common medication error: (3) giving a medication that is not indicated for the patient's genotype. It is not yet clear how the responsibility for avoiding these errors will be allocated, but it is quite likely that both physicians and pharmacists will be required to consider genetic information in prescribing and dispensing medications (see chapter by Brushwood, this volume). Dentists, nurses, and other health professionals also may need to consider genetic information before prescribing or administering medications.

By increasing the amount of medical information necessary for prescribing and dispensing, pharmacogenomics will demand that physicians
and pharmacists work together more closely than has often been the case.
For example, if pharmacists are to serve as a "safety net" for physician prescribing (Brushwood, this volume), then the pharmacist must have the
genetic information on which the prescription is based. Are physicians prepared to take the additional time to share this information and to include
nonphysicians in the prescribing function? Are pharmacists willing to accept
this additional duty, especially in some retail settings where productivity is
measured by the number of prescriptions filled? How are patient education
and warnings to be coordinated?

Regardless of the allocation of professional responsibilities, additional
educational programs in pharmacogenomics will be essential for all health
professionals. Genetics, genomics, and proteomics must be integrated into
the curricula of undergraduate medical students, training programs for residents, and continuing medical education programs for practicing physicians (Omenn and Motulsky, this volume). Pharmacogenomics must become
a standard part of the national medical boards and the examinations of the
medical specialty colleges. Similar efforts are necessary in schools of pharmacy and nursing, as well as in the continuing education courses of these
health professions.

B. LIABILITY ISSUES

As new medical advances become the "standard of care," there is the potential for liability of a health care provider if a patient sustains an injury
as a result of a provider's failure to meet the standard. In the context of
pharmacogenomics, provider liability may be based on, among other theories, failure to order genetic testing, improper interpretation of genetic test
results, failure to provide necessary genetic counseling, failure to prescribe
the proper medication and dosage, failure to warn the patient of possible
adverse events (and perhaps even failure to warn at-risk relatives), and
failure to dispense or administer the medication properly. It is unclear
whether these malpractice claims can be adjudicated without major changes
in the prevailing common law liability doctrines (see chapter by Palmer, this
volume). It is also unclear whether the fear of malpractice liability will drive
health care providers to prescribe unnecessary levels of pharmacogenomics-
based medications.

Manufacturers of products placed into commerce (including pharmaceutical manufacturers) are liable for harms caused by their defective products without the need for the plaintiff to prove that the manufacturer was
negligent or at fault. This approach, known as strict liability, is one of the
main approaches used in product liability cases. Strict liability is based on
the three-part rationale that manufacturers are in a better position to spread

the risk of loss among purchasers, manufacturers will be encouraged to make their products safer, and manufacturers at least implicitly represent their products as safe and consumers are entitled to rely on the representation (Dobbs, 2000). Lawsuits for injuries caused by defective products also may be brought under negligence and warranty theories. Product liability actions involving pharmacogenomics are likely to raise at least the following three issues: (1) failure to warn, (2) off-label uses, and (3) direct-to-consumer marketing.

First, manufacturers may be liable if they knew or should have known that their product may be harmful to some consumers based on their genotype and failed to issue appropriate warnings (McCormick, 1999). For example, one pending lawsuit involves the allegation that the manufacturer of a vaccine for Lyme disease failed to warn physicians and patients that individuals with a particular HLA marker were at risk of developing treatment-resistant arthritis (*Cassidy v. SmithKline Beecham*, 1999). Failure to warn cases could be brought in advance of the development of pharmacogenomics-based medications. An injured plaintiff would simply allege that the manufacturer had knowledge (and failed to warn) that a medication may be harmful to a subset of individuals defined by genotype. Cases such as these are likely to raise the issue of whether the manufacturer could reasonably rely on the prescribing physician or pharmacist to provide essential warnings. This is referred to as the "learned intermediary doctrine" (Castagnera et al., 2000). The cases also may raise the issue of the effect of package inserts in warning patients (Paytash, 1999).

Second, when a pharmaceutical product is approved by the FDA, the indications for use are listed in the package insert and the Physicians' Desk Reference (PDR). Although the FDA regulates drugs and devices, it does not regulate the practice of medicine. Accordingly, a physician is permitted to prescribe drugs for uses other than those approved by the FDA. These are known as "off-label" uses (Rheingold and Rheingold, 2001; Salbu, 1999). This issue could arise in pharmacogenomics when a medication approved for use by individuals with one genotype is prescribed for an individual with a different genotype (Gilhooley, 1999). The manufacturer's liability may depend on whether it knew about and failed to act against potentially harmful off-label uses, whether off-label uses were reasonably foreseeable, and whether the manufacturer encouraged off-label uses, such as by generating or promoting publications or advertisements touting the drug's value for unapproved uses.

Third, pharmaceutical advertisements increasingly are targeted directly to consumers. The major impetus for an increase in this technique was publication of a draft guidance by the FDA in 1997 and a final guidance in 1999 that permitted advertisements to include the purpose of the product (FDA, 1999). From 1997 to 2000, direct-to-consumer advertising in the U.S. increased from $860 million to $2.5 billion (Lancet, 2002). It must be assumed that new pharmacogenomic products would also be marketed directly to consumers. Such promotion, however, raises the possibility of a lawsuit based on negligent marketing, a theory being used increasingly to challenge harms allegedly caused by the overpromotion of pharmaceutical products to consumers (Ausness, 2002; Noah, 1997).

IV. Regulatory Issues

A. Food and Drug Law

There are at least four main ways in which the FDA is likely to become involved in the regulation of pharmacogenomics (see chapter by Feigal and Gutman, this volume). First, test kits used in the clinical setting for geno-typing are subject to regulation if they are considered commercial diagnostics. Second, pharmacogenomic testing may be an important part of a new drug application and therefore the scientific basis for subdividing the target population will be scrutinized by the FDA. Third, the FDA will evaluate the results of clinical trials, and, to the extent they are based on pharmacogenomic hypotheses, they will be analyzed to determine the safety, efficacy, and approved uses of the pharmaceutical product. Fourth, in reviewing post-marketing phase IV safety data the FDA will consider the effect of genetic factors in any reported adverse drug reactions. Although the FDA has jurisdiction in all of these areas under current law, it will be necessary to develop new regulatory guidelines and procedures to deal with the expected increase in products based on pharmacogenomic technologies.

B. Patent Protection

Many pharmaceutical products that were more effective than standard therapy for some individuals in traditional clinical trials have nevertheless failed to obtain FDA approval because there were unacceptably high adverse response rates. Without any information about the cause of the adverse responses, the FDA had no choice but to deny approval. It is possible that for some of these products the adverse responses are isolated among individuals with a certain genotype. Thus pharmacogenomics raises the possibility that drugs previously disapproved for general use could be "rescued" and approved for individuals with only certain genotypes. With FDA approval for these medications individuals with the indicated genotype may obtain a more effective product and the pharmaceutical companies could put into production a drug for which virtually all of the development costs already had been incurred.

Patent law, however, may be a problem for "rescued" drugs in that a substantial part of the 20-year period for patent protection may have elapsed before the drug's new safety profile is developed. Even without a modification in patent law, food and drug law could help to resolve the issue of "exclusivity." The Drug Price Competition and Patent Term Restoration Act of 1984 (Hatch–Waxman Act) grants the FDA the authority to award a pharmaceutical manufacturer "market exclusivity" for 5 years, or 7 years if the product qualifies under the Orphan Drug Act (Eisenberg, 2001). Although this is primarily designed to deal with rare disorders, it might be applicable to drugs designed for a narrow class of genotypes or, conceivably, could be extended to an even broader class of drugs. Where applicable, this additional time should be enough to make the production of "rescued" drugs economically viable.

V. Social Dimensions

A. Race and Ethnicity

Among the many findings of the HGP is that all humans are 99.9% identical at the level of DNA. Consequently, any two humans on the planet are likely to vary genetically only once in every one thousand base pairs. This finding reconfirms that the species *Homo sapiens* is relatively young (about 100,000 years old) and that there are no subspecies. Thus there is no scientific basis for racial classifications; race is a purely social construct.

The overall genetic homogeneity of human populations does not mean that there is no genetic diversity among individuals. At the molecular level there are, on average, about 3 million differences in the genomes of any two humans. The phenotypic diversity of humans is readily observable. Moreover, genetic variation is not random. Because of geographic and social isolation of populations over hundreds or thousands of years, patterns of genetic variation tend to be observed in individuals with a shared lineage. Thus, for example, certain monogenic disorders, such as Tay–Sachs disease and sickle-cell disease, are observed in higher frequencies in certain socially defined subpopulations.

In the U.S., race has been used as a proxy for a variety of physical, mental, and social dimensions. The legacy of racism still pervades society, including disparity in access to health care and in health care outcomes (see chapter by Nsiah-Jefferson, this volume). In attempting to eliminate these health disparities there have been efforts to include racial and ethnic minorities in research studies and to generate better data on racial and ethnic variations in health status. Although most policy makers and commentators believe that racial classifications in data collection and outcomes research are needed to eliminate health disparities, others believe that such policies prolong the use of a flawed concept and reify race as a pseudoscientific measure (Lee et al., 2001).

Issues surrounding when, if ever, it is appropriate to use race in research and clinical practice are certain to arise in the context of pharmacogenomics (Rothstein and Epps, 2001b). At every stage of research and development, from identifying drug targets through marketing and prescribing, it will be necessary to weigh the benefits and risks of using race as an imprecise proxy for genotype. In theory, all medical research and clinical care could become genome-specific; then, all variation would be individual variation. Yet, at present, it would be prohibitively expensive and impractical to recruit all research subjects by genotype. It would also be irresponsible for clinicians to ignore the prevalence of chronic diseases in certain populations (e.g., diabetes, hypertension) under the theory of avoiding "racial profiling" in medical practice.

In the U.S., pharmacogenomics will be introduced into a culture that continues to have a difficult time dealing with the issue of race (see chapter by Robertson, this volume), as well as a society in which many citizens lack scientific acumen. Most members of the public do not understand founder effect, genetic drift, and other principles of population genetics. Thus they would not understand why the same scientists who proclaim that race is not a scientific concept may choose to undertake their research in discrete ethnic populations. Scientists must appreciate the sensitivity of minority populations to targeted research; manufacturers must appreciate the risks of marketing pharmacogenomic products in a careless manner; clinicians must recognize the need to educate their patients about genetic testing and pharmacogenomic medications. By heedlessly equating race with genetic variation and genetic variation with genotype-based medications, we risk developing an oversimplified view of race-specific medications and a misleading view of the scientific significance of race (Braun, 2002; Epps and Rothstein, 2002). In the long run, genetics may help society eliminate the use of race in medicine and elsewhere, but in the short run those who work in genetics must be careful not to permit genetics to exacerbate the problem of race.

Although race is the primary social identity issue in the U.S., ethnicity is also a cause for concern in the U.S., and it is of even greater significance in other cultures. Thus there is a parallel risk that inappropriate conceptions of genetic variation could lead to ethnicity becoming reified as a biological category (Foster and Sharp, 2002). Indeed, the possibility that any social identity, such as language, religion, or geographic origin, could be accorded biological significance challenges researchers and policy makers to confront long-held conceptions of similarity and difference in individuals and populations.

B. PRIVACY, CONFIDENTIALITY, AND DISCRIMINATION

Privacy, confidentiality, and discrimination have received more attention from commentators and legislators than any other genome-related social issues (Rothstein, 1997). Thus it is important to consider how pharmacogenomics may affect these matters. In the usual situation, genetic information obtained from a family health history or genetic test indicates an individual's risk of illness. The disclosure of this genetic information to third parties may then cause embarrassment, stigmatization, or adverse treatment in such areas as employment and insurance. Genetic testing for purposes of prescribing pharmacogenomic medications differs from clinical genetic testing. Pharmacogenomics-related testing is likely to be of a much narrower scope, and the test, by itself, is likely to lack any independent, conclusive diagnostic or prognostic value. It will simply identify the polymorphic variation on which differential drug response is expected.

Notwithstanding this fundamental difference in the nature of genetic testing, it is possible that information derived from genetic tests performed solely for pharmacogenomic purposes could result in social harms to individuals. A pharmacogenomic test, for example, might reveal evidence of a genetic variation that could lead to individuals being classified as "difficult to treat" or "more expensive to treat." Such a finding could be used by any entity that has a possible financial obligation to pay for the individual's future health care. As a result, the fear that information of this nature might be generated could discourage some individuals from genetic analysis (depending on the importance of the pharmaceutical intervention), but, in any event, it is likely to contribute to an already-existing sense of public concern over the uses of genetic information. Consequently, the likelihood of developing pharmacogenomic tests increases the importance of enacting meaningful and comprehensive medical privacy and nondiscrimination protections.

C. ACCESS TO HEALTH CARE

Traditionally, pharmaceutical companies have sought to develop products with widespread applicability and large potential markets. Because pharmacogenomics is based on market segmentation, the effects on the economy and pharmaceutical pricing are difficult to predict (see chapter by Reeder and Dickson, this volume). It is quite possible, however, that allowing drug development, distribution, and pricing to be left entirely to the market could result in even greater patterns of skewed access to these new pharmaceutical products than already exist with current drugs.

As for-profit enterprises, pharmaceutical companies must consider whether they will be able to recoup their substantial investments in drug discovery. Thus it is reasonable to assume that pharmaceutical companies would consider not only the number of individuals for whom a particular drug would be efficacious but also whether the individuals would be able to purchase the medication, either personally or through their insurer or other third-party payer. Pharmacogenomics adds another dimension. Once a genetic variation that forms the drug target has been identified, biotech and pharmaceutical companies then could conduct a demographic analysis to determine the socioeconomic profile of individuals with the genotype. This will help analyze whether it would be profitable to invest the millions of dollars necessary to develop the product.

Rational economic calculations on the part of pharmaceutical companies do not always translate into sound social policy. That is one reason why, in 1983, Congress enacted the Orphan Drug Act (Orphan Drug Act, 1983). An orphan disease is defined as a rare disease or condition that "affects less than 200,000 persons in the United States" or one that "affects more than 200,000 in the United States and for which there is no reasonable expectation that the cost of developing and making available in the United States a drug for such disease or condition will be recovered from sales in the United States of such drug" (21 U.S.C. § 360bb(a)(2)). To encourage development of drugs for orphan diseases, the law streamlines the FDA development process, provides tax incentives, authorizes the FDA to help fund clinical trials, and grants the manufacturer 7 years of exclusive marketing rights (Bohrer and Prince, 1999; Rohde, 2000).

The development of the pharmacogenomic paradigm of drug development suggests that it is appropriate for Congress to reconsider the Orphan Drug Act. The law must be sufficiently flexible and its implementation sufficiently well funded through the FDA to support the development of pharmaceutical products for rare ("orphan") genotypes as well as for more common genotypes where drug development is not economically viable.

Because a substantial percentage of increased health care costs are attributable to pharmaceutical products, new and possibly expensive pharmacogenomics-based medications need to be considered carefully by individuals, health care providers, and third-party payers. A drug with 10% increased effectiveness and twice the cost may not be worth the additional cost for many individuals and payers, depending on various medical and economic factors. For public payers and society in general there is also an opportunity cost in committing a higher percentage of health care dollars to pharmaceutical products rather than other health care products and services.

The economics of pharmacogenomic medications must be considered in the overall context of a health care system that does not provide equal access to medications or any other health services and that produces widely differing outcomes based on various social factors. Although public policy recently has tended to focus on providing medications for low-income and elderly Americans, the issue of access is more complex. As noted in Chapter 1, individuals in the $20,000–25,000 per year income range ("the working poor") are the most concerned about the affordability of prescription medications and the most likely to have not had a prescription filled because of cost. Clearly, additional efforts are needed to extend health care coverage, with prescription medicine benefits, to this segment of the population.

VI. CONCLUSION

For science, molecular genetics represents a challenge to observe, understand, and manipulate ever-smaller units of biological materials. For ethics,
law, and policy, the opposite is true. Genetics challenges us to see the big
picture and to adopt thoughtful and responsible policies of a broad nature.
In resolving the seemingly new issues raised by genetics we invariably
return to familiar terrain. Thus policies on pharmacogenomics must address
the funding levels and direction of basic and translational research, the regulatory approval process for pharmaceutical products, the system of patent
protection for new drugs, the role of health care professionals in providing
optimum patient care, the responsibility for harms caused in the health care
system, the conflict between scientific and social conceptions of human variation, and societal notions of equality and inequality.

REFERENCES

Agency for Healthcare Research and Quality (AHRQ) (2002a) Translating Research
    Into Practice (TRIP)-II, *www.ahrq.gov/research/trip2fac.htm* (accessed May 24,
    2002).

Agency for Healthcare Research and Quality (AHRQ) (2002b) AHRQ Fiscal Year
    2003 Budget in Brief, *www.ahrq.gov/about/cj2003/budbrf03.htm* (accessed May 24,
    2002).

Ausness, R.C., "Will More Aggressive Marketing Practices Lead to Greater Tort Lia-
    bility for Prescription Drug Manufacturers?," *Wake Forest L. Rev.,* **37**, 97–139
    (2002).

Bohrer, R.A. and J.T. Prince, "A Tale of Two Proteins: The FDA's Uncertain Interpre-
    tation of the Orphan Drug Act," *Harvard J. Law Technol.,* **12**, 365–417 (1999).

Braun, L., "Race, Ethnicity, and Health: Can Genetics Explain Disparities? *Perspect.
    Biol. Med.,* **45(2)**, 159–174 (2002).

Buchanan, A., et al., "Pharmacogenetics: Ethical Issues and Policy Options," *Kennedy
    Inst. Ethics J.,* **12**, 1–17 (2002).

Cardon, L.R., et al., "Testing Drug Response in the Presence of Genetic Information,"
    Pharmacogenetics, **10**, 503–510 (2000).

Cassidy v. SmithKline Beecham, No. 99-10423 (Pa. Ct. Common Pleas, Chester County,
    filed Dec. 14, 1999).

Castagnera, J.O., et al., "The Gradual Enfeeblement of the Learned Intermediary Rule
    and the Argument in Favor of Abandoning it Entirely," *Tort Insurance L. J.,* **36**,
    119–146 (2000).

Center for Drug Evaluation and Research (CEDR) Postmarketing Surveillance
    Programs, *www.fda.gov/cder/regulatory/applications/postmarketing/
    surveillancepost.htm* (accessed May 24, 2002).

Centers for Disease Control and Prevention, "Policy on the Inclusion of Women and
    Racial and Ethnic Minorities in Externally Awarded Research," 60 Fed. Reg.
    47,947 (1995).

Collins, F.S. and V.A. McKusick, "Implications of the Human Genome Project for Medical Science," *JAMA*, **285**, 540–544 (2001).

Dobbs, D.B., *The Law of Torts*. St. Paul, MN: West Group, pp. 975–976 (2000).

Eisenberg, R.S., "The Shifting Functional Balance of Patents and Drug Regulation," *Hlth. Affairs*, **20(5)**, 119–135 (2001).

Epps, P.G. and M.A. Rothstein, "Pharmacogenomics: Ensuring Equity Regarding Drugs Based on Genetic Difference," In: J. Licinio and M.-L. Wong (eds.), *Pharmacogenomics: The Search for Individualized Therapies*. Weinheim, Germany: Wiley-VCH (2002).

Evans, W.E. and M.V. Relling, "Pharmacogenomics: Translating Functional Genomics Into Rational Therapeutics," *Science*, **286**, 5439–5491 (1999).

Food and Drug Administration (FDA) (1999) Final Guidance on Direct-to-Consumer Advertising. Federal Register 64:43,197–198, available at *www.fda.gov/cder/guidance/1804fnl.htm* (accessed May 26, 2002).

Food and Drug Administration (FDA) (2002) Report to Congress on Post-Marketing Surveillance of Drugs, March 13, 2002, available at *www.fda.gov/cber/fdama/pstmrktfdama130.htm* (accessed May 30, 2002.)

Foster, M.W. and R.R. Sharp, "Race, Ethnicity, and Genomics: Social Classifications as Proxies of Biological Heterogeneity," *Genome Res.*, **12**, 844–850 (2002).

Gilhooley, M., "When Drugs Are Safer for Some But Not Others: The FDA Experience and Alternatives for Products Liability," *Houston L. Rev.*, **36**, 927–949 (1999).

Holtzman, N.A., "Putting the Search for Genes in Perspective," *Intl. J. Hlth. Serv.*, **31**, 445–458 (2001).

International Human Genome Sequencing Consortium, "Initial Sequencing and Analysis of the Human Genome," *Nature*, **409**, 860–921, (2001).

Knoppers, B.M., "Of Populations, Genetics and Banks," *Genet. Law Monit.*, Jan.–Feb., 3–6 (2001).

The Lancet, "Europe on the Brink of Direct-to-Consumer Drug Advertising," *Lancet*, **359**, 1709 (2002).

Lazarou, J., et al., "Incidence of Adverse Drug Reactions in Hospitalized Patients: A Meta-Analysis of Prospective Studies," *JAMA*, **279**, 1200–1205 (1998).

Lee, S.S.-J., et al., "The Meanings of 'Race' in the New Genomics: Implications for Health Disparities Research," *Yale J. Hlth. Policy L. Ethics*, **1**, 33–75 (2001).

McCormick, R., "Pharmaceutical Manufacturer's Duty to Warn of Adverse Drug Interactions," *Defense Counsel J.*, **66**, 59–68 (1999) .

Moldrup, C., "When Pharmacogenomics Goes Public," *New Genet. Soc.*, **21**, 29–37 (2002).

National Institute of Environmental Health Sciences (NIEHS) (2001) Environmental Genome Project Overview, *www.niehs.gov/envgenom/concept.htm* (accessed May 27, 2002).

National Institutes of Health, "Guidelines on the Inclusion of Women and Minorities as Subjects in Clinical Research," 59 Fed. Reg. 14,508, (1994).

Noah, L., "Advertising Prescription Drugs to Consumers: Assessing the Regulatory and Liability Issues," *Ga. L. Rev.*, **32**, 141–180 (1997).

Omenn, G.S., "Prospects for Pharmacogenetics and Ecogenetics in the New Millennium," *Drug Metab. Dispos.*, **29**, 611–614 (2001).

Orphan Drug Act (1983) 21 U.S.C. §§ 360aa–360ee.

Paytash, C.A., "Note, The Learned Intermediary Doctrine and Patient Package Inserts: A Balanced Approach to Preventing Drug-Related Injury," *Stanford L. Rev.*, **51**, 1343–1371 (1999).

Phillips, K.A., et al., "Potential Role of Pharmacogenomics in Reducing Adverse Drug Reactions: A Systematic Review," *JAMA*, **286**, 2270–2279 (2001).

Rheingold, P.D. and D.B. Rheingold, "Offense or Defense? Managing the Off-Label Use Claim," *Trial*, **37**, 52–56 (March) (2001).

Robertson, J.A., et al., "Pharmacogenetic Challenges for the Health Care System," *Hlth. Affairs*, **21(4)**, 1–13 (2002).

Rohde, D.D., "The Orphan Drug Act: An Engine of Innovation? At What Cost?" *Food Drug L. J.*, **55**, 125–143 (2000).

Rothenberg, K., "Gender Matters: Implications for Clinical Research and Women's Health Care," *Houston L. Rev.*, **32**, 1201–1272 (1996).

Rothstein, M.A., "The Role of IRBs in Research Involving Biobanks," *J. L. Med. Ethics,* **30**, 105–108 (2002).

Rothstein, M.A. and P.G. Epps, "Ethical and Legal Implications of Pharmacogenomics," *Nat. Rev. Genet.*, **2**, 228–231 (2001a).

Rothstein, M.A. and P.G. Epps, "Pharmacogenomics and the (Ir)relevance of Race," *Pharmacogenomics J.*, **1**, 104–108 (2001b).

Salbu, S., "Off-Label Use, Prescription, and Marketing of FDA-Approved Drugs: An Assessment of Legislative and Regulatory Policy," *Fla. L. Rev.*, **51**, 181–226 (1999).

United States Dept. of Health and Human Services, The Initiative to Eliminate Racial and Ethnic Disparities in Health; *http://raceandhealth.hhs.gov.*

Venter, J.C., et al., "The Sequence of the Human Genome," *Science*, **291**, 1304–1351 (2001).

PHARMACOGENOMICS AND MINORITY POPULATIONS

GENERAL POPULATION SURVEY QUESTIONNAIRE

Good (morning/afternoon/evening). This is (FIRST & LAST NAME) calling for the University of Louisville School of Medicine. We are conducting a national survey funded by the National Institutes of Health about important health care issues, and your household was randomly selected to represent people living in your area. I would like to invite you to participate.

S1. First, for this survey, I need to speak with the adult (male/female) member of your household who had the most recent birthday. (Would that be you/Is [he/she] available)?
    <1> Yes (CONTINUE)
    <2> No (ASK TO SPEAK WITH SELECTED RESPONDENT)
    <9> RF/DK

UPON REACHING RESPONDENT:
The research we are conducting asks your opinions about new discoveries in medicine. The interview takes about 20 minutes, and your participation is completely voluntary. You may decline to participate or end your participation at any time. In addition, you may decline to answer any specific item. All answers you provide will be confidential, meaning that we will not ask you for your name, and a record of the telephone numbers called will not be linked with the responses. In all other respects, confidentiality will be protected to the extent permitted by law.

Pharmacogenomics: Social, Ethical, and Clinical Dimensions, Edited by Mark A. Rothstein.
ISBN 0-471-22769-2   Copyright © 2003 Wiley-Liss, Inc.

Although there is the potential for some scientific benefit from this research, you may not derive any personal benefit from your participation. However, your participation in this study poses no risk to you whatsoever. By agreeing to this interview, you acknowledge that the questions you have about the research at this time have been answered in a language you understand. If you have any future questions about this research, please contact Professor Mark Rothstein at (502) 852-4980 or the University of Louisville Human Subjects Committee at (502) 852-5188.

S2. Would you be willing to participate in this important study?
    <1> Yes (CONTINUE)
    <2> No (THANK & TERMINATE)

Some differences among people are inherited from their parents through genes. Some genes determine things like hair color and height. Other genes determine things like how likely we are to develop specific diseases. New research involves finding genes that determine how we respond to specific medicines. For the most part, this interview will focus on the new research on genetic differences among individuals as they relate to medicines.

1. First, do you feel the study of genetic differences in the way individuals respond to specific medicines is . . . (READ LIST)
    <1> A Good Thing
    <2> More Good Than Bad
    <3> Equally Good and Bad
    <4> More Bad Than Good, or
    <5> A Bad Thing
    <7> RF
    <8> DK

2. Genetic testing is a quick and painless procedure, such as having a blood test or simply brushing the inside of your mouth with a cotton swab. Genetic tests may be done for different reasons and may give different kinds of information. I am going to read a list of potential reasons to have a genetic test and ask you to tell me how likely you would be to have a genetic test for each reason. First, . . .

    A. How likely would you be to have a genetic test to help diagnose a <u>serious</u> disease? Would you be . . . (READ CATEGORIES)
        <1> Very Likely
        <2> Somewhat Likely
        <3> Somewhat Unlikely
        <4> Very Unlikely
        <7> RF
        <8> DK

B. How likely would you be to have a genetic test to help diagnose a non-serious disease? Would you be . . . (READ CATEGORIES)
  <1> Very Likely
  <2> Somewhat Likely
  <3> Somewhat Unlikely
  <4> Very Unlikely
  <7> RF
  <8> DK

C. How likely would you be to have a genetic test to learn whether you were at increased risk of getting a serious disease in the future? Would you be . . . (READ CATEGORIES)
  <1> Very Likely
  <2> Somewhat Likely
  <3> Somewhat Unlikely
  <4> Very Unlikely
  <7> RF
  <8> DK

D. How likely would you be to have a genetic test to learn whether you were at increased risk of getting a serious disease if there were no treatment available for the disease? Would you be . . . (READ CATEGORIES)
  <1> Very Likely
  <2> Somewhat Likely
  <3> Somewhat Unlikely
  <4> Very Unlikely
  <7> RF
  <8> DK

3. Genetic testing also may be used to identify which medications are best for you based on your genetic makeup. For example, a medication that is very effective for one person may be less effective or even harmful to someone with a different genetic makeup. By studying genetic differences, scientists hope to develop medications that are matched to genetic makeup.

A. How likely would you be to have a genetic test to learn which medicine is most effective for you based on your genetic makeup? Would you be . . . (READ CATEGORIES)
  <1> Very Likely
  <2> Somewhat Likely
  <3> Somewhat Unlikely
  <4> Very Unlikely
  <7> RF
  <8> DK

B.   How likely would you be to have a genetic test to learn what <u>dose</u> of medicine you should take based on your genetic makeup? Would you be . . . (READ CATEGORIES)

    <1> Very Likely
    <2> Somewhat Likely
    <3> Somewhat Unlikely
    <4> Very Unlikely
    <7> RF
    <8> DK

4. A great deal of research is needed to develop medicines based on genetic makeup. How likely would you be to participate in genetic research under each of the following conditions?

A.   First, how likely would you be to participate in research if you knew your genetic test results would be anonymous, meaning that nobody—not even the researchers—would know which test results came from which person? Would you be . . . (READ CATEGORIES)

    <1> Very Likely
    <2> Somewhat Likely
    <3> Somewhat Unlikely
    <4> Very Unlikely
    <7> RF
    <8> DK

B. How likely would you be to participate in research if it involved researchers reviewing your medical records in addition to a genetic test in order to explore possible links between genetic test results and health history? Would you be . . . (READ CATEGORIES)

    <1> Very Likely
    <2> Somewhat Likely
    <3> Somewhat Unlikely
    <4> Very Unlikely
    <7> RF
    <8> DK

C. How likely would you be to participate in research if you knew your genetic test results, without your name or other identifying information, would be shared with other scientists nationwide for use in additional genetic research? Would you be . . . (READ CATEGORIES)

    <1> Very Likely
    <2> Somewhat Likely
    <3> Somewhat Unlikely
    <4> Very Unlikely
    <7> RF
    <8> DK

D. How likely would you be to have a genetic test if the possibility existed to develop a treatment in the future based on your participation in the research study?
<1> Very Likely
<2> Somewhat Likely
<3> Somewhat Unlikely
<4> Very Unlikely
<7> RF
<8> DK

5. If you were paid $50.00 to have a genetic test as part of a medical research project, what impact, if any, would the money have on your likelihood of participating? Would it make you . . .
<1> Much More Likely
<2> Somewhat More Likely
<3> Have No Impact
<4> Somewhat Less Likely, or
<5> Much Less Likely to Participate?
<7> RF
<8> DK

6. Genetic research could be conducted by several different groups. How much trust would you have in each of the following groups to carry out genetic research? First, . . . (ROTATE):

A. How much trust would you have in genetic research conducted by universities and Medical Schools? Would you have . . . (READ CATEGORIES)
<1> A Great Deal of Trust
<2> Some Trust
<3> Some Lack of Trust
<4> No Trust
<7> RF
<8> DK

B. How much trust would you have in genetic research conducted by the Federal Government? Would you have . . . (READ CATEGORIES)
<1> A Great Deal of Trust
<2> Some Trust
<3> Some Lack of Trust
<4> No Trust
<7> RF
<8> DK

C. How much trust would you have in genetic research conducted by drug companies? Would you have . . . (READ CATEGORIES)
<1> A Great Deal of Trust
<2> Some Trust
<3> Some Lack of Trust
<4> No Trust
<7> RF
<8> DK

D. How much trust would you have in genetic research conducted by organizations like the American Cancer Society or the March of Dimes? Would you have . . . (READ CATEGORIES)
<1> A Great Deal of Trust
<2> Some Trust
<3> Some Lack of Trust
<4> No Trust
<7> RF
<8> DK

7. With the exception of identical twins, every individual has a unique genetic makeup. Because genetic factors are inherited from our parents, family members tend to have genetic similarities. Sometimes, members of the same racial or ethnic group are also more likely to have some similar genetic characteristics.

A. How likely would you be to participate in research if it might determine whether family members have some genetic similarities in their response to certain medications? Would you be . . . (READ CATEGORIES)
<1> Very Likely
<2> Somewhat Likely
<3> Somewhat Unlikely
<4> Very Unlikely to Participate?
<7> RF
<8> DK

B. How likely would you be to participate in research if it might determine whether there are genetic similarities and differences in the way people of the same racial or ethnic group respond to certain medications? Would you be . . . (READ CATEGORIES)
<1> Very Likely
<2> Somewhat Likely
<3> Somewhat Unlikely
<4> Very Unlikely to Participate?
<7> RF
<8> DK

8. Because of possible genetic similarities, some medications based on genetic studies could be prescribed more often for people belonging to certain racial or ethnic groups. What effect, if any, would this have on your willingness to take this kind of medicine if it was prescribed for you? Would this make you . . . (READ CATEGORIES)

    <1> Much More Willing
    <2> Somewhat More Willing
    <3> Have No Impact
    <4> Somewhat Less Willing, or
    <5> Much Less Willing to Take Such A Medicine?
    <7> RF
    <8> DK

9. If medicines were developed that were matched to the genetic makeup of individuals, do you think that people of your income level could afford them?

    <1> Yes
    <2> No
    <7> RF
    <8> DK

10. Some genetic tests can be used to predict whether certain individuals are more likely to get sick in the future. The next questions ask your opinions about who should have access to these genetic test results. (ROTATE):

    A. If your employer could get the results of a genetic test that showed whether you were more likely to get sick in the future, what impact, if any, would this have on your willingness to take the test? Would it make you . . . (READ CATEGORIES)

    <1> Much More Likely
    <2> Somewhat More Likely
    <3> Have No Impact
    <4> Somewhat Less Likely, or
    <5> Much Less Likely to be Willing to Take a Genetic Test?
    <7> RF
    <8> DK

    B. If your health insurance company could get the results of a genetic test that showed whether you were more likely to get sick in the future, what impact, if any, would this have on your willingness to take the test? Would it make you . . . (READ CATEGORIES)

    <1> Much More Likely
    <2> Somewhat More Likely
    <3> Have No Impact
    <4> Somewhat Less Likely, or
    <5> Much Less Likely to be Willing to Take a Genetic Test?
    <7> RF
    <8> DK

C. If your life insurance company could get the results of a genetic test that showed whether you were more likely to get sick in the future, what impact, if any, would this have on your willingness to take the test? Would it make you . . . (READ CATEGORIES)

<1> Much More Likely
<2> Somewhat More Likely
<3> Have No Impact
<4> Somewhat Less Likely, or
<5> Much Less Likely to be Willing to Take a Genetic Test?
<7> RF
<8> DK

11. How optimistic are you about the possibility of medical advances as a result of genetic research? Would you say you are . . . (READ CATEGORIES)

<1> Very Optimistic
<2> Somewhat Optimistic
<3> Not Too Optimistic
<4> Not At All Optimistic
<7> RF
<8> DK

12. Based on your prior knowledge and everything we have discussed in this interview, do you feel the study of genetic differences in the way individuals respond to specific medicines is . . . (READ LIST)

<1> A Good Thing
<2> More Good Than Bad
<3> Equally Good and Bad
<4> More Bad Than Good, or
<5> A Bad Thing
<7> RF
<8> DK

D1. Now for some background questions, and we will be finished. First, would you say your health is . . . (READ LIST)

<1> Excellent	435	24.2%
<2> Very Good	566	31.5
<3> Good	473	26.3
<4> Fair, or	243	13.5
<> Poor?	69	3.8
<7> Refused	2	.1
<8> Don't Know	8	.4

D2. Do you, yourself, ever take any prescription or over-the-counter medicines?

         <1> Yes (ASK A)   1206   67.1%
         <2> No (SKIP TO D3)   587   32.7
         <7> RF (SKIP TO D3)   3   .2
         <8> DK (SKIP TO D3)   —   —
            IF YES, ASK:

D2A. Do you take prescription or over-the-counter medicines . . .

         <1> On a daily basis, or   608   33.9%
         <2> Occasionally, as needed?   595   33.1
         <7> RF   3   .2
         <8> DK   —   —

D3. Do you have a close family member whose health is poor?

         <1> Yes   606   33.7%
         <2> No   1175   65.4
         <7> RF   2   .1
         <8> DK   13   .7

D4. Including yourself, how many people live in your household full time?

                                              NUMBER: ___ ___
                                                  <97> RF
                                                  <98> DK

          IF ONE, SKIP TO D5. IF MORE THAN ONE, ASK D4:

D4. How many of these are age 18 and older?

                                            NUMBER: ___ ___
                                                    <97> RF
                                                  <98> DK

D5. What is your age?

                                            AGE: ___ ___
                                                     <97> RF
                                                  <98> DK

D6. What is the highest grade of school or college you completed? (DO NOT READ LIST)

         <01> 8th Grade or Less   90   5.0%
         <02> Some High School   133   7.4
         <03> High School Graduate   400   22.3
         <04> Trade or Technical School 49   2.7
         <05> Some College   427   23.8
         <06> Bachelor's Degree   407   22.7
         <07> Graduate Degree   283   15.8
         <97> RF   5   .3
         <98> DK   2   .1

D7. Do you regularly use a computer at home?
       <1> Yes   1020   56.8%
       <2> No   773   43.0
       <7> RF   2   .1
       <8> DK   1   .1

D8. Are you currently . . . (READ LIST)
       <1> Married   778   43.9%
       <2> Widowed   133   7.4
       <3> Separated or Divorced   215   12.0
       <4> In a Domestic Partnership, or   77   4.3
       <5> Single/Never Married?   571   31.8
       <7> RF   10   .6
       <8> DK   2   .1

D9. Are you currently . . . (READ LIST)
       <01> Employed Full Time   912   50.8%
       <02> Employed Part Time   181   10.1
       <03> Unemployed   92   5.1
       <04> A Homemaker   102   5.7
       <05> Retired   257   14.3
       <06> Disabled   56   3.1
       <07> A Student, or   153   8.5
       <08> Something else? *(Specify)*   34   1.9

       <97> RF   9   .5
       <98> DK   —   —

D10. Have you ever worked in the health care field?
       <1> Yes   490   27.3%
       <2> No   1302   72.5
       <7> RF   2   .1
       <8> DK   2   .1

D11. Do you live . . . (READ LIST)
       <1> In or near a large city with a population over 100,000   1360   75.7%
       <2> In a small city with a population of less than 100,000, or   241   13.4
       <3> In a rural or farm area?   175   9.7
       <7> RF   2   .1
       <8> DK   18   1.0

D12. Do you consider yourself . . . (READ LIST)
       <1> Caucasian or White, Non-Hispanic   848   47.2%
       <2> Black or African-American   310   17.3
       <3> Hispanic   300   16.7
       <4> Asian or Asian-American   298   16.6
       <5> Native American, or   13   .7

<6> Something else? *(Specify)*	27	1.5
<7> RF	—	—
<8> DK	—	—

D13. What language do you speak most often at home? (DO NOT READ LIST)

<01> English	1345	74.9%
<02> Spanish	237	13.2
<03> Chinese (Mandarin or Cantonese)	110	6.1
<04> Vietnamese	29	1.6
<05> Korean	32	1.8
<06> Other Asian Language *(Specify)*	21	1.2
<07> Other Language (Non-Asian) *(Specify)*	20	1.1
<97> RF	1	.1
<98> DK	1	.1

D14. In what country were you born? (DO NOT READ LIST)

<01> United States (including Alaska and Hawaii)	1231	68.5%
<02> U.S. Territory (Puerto Rico, Guam, etc.)	26	1.4
<02> Mexico	90	5.0
<03> Cuba	34	1.9
<04> China	100	5.6
<05> Taiwan	9	.5
<06> Vietnam	38	2.1
<07> Korea	38	2.1
<08> Other *(Specify)*	222	12.4
<97> RF	6	.3
<98> DK	2	.1

D15. Do you currently have any type of health insurance coverage, including Medicare or Medicaid? (DO NOT READ LIST)

<1> Yes (ASK A & B)	1391	77.4%
<2> No (SKIP TO B)	385	21.4
<7> RF (SKIP TO B)	12	.7
<8> DK (SKIP TO B)	8	.4

IF YES, ASK:

A. Does your health insurance have full coverage, partial coverage or no coverage for prescription drugs? (DO NOT READ LIST)

<1> Full Coverage	532	29.6%
<2> Partial Coverage	694	38.6
<3> No Coverage	112	6.2
<7> RF	2	.1
<8> DK	51	2.8

B. Have you ever had a prescription that you did not have filled because of the cost? (DO NOT READ LIST)

<1> Yes	384	21.49%
<2> No	1394	77.6
<7> RF	—	—
<8> DK	18	1.0

D16. What is your religious preference, if any? Is it . . . (READ LIST)

<01> Protestant,	613	34.1%
<02> Catholic,	458	25.5
<03> Jewish,	36	2.0
<04> Mormon,	15	.8
<05> Islam,	11	.6
<06> Buddhism,	44	2.4
<07> Other Religion (Specify)	181	10.1
<08> No Religious Preference?	406	22.6
<97> RF	25	1.4
<98> DK	7	.4

D17. Finally, what was your total household income in 2000? That is, income for all members of your household during 2000. Was it . . . (READ LIST)

<01> Under $10,000	163	9.1%
<02> $10,000 to $14,999	136	7.6
<03> $15,000 to $19,999	134	7.5
<04> $20,000 to $24,999	141	7.9
<05> $25,000 to $34,999	193	10.7
<06> $35,000 to $49,999	245	13.6
<07> $50,000 to $74,999	233	13.0
<08> $75,000 to $99,999, or	151	8.4
<09> $100,000 or more?	156	8.7
<97> RF	130	7.2
<98> DK	114	6.3

That concludes our interview. Thank you for your time and help with this important research effort.

D18. Interviewer Record:

<1> Male	851	47.4%
<2> Female	945	52.6

D19. Language of Interview:

<1> English	1445	80.5%
<2> Spanish	213	11.9
<3> Chinese	86	4.8
<4> Vietnamese	29	1.6
<5> Korean	23	1.3

Index